Essential Oils
Contributions to the
Chemical–Biological Knowledge

Editors

Valdir Florêncio da Veiga Jr.
Military Institute of Engineering
Rio de Janeiro, Brazil

Isiaka Ajani Ogunwande
Foresight Institute of Research and Translation
Ibadan, Oyo, Nigeria

José L. Martinez
Adjunct to the Vice-Rectory for Research, Innovation and Creation
Universidad de Santiago Chile
Santiago, Chile

CRC Press
Taylor & Francis Group
Boca Raton London New York

CRC Press is an imprint of the
Taylor & Francis Group, an **informa** business

A SCIENCE PUBLISHERS BOOK

Cover credit:
Psidium guajava L. (Myrtaceae)
Common name: Guava or Guayaba
Image captured by Valdir Florêncio da Veiga Jr. in Cachoeiras de Macacu, Rio de Janeiro, Brazil
Essential oil extractor, Laboratory of Dr. Valdir Florêncio da Veiga Jr.

First edition published 2024
by CRC Press
2385 NW Executive Center Drive, Suite 320, Boca Raton FL 33431

and by CRC Press
4 Park Square, Milton Park, Abingdon, Oxon, OX14 4RN

© 2024 Valdir Florêncio da Veiga Jr., Ogunwande Isiaka Ajani and José L. Martinez

CRC Press is an imprint of Taylor & Francis Group, LLC

Library of Congress Cataloging-in-Publication Data (applied for)

ISBN: 978-1-032-12810-8 (hbk)
ISBN: 978-1-032-12816-0 (pbk)
ISBN: 978-1-003-22634-5 (ebk)

DOI: 10.1201/9781003226345

Typeset in Times New Roman
by Radiant Productions

Preface

There are hundreds of extracts and molecules of clinical value waiting to be discovered around the world. And this could have been responsible for the continued prevalence of disease and infection. For this to happen, a large number of researchers must be well versed in the interconnection and relationship between different types of knowledge, ethnomedicine, pharmacology, and natural product chemistry.

Essential oil is an emerging area in the supply of extracts and compounds of pharmacological importance. Each of the chapters of the book is designed to provide accurate and reliable information regarding the subject matter in it. The authors determine the chemical compositions and in most cases evaluate the biological activities. Essential oils are explored in the context of their ethnobotanical study in the treatment of one or more ailments.

The book presents 6 chapters by authors from South America from Colombia, Brazil, Peru and Venezuela and three chapters from other continents, being from Turkey, India and Pakistan.

I hope readers find useful and beneficial information in this book.

Valdir Florêncio da Veiga Jr.
Isiaka Ajani Ogunwande
José L. Martínez

Contents

1

Purification of Essential Oil Compounds by Countercurrent Chromatography

André Mesquita Marques,[1] *Shaft Corrêa Pinto,*[2]
Suzana Guimarães Leitão[3] *and Gilda Guimarães Leitão*[4,*]

Introduction

General Principle of Countercurrent Chromatography

Countercurrent chromatography (CCC) is a form of liquid-liquid partition chromatography in which the stationary liquid phase is retained inside the column without solid support (Conway 1990, Vetter et al. 2020). The solvent systems used in this technique are mixtures of solvents that form a biphasic liquid system wherein one of the phases acts as stationary and the other as mobile. The principle of separation involves the partitioning of a solute between these two immiscible liquid phases, according to its distribution coefficient (K_D) or partition coefficient (K) if only one solute form is involved (Bojczuk et al. 2017, Vetter et al. 2020). The absence of solid support and the nature of its stationary liquid phase brings many advantages to this technique over others based on liquid-liquid partition mechanisms; one of the most important advantages is the high loading capacity that turns it most useful in preparative separations and total sample recovery as there are no losses by adsorption (Conway 1990).

[1] Laboratório de Produtos Naturais, TecBio, FarManguinhos, Fiocruz, 21.045-900, Rio de Janeiro, RJ, Brazil.

[2] Instituto de Ciências Farmacêuticas, Universidade Federal do Rio de Janeiro, Centro Multidisciplinar UFRJ-Macaé, Av. Aluizio da Silva Gomes, 50, Novo Cavaleiros, 27.930-560 Macaé, Brazil.

[3] Faculdade de Farmácia, Universidade Federal do Rio de Janeiro, Av. Carlos Chagas Filho, 373, Centro de Ciências da Saúde, Bloco A segundo andar, Ilha do Fundão, 21941-902, Rio de Janeiro, RJ, Brazil.

[4] Instituto de Pesquisas de Produtos Naturais, Universidade Federal do Rio de Janeiro, Av. Carlos Chagas Filho, 373, Centro de Ciências da Saúde, Bloco H, Ilha do Fundão, 21.941-902, Rio de Janeiro, RJ, Brazil.

Emails: andre.marques@far.fiocruz.br; shaft@macae.ufrj.br; sgleitao@pharma.ufrj.br

* Corresponding author: ggleitao@ippn.ufrj.br

In modern CCC equipment operate in a centrifugal way, which generates a centrifugal force field that retains the stationary phase inside the column. The rotational movement of the column can be of two types. Either it can rotate along one axis in a constant field generating the hydrostatic equilibrium between the two immiscible liquid phases or it can rotate along the two axes (its axis; generally called the planetary axis, central or solar axis) in a highly variable field generating a hydrodynamic equilibrium. To make it simple, the equipment can be either called a hydrostatic machine or CPC (from *Centrifugal Partition Chromatography*) or a hydrodynamic machine (or HSCCC from *High-Speed Countercurrent Chromatography*). In CPC, the chromatographic column is formed by channels interconnected by small ducts that are engraved on a disk, which when pilled can form chambers where the stationary phase is held while the mobile phase passes through them (Berthod 2009). In HSCCC, the chromatographic column is instead formed by a tubing coiled around a bobbin, and due to the varying centrifugal field, there is the formation of mixing and settling zones (Berthod 2009). Most solvent systems are very well retained in hydrodynamic columns, whereas in CPC instruments all solvent systems are well retained.

In modern CCC, the mobile phase can be either one of the solvent systems (upper or lower). Thus, in hydrostatic instruments, when the heavier (lower) phase is used as the mobile phase, the instrument is operating in the descending mode, whereas when the lighter (upper) phase is the mobile phase, the instrument is operating in the ascending mode. In hydrodynamic instruments, due to Archimedean forces, zones move to the high-pressure side of the coil, which is called the "head" (Berthod 200). The terminology used for operations in HSCCC machines is "head-to-tail" (H → T) when the lower phase is used as mobile and "tail-to-head" (T → H) when the upper phase is employed.

Advantages of CCC on the Purification of Essential Oil Compounds

CCC is extremely useful in the purification of compounds from complex matrices, such as plant extracts. During the past few decades, CCC has been applied to both analytical and preparative separations of a wide variety of natural compounds (Skalicka- Woźniak and Garrard 2014, Costa et al. 2010) and essential oils (Marques and Kaplan 2013). Essential oils (EO) are complex hydrophobic mixtures composed of volatile compounds. They are normally described as oily liquids at room temperature, which may have a characteristic aroma and flavor and insolubility in water, but they are soluble in organic solvents (Maurya et al. 2021). EO can be found in aromatic plants, being constituted by a complex mixture of volatile metabolites from mevalonate, such as terpenes (mono and sesquiterpenes) and shikimate pathways (phenolic compounds, such as phenylpropanoids). They can be composed of solid and liquid volatile compounds and considering their numerous biological activities and pleasant scent, the EO can be considered an important raw material with economic value for many pharmaceutical, perfumery and food industries (Silva-Santos et al. 2006).

Different plant structures can store essential oils, such as glandular trichomes (as in Asteraceae and Lamiaceae) and secretory bags (as in Rutaceae and Myrtaceae) that are found in all plant organs. They can be used as raw material with application in industry as adjuvants in medicines mainly used as aromas, fragrances and fragrance fixatives in pharmaceutical and oral compositions and are marketed in their raw or processed form. EO is also a source of compounds of economic interest, such as limonene, citral, citronellal, eugenol, menthol and safrole (Bizzo et al. 2009).

The components of essential oils can be grouped according to their chemical groups that are constituted by a complex mixture of hydrocarbons, alcohols and carbonyl compounds. The most frequently found hydrocarbons belong to the class of terpenes, mainly mono- and sesquiterpenes in a wide variety of chemical groups containing alcohol, aldehyde or ketone functions (Burčul et al. 2020). Important phenylpropanoids can also be found in some EO. For example, safrole metabolites from *Piper hispidinervum* present economic importance in the chemical industry as being a raw material for the synthesis of pyrethroid insecticides (Andrés et al. 2017).

The obtention of pure compounds in quantity from natural sources is frequently necessary for pharmacological investigations (Carini et al. 2015). The development of a purification protocol to obtain a single pure compound from complex natural matrices is always laborious work and time-consuming activity. The choice of an efficient purification technique can be determinant for the natural product research success (Skalicka-Wosniak and Garrard 2014).

Due to the complex composition, poor stability and structural similarity of many terpenes from volatile oils, the preparative separation of individual constituents from essential oils is a challenging task, which is yet more difficult when the target compound demands high productivity and purity levels in a single and fast separation process (Beek and Joulain 2018).

As major EO constituents, the terpenoid/phenylpropanoid derivatives are low to intermediate polarity metabolites that usually vary at the number/position of double bonds or oxygenated groups attached to their carbon skeleton. For instance, considering their high volatility and low stability characteristics, these types of compounds are difficult to separate through conventional normal-phase silica gel column chromatography or even preparative HPLC since many volatile terpenes do not present suitable chromophores to be detected by UV (Zhang et al. 2015).

Regarding efficient preparative scale techniques for the separation/isolation of volatile compounds from essential oil, the spinning band distillation is frequently reported as being successful for this purpose (Kurniawan et al. 2019). It is a useful tool in the fractionation of essential oils as a time-saving technique that does not require the use of solvents in the fractionation, making this process less expensive and more environmentally friendly. However, it is known that in many cases, it is not able to furnish single pure compounds in one-step processed samples, making it necessary to use other chromatographic techniques (Ribeiro et al. 2018). Considering that high-purity compounds are essential for validation of analytical methods and for undertaking biological assays, the choice of an efficient and fast separation technique in one step process should be prioritized.

In this context, countercurrent chromatography, including high-speed counter-current chromatography (HSCCC) and centrifugal partition chromatography (CPC) is presented as a universal preparative chromatographic technique that permits both normal and reversed-phase operation in both isocratic or gradient elution modes, which is a useful separation technique for EO fractionation and obtention of pure isolated compounds in high purity level and low solvent consumption. CCC is ideal for the fast preparative separation of volatile compounds from essential oils once there is no solid support for stationary phase retention, thus avoiding the risk of sample denaturation. In addition, unlike plant extracts, once EO is easily soluble in organic phases, the loading capacity is much higher than in other chromatographic techniques, which is possible to inject, such as up to 2 g or more (Urbain et al. 2010) of raw EO sample in a single preparative separation run without the risk of precipitation or clogging of columns. This technique is useful in the isolation of compounds that normally have limited thermal and chemical stability, such as volatile compounds (Zhang et al. 2018). In addition, it allows performing isolation using low solvent volumes in only one separation step.

Although still sparse when compared to non-volatile metabolites (only 29 papers have been published on the isolation of compounds from EO in the last years). CCC became recognized as an efficient chromatographic tool for the short-time separation of EO compounds. Many researchers all over the world have developed different CCC methodologies and a wide range of solvent systems, including some environmentally friendly ones that have been successfully used in essential oil separations leading to the isolation of many volatile compounds (Marques et al. 2017, 2018a) and increasing the number of publications with impact in the field (Figure 1).

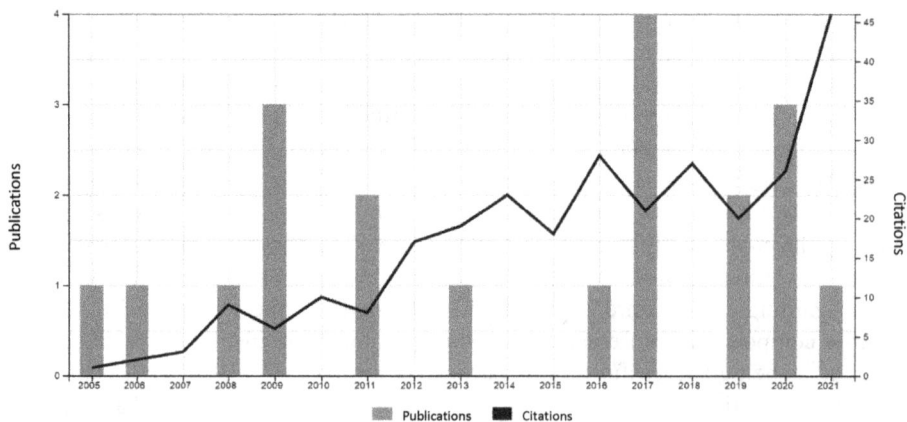

Figure 1: Number of publications X citation by year according to web of science clarivate database.

Solvent Systems Used in CCC for the Purification of Essential Oil Compounds

A successful liquid-liquid separation using countercurrent chromatography requires a careful evaluation of a suitable two-phase solvent system to provide an ideal range of partition coefficient (K) for the target substances and obtain good resolution. The simplest and most frequently used test for guiding the choice of the solvent system is the shake-flask experiment, which consists of dissolving the sample in the biphasic solvent system and shaking the flask to allow the compounds to partition between the two phases. After that, equal amounts of both phases are spotted side by side on thin-layer chromatography (TLC) plate, and K can be visually estimated after elution of the TLC with an appropriate solvent system. K can also be calculated by other chromatographic methods, such as GC or HPLC.

Some basic requirements for a suitable two-phase solvent system (Marston and Hosttetmann 1994, Marston and Hosttetmann 2006) include (a) the settling time of the solvent system should be shorter than the 30 s (to ensure satisfactory retention of the stationary phase); (b) the partition coefficient (K) of the target compounds should lie in the range $0.5 < K < 2.0$, and the separation factor between any two components $(\alpha = K2/K1, K2 > K1)$ should be greater than 1.5; (c) to avoid wasting solvents, the two-phase solvent system under consideration should produce more mobile phase than a stationary phase, especially when multiple injections are made without changing the stationary phase.

Being an all-liquid chromatographic system, CCC can select from an almost infinite range of possible two-phase solvent systems. Most of the work on the separation and isolation of volatile compounds from essential oils by CCC has focused on terpenoids and phenylpropanoids, which are the most diverse and biologically important classes of natural products present in EO. Literature shows a great number of possible combinations of solvents for this purpose, as can be seen in Table 1, which have demonstrated to be efficient in the isolation of different classes of compounds, such as monoterpene aldehydes (cuminaldehyde, neral and geranial), mono and sesquiterpene alcohols (terpineol; linalool; geraniol; nerolidol; patchoulol), mono and sesquiterpene ketones (camphor; cyperone; selina-1,3,7(11)-trien-8-one), sesquiterpene lactones (senkyunolide I; senkyunolide H), phenylpropanoids (chavibetol; methyleugenol; eugenol; thymol; carvacrol), coumarins (coriandrone A; coriandrone B), etc.

As most volatile compounds have strong hydrophobic properties, non-aqueous solvent systems are often used in the CCC purifications of essential oil compounds as shown in Table 1. Nevertheless, aqueous solvent systems have also been used, especially in the case of purification of oxygenated derivatives, such as monoterpene and sesquiterpene alcohols/ketones, coumarins and phenylpropanoids. The following non-aqueous solvents systems are the most frequently reported and may be used as a starting point for further optimization of solvent ratios and/or upper or lower phase modifiers in the biphasic systems; hexane (or heptane, light petroleum ether)-acetonitrile and hexane-methanol (or ethanol) are modified by chlorinated solvents like chloroform or dichloromethane, acetone or ethyl acetate. One example of solvent system optimization is the obtention of β-caryophyllene from the leaf essential oil of *Vitex negundo* L. var. *heterophylla* (Franch.) Rhed. (Xie et al. 2008). The optimization of the solvent ratios of *n*-hexane-chloroform-acetonitrile from 6:2:5 to 10:3:7 v/v, and replacement of chloroform for dichloromethane has resulted in the purification of β-caryophyllene from the other essential oil components in higher purity and recovery rate.

Table 1: Solvent systems used for the purification of essential oil compounds by countercurrent chromatography (CCC).

Class of Compound	Isolated Compound	Plant Species	Solvent System	Reference
Monoterpenes	Linalool	*Pimpinella anisum*	Hep-EtOAc-MeOH-H$_2$O (5:2:5:2)	Skalicka-Woźniak et al. 2014
		Piper claussenianum	Hex-ACN (1:1)	Marques et al. 2017
	Camphor	*Piper molicomum*	Hex-ACN-EtOAc (1:1:0.4)	Marques et al. 2018a
		Curcuma wenyujin	Hex-ACN-EtOH (5:3:2)	Wang et al. 2020
	Camphene	*Piper molicomum*	Hex-ACN-EtOAc (1:1:0.4)	Marques et al. 2018a
	Bornyl acetate	*Piper molicomum*	Hex-ACN-EtOAc (1:1:0.4)	Marques et al. 2018a
	1,8-Cineole	*Curcuma wenyujin*	PET-ACN-ACE (4:3:1)	Dang et al. 2010
		Curcuma wenyujin	Hex-ACN-EtOH (5:3:2)	Wang et al. 2020
	Neral + geranial (citral)	*Pectis brevipedunculata*	Hex-ACN (1:1)	Marques and Kaplan 2013
	Geraniol	*Pectis brevipedunculata*	Hex-ACN (1:1)	Marques and Kaplan 2013
	Thymol	*Origanum vulgare*	Hex–MTBE-ACN (1:0.1:1)	Puertas et al. 2002
	Carvacrol	*Origanum vulgare*	Hex–MTBE-ACN (1:0.1:1)	Puertas et al. 2002
	Cuminaldehyde	*Cuminum cyminum*	Hex-MeOH-H$_2$O (5:4:1)	Chen et al. 2011
	Terpinen-4-ol	*Pimpinella anisum*	Hep-EtOAc-MeOH-H$_2$O (5:2:5:2)	Skalicka-Woźniak et al. 2014
	P-Menta-1,4-dien-7-al	*Cuminum cyminum*	Hex-MeOH-H$_2$O (5:4:1)	Chen et al. 2011
	Terpinen-4-ol	*Pimpinella anisum*	Hep-EtOAc-MeOH-H$_2$O (5:2:5:2)	Skalicka-Woźniak et al. 2014
	α-Terpineol	*Pimpinella anisum*	Hep-EtOAc-MeOH-H$_2$O (5:2:5:2)	Skalicka-Woźniak et al. 2014
Phenylpropanoids	Zingerone	*Zingiber officinale*	Hex-MeOH-H$_2$O (3:2:1)	Wang et al. 2020
	6-Gingerol	*Zingiber officinale*	Hex-EtOAc-MeOH-H$_2$O (7:3:5:5)	Wang et al. 2020
		Zingiber officinale	Hex-EtOAc-MeOH-H$_2$O (10:2:5:7)	Gan et al. 2016
	Eugenol	*Eugenia caryophyllata*	Hex-EtOAc-MeOH-H$_2$O (1:0.5:1:0.5)	Geng et al. 2007

	Chavibetol	Hex-BuOH-MeOH-H$_2$O (12:4:4:3)	Dos Santos et al. 2009
	Anethole	Hep-MeOH (1:1)	Skalicka-Woźniak et al. 2014
		Pimpinella anisum	
	Anethole	Illicium verum	Wang et al. 2011b
		Hex-EtOAc-MeOH-H$_2$O (1:0.2:1:0.1)	
		Foeniculum vulgare	Xiao et al. 2007
		Hex-EtOAc-MeOH-H$_2$O (5:2:5:2)	
	Methyleugenol	Pimenta pseudocaryophyllus	Dos Santos et al. 2009
		Hex-BuOH-MeOH-H$_2$O (12:4:4:3)	
		Illicium verum	Wang et al. 2011b
		Hex-EtOAc-MeOH-H$_2$O (1:0.2:1:0.1)	
	α-Asarone	Acorus tatarinowii	Wang et al. 2011a
		Hex-EtOAc-MeOH-H$_2$O (1:0.2:1:0.3)	
	β-Asarone	Acorus tatarinowii	Wang et al. 2011a
		Hex-EtOAc-MeOH-H$_2$O (1:0.2:1:0.3)	
	Foeniculin	Pimpinella anisum	Skalicka-Woźniak et al. 2014
		Hep-EtOAc-MeOH-H$_2$O (5:2:5:2)	
Benzyl metabolites	Anisaldehyde	Illicium verum	Wang et al. 2011b
		Hex-EtOAc-MeOH-H$_2$O (1:0.2:1:0.1)	
		Hex-EtOAc-MeOH-H$_2$O (5:2:5:2)	Xiao et al. 2007
Sesquiterpene lactones	Senkyunolide I	Rhizoma chuanxiong	Xiong et al. 2014
		BuOH-ACOH-H$_2$O (4:1:5)	
	Senkyunolide H	Rhizoma chuanxiong	Xiong et al. 2014
		BuOH-ACOH-H$_2$O (4:1:5)	
	Senkyunolide A	Ligusticum chuanxiong	Liu et al. 2010
		Hex-EtOAc-EtOH-H$_2$O (1:1:1:1)	
		Ligusticum chuanxiong	Zhang et al. 2006
		Hex-EtOAc-MeOH-H$_2$O-ACN (8:2:5:5:3)	
	(Z)-Ligustilide	Ligusticum chuanxiong	Liu et al. 2010
		Hex-EtAOc-EtOH-H$_2$O (1:1:1:1)	
		Ligusticum chuanxiong	Zhang et al. 2006
		Hex-EtOAc-MeOH-H$_2$O-ACN (8:2:5:5:3)	
Alkyl lactones (α-Pyrones)	(R)–(−)–C-10 Massoia Lactone	Cryptocarya massoy	Urbain et al. 2010
		Hex-MeOH-H$_2$O (10:9:1)	
	(R)–(−)–C-12 Massoia Lactone	Cryptocarya massoy	Urbain et al. 2010
		Hex-MeOH-H$_2$O (10:9:1)	
	(R)–(−)–C-14 Massoia Lactone	Cryptocarya massoy	Urbain et al. 2010
		Hex-MeOH-H$_2$O (10:9:1)	
Coumarins	Coriandrone B	Coriandrum sativum	Chen et al. 2009
		Hex-EtOAc-MeOH-H$_2$O (3:7:5:5)	
	Coriandrin	Coriandrum sativum	Chen et al. 2009
		Hex-EtOAc-MeOH-H$_2$O (3:7:5:5)	

Table 1 contd...

...Table 1 contd.

Class of Compound	Isolated Compound	Plant Species	Solvent System	Reference
	Dihydrocoriandrin	*Coriandrum sativum*	Hex-EtOAc-MeOH-H$_2$O (3:7:5:5)	Chen et al. 2009
	Coriandrone A	*Coriandrum sativum*	Hex-EtOAc-MeOH-H$_2$O (3:7:5:5)	Chen et al. 2009
Sesquiterpenes	δ-Elemene	*Curcuma wenyujin*	Hex-ACN-EtOH (5:3:2)	Wang et al. 2020
	β-Elemene	*Curcuma wenyujin*	Hex-ACN-EtOH (5:3:2)	Wang et al. 2020
		Curcuma wenyujin	PET-ACN-ACE (4:3:1)	Dang et al. 2010
		Nigella damascena	PET-ACN-ACE (2:1.5:0.5)	Sieniawska et al. 2018
	Curzerene	*Curcuma wenyujin*	Hex-ACN-EtOH (5:3:2)	Wang et al. 2020
		Curcuma wenyujin	PET-ACN-ACE (4:3:1)	Dang et al. 2010
	Nerolidol	*Piper clausenianum*	Hex-ACN (1:1)	Marques et al. 2017
		Piper molicomum	Hex-ACN-EtOAc (1:1:0.4)	Marques et al. 2018a
		Baccharis dracunculifolia	Hex-MeOH-H$_2$O (5:4:1)	Queiroga et al. 2014
	Selina-1,3,7(11)-trien-8-One	*Eugenia uniflora*	Hex-ACN (1:1)	Marques et al. 2018b
	Selina-1,3,7(11)-trien-8-One epoxide	*Eugenia uniflora*	Hex-ACN (1:1)	Marques et al. 2018b
	α-Cyperone	*Cyperus rotundus*	Hex-EtOAc-MeOH-H$_2$O (1:0.2:1.1:0.2)	Shi et al. 2009
	ar-Turmerone	*Curcuma longa*	Hep-EtOAc-ACN-H$_2$O (9.5/0.5/9/1)	Zhou et al. 2017
	b-Turmerone	*Curcuma longa*	Hep-EtOAc-ACN-H$_2$O (9.5/0.5/9/1)	Zhou et al. 2017
	(*E*)-α-Atlantone	*Curcuma longa*	Hep-EtOAc-ACN-H$_2$O (9.5/0.5/9/1)	Zhou al. 2017
	α-Turmerone	*Curcuma longa*	Hep-EtOAc-ACN-H$_2$O (9.5/0.5/9/1)	Zhou et al. 2017
	Nootkatone	*Alpinia oxyphylla*	Hex-MeOH-H$_2$O (5:4:1)	Xie et al. 2009

Compound	Species	Solvent system	Reference
b-Caryophyllene	*Vitex negundo*	Hex-CH$_2$Cl$_2$-ACN (10:3:7)	Xie et al. 2008
	Flaveria bidentis	Hex-ACN-EtOH (5:4:3)	Wei et al. 2008
Caryophyllene oxide	*Flaveria bidentis*	Hex-ACN-EtOH (5:4:3)	Wei et al. 2008
β-Farnesene	*Flaveria bidentis*	Hex-ACN-EtOH (5:4:3)	Wei et al. 2008
Germacrone	*Curcuma wenyujin*	PET-EtOH-Et2O-H$_2$O (5/4/0.5/1)	Yan et al. 2005
	Curcuma wenyujin	PET-ACN-ACE (4:3:1)	Dang et al. 2010
Curdione	*Curcuma wenyujin*	PET-EtOH-Et2O-H$_2$O (5/4/0.5/1)	Yan et al. 2005
	Curcuma wenyujin	PET-ACN-ACE (4:3:1)	Dang et al. 2010
Curcumol	*Curcuma wenyujin*	PET-ACN-ACE (4:3:1)	Dang et al. 2010
Patchoulol	*Pogostemom cabin*	PET-ACN (1:1)	Li et al. 2011

ACE = Acetone; ACN = Acetonitrile; ACOH = Acetic Acid; BuOH = Butanol; EtOAc = Ethyl Acetate; Et$_2$O = Ethyl Ether; EtOH = Ethanol; Hep = Heptane; Hex = Hexane; MeOH = Methanol; MTBE = Methyl *t*-Butyl Ether; PET = Light Petroleum Ether.

Examples of aqueous solvent systems are hexane (or petroleum ether)-ethanol (or methanol)-H$_2$O, hexane (or petroleum ether)-ethyl acetate-methanol-H$_2$O and hexane-butanol-methanol-H$_2$O (Table 1). However, most solvent systems are composed of protic solvents including water, which normally impacts the volatile sample recovery when we focus on the isolation of volatile metabolites. The choice of some non-protic volatile solvents may represent an advantage in the isolation and recovery process. Once both upper and lower phases can be fast and easily evaporated, a minimum loss would occur during the purification process and sample drying. Thus, the combination of non-protic solvents could be a post benefit for CCC users to ensure a high recovery rate compared to solvent systems containing protic and aqueous phases.

Selected Examples

Many solvent systems, including aqueous and non-aqueous systems, can be used in CCC in different elution modes. Strategies for using CCC in a rational and targeted manner to achieve the isolation and purification of economically or therapeutically important natural products from volatile oils are discussed.

Purification of Essential Oil Compounds from Plants of the Brazilian Native Flora with Non-Aqueous Solvent Systems

Non-aqueous systems may represent a suitable alternative for the purification of volatile and low to intermediate polarity compounds because of their higher volatility when compared to aqueous systems. In previous work from our group, five different non-aqueous solvent systems, such as *n*-hexane-acetonitrile (1:1, v/v), *n*-hexane-methanol (1:1, v/v), *n*-hexane-acetonitrile-ethyl acetate (1:1:0.4, v/v), *n*-hexane-acetonitrile-methanol (1:1:0.5, v/v) and petroleum ether-acetonitrile-acetone (4:3:1, v/v), were successfully used and proved to be effective in the isolation of several different compounds from essential oils of Brazilian native plants (Marques et al. 2013, Marques et al. 2017, 2018a, Marques et al. 2018b). As the target compounds have considerable volatility, the choice of those solvents was thought to obtain not only the best resolution but also the maximum recovery rate after separation. In all separation experiments, the two solvent systems *n*-hexane-acetonitrile (Marques et al. 2013, Marques et al. 2017, 2018a) and *n*-hexane-acetonitrile-ethyl acetate (1:1:0.4) (Marques et al. 2018b) presented the best resolution of compounds and mass recovery when compared to other systems containing methanol and water. Hexane-acetonitrile-ethyl acetate 1:1:0.4 was also effectively used in the fractionation of the oleoresin of *Copaifera glycycarpa* Ducke (de Souza et al. 2010), and the so-called Copaiba oil was used in Brazilian folk medicine as a healing, antiseptic and anti-inflammatory agent. By using the mentioned solvent system, de Souza and co-workers described the isolation of two bioactive diterpenes, kaurenoic and polyalthic acids, which were separated from the sesquiterpene fraction and obtained in high purity and recovery rates.

The binary solvent system *n*-hexane-acetonitrile has proven to be quite good for the CCC separation of mono- and sesquiterpenes, showing a good distribution of compounds between the two phases (K close to 1). In addition, this biphasic system presents similar volumes of upper and lower phases after equilibrium and a short settling time of the two phases, which is imperative in CCC. This similar distribution volume can also be beneficial when both elution modes (tail-to-head and head-to-tail) are performed as both phases can be used as mobile phases. In addition, the stationary phase retention, which is directly related to the resolution and efficiency of the separation, is normally up to 85% in a tail-to-head elution mode with the *n*-hexane-rich phase as mobile. On the other hand, when the *n*-hexane-rich phase acts as stationary, in head-to-tail elution mode, up to 77% of the stationary phase is retained. It is known that a successful separation process depends on the stationary phase retention in the column, which should be preferred over 50% to achieve the best resolution among the target metabolites (Berthod and Hassoun 2006).

To illustrate the above findings, we present the HSCCC preparative separation in both elution modes of the major *Eugenia uniflora* L. chemical markers in one step. *E. uniflora* L. (Myrtaceae) is commonly known in Brazil as *pitangueira*, and its fruits are important in the food industry. Two sesquiterpenes have been described as the main compounds in *E. uniflora* leaf essential oil from the red-orange fruit biotype, which was characterized by a high content of selina-1,3,7(11)-trien-8-one (32.6 ± 2.1%) (Compound A in Figure 2) and selina-1,3,7(11)-trien-8-one epoxide (24.5 ± 1.7%) (Compound B in Figure 2) among other 62 compounds (Marques et al. 2018a).

According to Marques and co-workers, all the initially tested solvent systems composed of *n*-hexane-acetonitrile modified by either ethyl acetate or methanol, and *n*-hexane-methanol presented K between 0.5 and 2.0, which is suitable for a CCC separation. However, the binary system *n*-hexane-acetonitrile (1:1) was chosen for its simpler composition as well as for the considerable difference between the K values of the targeted sesquiterpenes (meaning a good selectivity; $\alpha = K2/K1$ greater than 1.5), which were 0.91 and 1.55 in the tail-to-head mode (normal elution mode) and 1.09 and 0.65 in the head-to-tail mode (reversed-phase elution mode) for selina-1,3,7(11)-trien-8-one and selina-1,3,7(11)-trien-8-one epoxide, respectively. It also presented the best volume phase ratio and settling time among all the solvent systems tested. By using the acetonitrile-rich phase as stationary in an isocratic tail-to-head normal elution mode, it was possible to obtain 78.2 mg of (+/–)-selina-1,3,7(11)-trien-8-one epoxide (97.5% purity) from 800 mg of crude EO after a single 80 min run. The other isomer instead was obtained with a purity lower than 80%. On the other hand, when using the hexane-rich phase as stationary in a head-to-tail reversed elution mode, only 46.3 mg of racemic (+/–)-selina-1,3,7(11)-trien-8-one epoxide (93.1% purity) and 24.7 mg of racemic (+/–)-selina-1,3,7(11)-trien-8-one (92.0% purity) were obtained after a 40 min run (Figure 2). Even with lower recovery rates in the reversed-phase mode, the purity for the ketone isomer was higher than that obtained in the other elution mode.

In another study, the preparative purification of monoterpene isomers from the citral-rich essential oil (with a lemon-grass odor) from a Brazilian ornamental and aromatic grass, *Pectis brevipedunculata* (Gardner) Sch. Bip. (Asteraceae), was successfully achieved by HSCCC again with the binary hexane-acetonitrile solvent

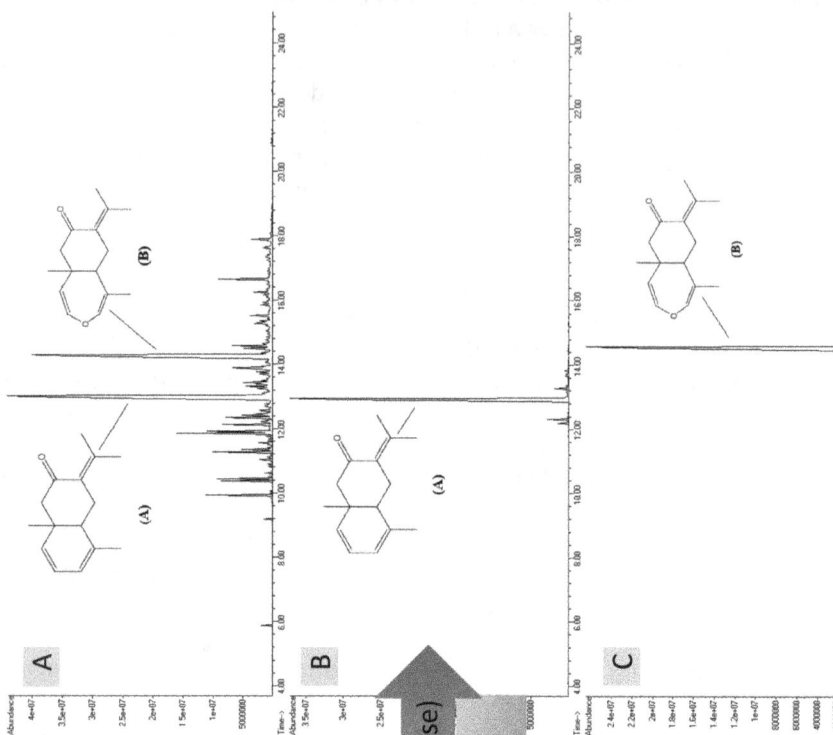

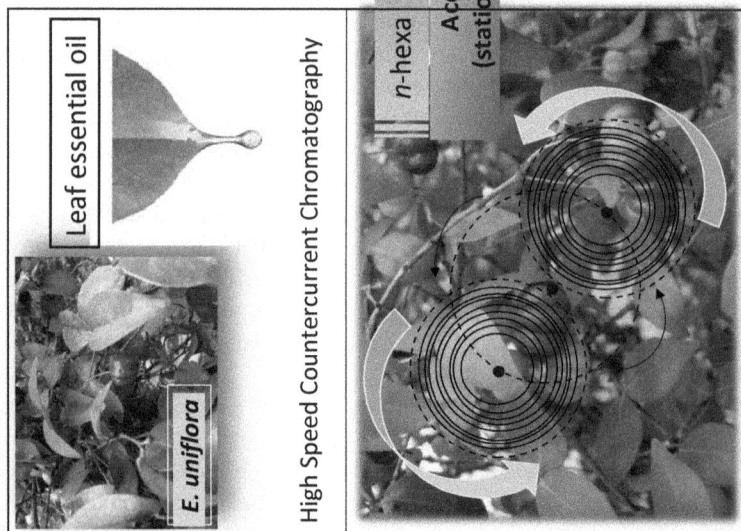

Figure 2: GC-FID chromatograms of *Eugenia uniflora* L. leaf essential oil showing the metabolites separation by HSCCC. (A) Crude *E. uniflora* leaf essential oil. (B) Fractions (33–35) containing (+/–)-Selin-1,3,7(11)-Trien-8-One. (C) Fractions (72–80) containing (+/–)-Selin-1,3,7(11)-Trien-8-One epoxide.

system (Marques et al. 2013). The high percentage of citral (mixture of the isomers neral and geranial) in the EO of *P. brevipedunculata* (up to 80% throughout the whole year) motivated further purification investigations. Chemical analysis of the EO contents revealed a high percentage of neral (32.7%) and geranial (49.2%), followed by the monoterpenes limonene (4.7%), α-pinene (3.4%) and by the cis- and trans-alcohol derivatives nerol and geraniol (1.49% and 5.1%, respectively). The preparative purification of 1.0 g of the crude essential oil containing 81.9% of citral by isocratic elution using the acetonitrile-rich phase as stationary afforded 780 mg of citral with a purity higher than 98.7% and a recovery rate of 84.9% in a single two hours chromatographic run (Figure 3C), along with geraniol (8.0 mg, 86% purity). Once again, the *n*-hexane-acetonitrile solvent system showed to be selective for citral (target compound) instead of all other minor essential oil constituents, even the monoterpene alcohol geraniol which was completely removed from the mixture. In this case, despite terpene structural and polarity similarities, HSCCC proved to be effective in separating citral isomers from the EO in a short time. This high percentage of mass recovery would probably not be possible using solid-liquid techniques, avoiding possible reactions with silica gel.

A second CCC run was intended to propose the separation of the *cis/trans* aldehyde isomers of citral (citral/geranial) using the same binary phases. To improve resolution between the target compounds the flow rate was reduced from 2.0 ml/min to 1.0 ml/min and the number of collected fractions was doubled. Unfortunately, the liquid-liquid partition was not able to completely separate the geometric isomers at a high selectivity level. Although in good purities, only enriched samples of 7.0 mg of neral and 12.0 mg of geranial were recovered (87.5% and 91.0% respectively; Figure 3D).

The following examples show the useful, efficient and fast preparative purification of compounds from *Piper* essential oils (Marques et al. 2017, Marques et al. 2018b). Species of this genus produce a variety of bioactive EO, which is being recognized as a source of many natural compounds with promising biological activities and economic importance (Marques et al. 2015, Peixoto et al. 2021).

HSCCC was used for the preparative isolation of the major biological markers from the leishmanicidal EO of the inflorescences of *P. claussenianum* (Miq.) C. DC.; Piperaceae is mainly composed of two terpene alcohols that are the monoterpene linalool (racemic mixture, 53.5%) and the sesquiterpene (+/–)-(*E*)-nerolidol (24.3%), representing 77.8% of the crude oil among other minor compounds (Marques et al. 2017, Marques et al. 2018b). These two terpene alcohols are economically important in the industry of flavors and fragrance due to their flower aroma in which nerolidol is appreciated as a base note in perfumery for its long-lasting odor (Marques et al. 2017). The *K* values of these compounds were calculated by GC-MS in the systems hexane-acetonitrile, hexane-methanol and hexane-acetonitrile modified by ethyl acetate or methanol aiming their purification by either normal (tail-to-head) or reversed (head-to-tail) elution mode. The two terpene alcohols showed a high affinity for the methanol-rich phase in the systems containing methanol, providing either high or low *K* value and low selectivity and are thus not suitable for the purification of the targets. Again, as in the case of *E. uniflora* EO, both hexane-acetonitrile and

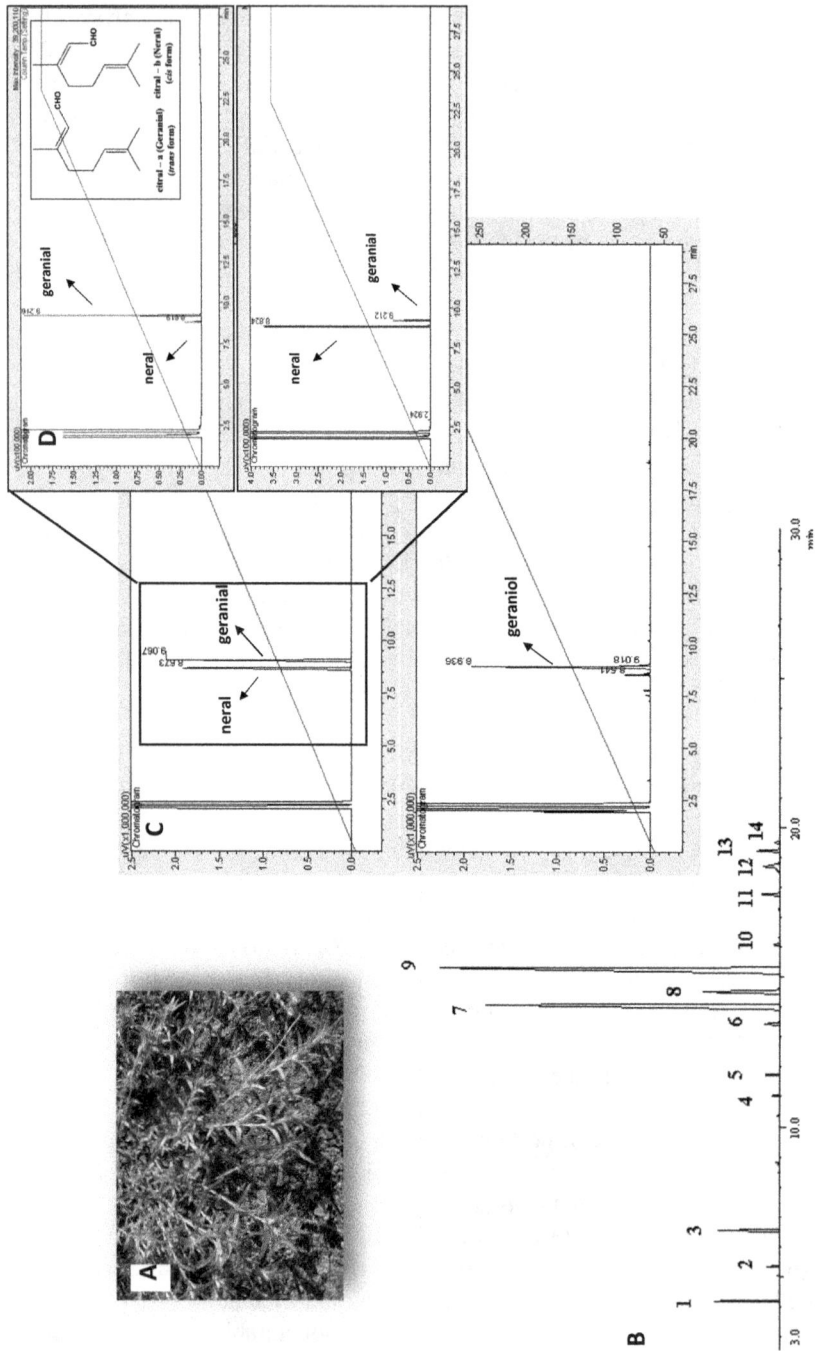

Figure 3: (A) *Pectis brevipedunculata* (Gardner) Sch. Bip. (B) GC-FID chromatographic profile of *P. brevipedunculata* crude essential oil. Major identified volatile compounds present in the essential oil: (1) α-pinene; (2) Limonene; (4) α-pinene epoxide; (6) Nerol; (7) Neral; (8) Geranial; (9) Geraniol; (11) Neryl acetate; (12) 4-Isopropylcyclohexanol; (14) Geranyl acetate. (C) Citral (Neral + Geranial) and Geraniol separation. (D) Fractionation of citral isomers by HSCCC.

hexane-acetonitrile-ethyl acetate (1:1:0.4) presented good K values for the targets, and their purification was then successfully performed using hexane-acetonitrile as the phase ratio for this system was better suitable for the two elution modes (K values of 0.56 and 1.20 in the tail-to-head mode, and 1.76 and 0.82 in head to the tail mode for linalool and (E)-nerolidol respectively). By injecting 1.0 g of the crude essential oil after optimization of chromatographic conditions, the one-step CCC separation in normal elution mode afforded two main fractions, one containing linalool (racemic mixture, 320.0 mg and 96.2% purity) and the second containing (E)-nerolidol (racemic mixture, 95.0 mg and 92.0% purity) (Figure 4). The obtention of (E)-nerolidol in high purity and recovery rate (92.5% and 84%, respectively) has also been reported from the essential oil of *Baccahris dracunculifolia* DC (Asteraceae) by using the aqueous solvent system hexane-methanol-water in normal phase elution mode (Queiroga et al. 2014).

In another *Piper* study, the essential oil from leaves of *P. mollicomum* Kunth was successfully fractionated by HSCCC (Marques et al. 2018b). The leaf EO was characterized by its high percentage of camphor derivatives, such as camphor itself

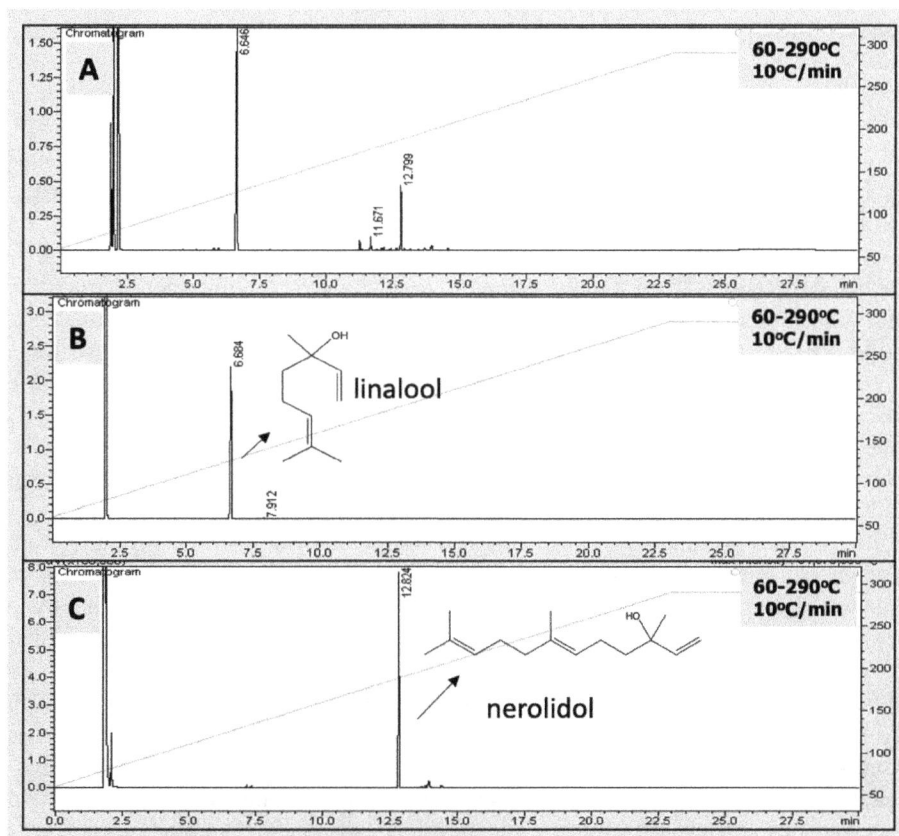

Figure 4: GC-FID chromatograms of *P. claussenianum* inflorescence essential oil and HSCCC fractions containing the target compounds. (A) Crude essential oil, (B) HSCCC fractions containing (+/–)-linalool and (C) HSCCC fractions containing (+/–)-(E)-nerolidol.

(39.9 ± 1.9%) and camphene (25.3 ± 2.1%) as major components followed by bornyl acetate (2.4 ± 0.2%) and nerolidol (7.5 ± 1.1%); all of them have economic value. The same systems tested for the EO from the inflorescences of *P. claussenianum* were tested here as a start point added by a system with acetone as a modifier, such as petroleum ether-acetonitrile-acetone 4:3:1. Some of the biphasic systems afforded $K \ll 0.5$, for instance, for camphene (0.09) in *n*-hexane-methanol (1:1), which means that the target compounds would be eluted too fast by the mobile phase instead of being retained in the stationary phase. In some cases, K values between the target compounds were very close, compromising the resolution of camphor and bornyl acetate, such as in *n*-hexane-acetonitrile-ethyl acetate 1:1:0.4 (K values 0.87 and 0.70, respectively). Based on K values for the target compounds, the best separation results were achieved by using *n*-hexane-acetonitrile 1:1 and *n*-hexane-acetonitrile-ethyl acetate 1:1:0.4. The insertion of ethyl acetate in the *n*-hexane-acetonitrile system increased the partition coefficient of camphene from 0.07 to 0.37, improving its retention in the coil and thus enabling a better separation and its obtention in a purity level of 82% in one step. Also, both K of camphor (1.13 to 1.40) and bornyl acetate (0.85 to 1.20) were increased, improving the distribution of these metabolites between the two solvent phases. Thus, under the optimized condition, an isocratic elution was conducted with 1.4 g of the EO using *n*-hexane-acetonitrile-ethyl acetate 1:1:0.4 in a one-step 120 minutes process in tail-to-head (normal) elution mode, which afforded 150.0 mg of camphene (82.0% purity), 85.0 mg of camphor (98.5% purity), 16.2 mg of bornyl acetate (91.2% purity), which is one of the minor compounds in the mixture, and 100.0 mg of (*E*)-nerolidol (92.0% purity). See the following Figure 5.

The last example of purification of essential oil compounds with non-aqueous solvent systems is the isolation of triquinane sesquiterpenes from the fern *Anemia tomentosa* var. *anthriscifolia* (Schrader) Mickel (Anemiaceae). Triquinane sesquiterpenes are characterized by a tricyclic structure with either three cyclopentane fused rings or two cyclopentane and one cyclohexane fused ring. These fused rings give rise to a series of stereocenters in the molecule, providing epimeric, enantiomeric and regioisomeric relationships (Kutateladze and Kuznetsov 2017). These compounds have aroused the interest of researchers to synthesize them (Hong and Stoltz 2014, 2012) and their derivatives (Syntrivanis et al. 2020).

In a previous study by our group, the composition of the bioactive EO from *A. tomentosa* var. *anthriscifolia* was investigated, showing a diverse and large amount of triquinane sesquiterpenes, presilphiperfolane and silphiperfolane-type sesquiterpenes that are most abundant in its composition (Pinto et al. 2009a). In that study, we performed the fractionation of the EO by silica-gel column chromatography to isolate its major constituent, which was not readily identified by comparison with the literature mass spectral data nor by comparison with the reference mass spectra from the Wiley library. The major constituent of the EO, 9-*epi*-presilphiperfolan-1β-ol (Pinto et al. 2009b, Joseph-Nathan et al. 2010) was obtained in high purity in two fractions (91 and 99% purities by GC-FID), but none of the other constituents could be purified. Thus, the HSCCC fractionation of 500 mg of the EO from the aerial parts of this fern was performed with the solvent system *n*-hexane-acetonitrile

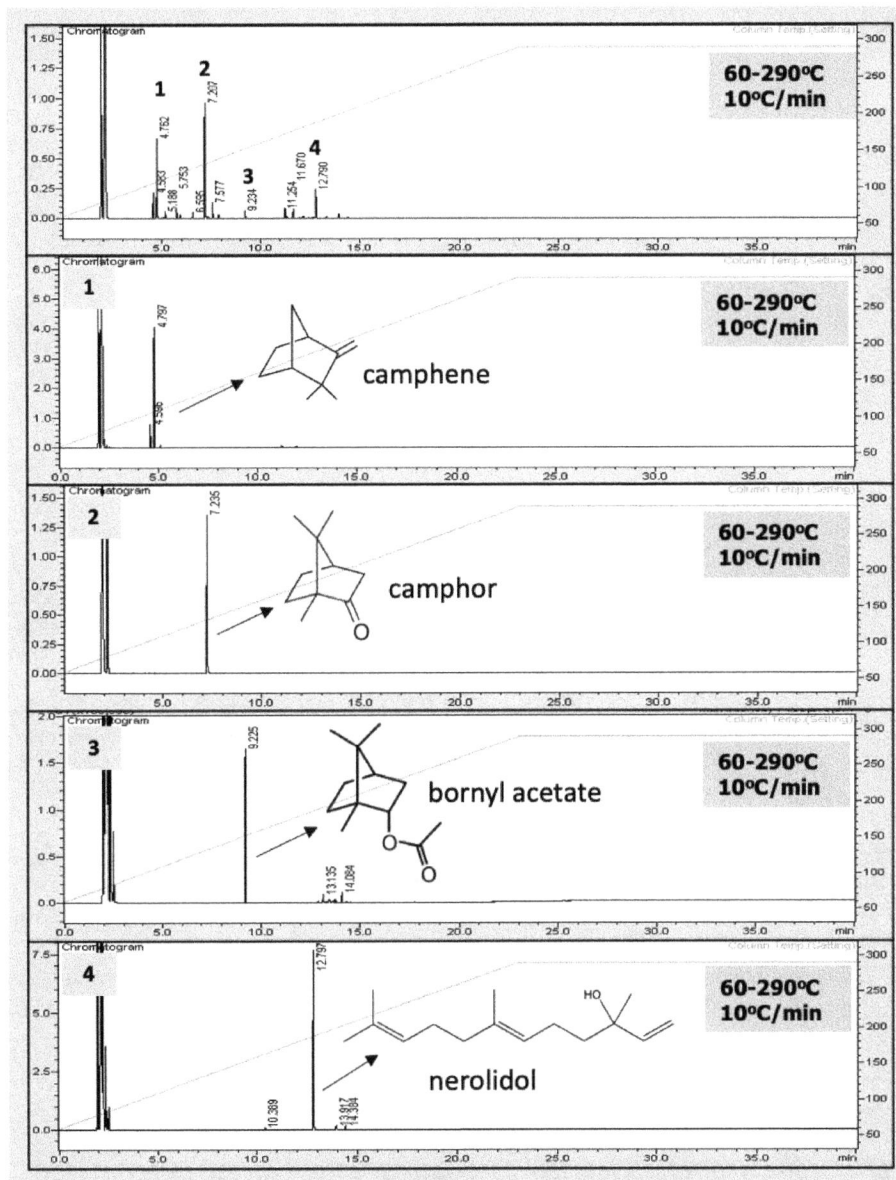

Figure 5: GC-FID chromatograms of the essential oil of leaves of *Piper mollicomum* Kunth and HSCCC fractions containing the target compounds. (1) HSCCC fractions containing camphene. (2) HSCCC fractions containing camphor. (3) HSCCC fractions containing bornyl acetate. (4) HSCCC fractions containing nerolidol.

in the head-to-tail mode (reversed-phase elution mode), affording 9 fractions (Figure 6).

The major constituent, 9-*epi*-presilphiperfolan-1β-ol was isolated in fraction AT6 with 95.6% purity by GC/MS (LRI$_{calc}$ = 1514) along with its isomer presilphiperfolan-

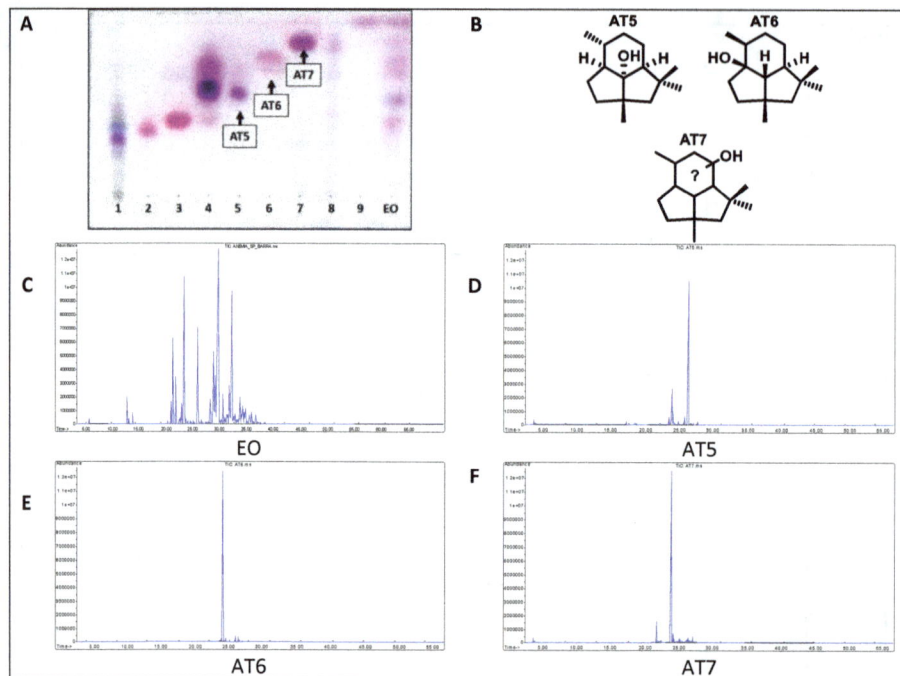

Figure 6: Chromatographic profile of the fractions obtained by HSCCC purification of the EO from *Anemia tomentosa* var. anthriscifolia (Schrader) Mickel. A: TLC profile of HSCCC fractions (silica-gel; Hex/EtOAc, 9.5:0.5, v/v); B: Structures of the isolated compounds; C: GC-MS profile of the EO; D: GC-MS profile of fraction AT5; E: GC-MS profile of fraction AT6; F: GC-MS profile of fraction AT7.

8-ol in fraction AT5 (76.9%; LRI_{calc} = 1579). These presilphiperfolane type sesquiterpenes were obtained in a single step with a recovery rate of 80% and 67%, respectively. Additionally, another sesquiterpenoid was isolated in fraction AT7 (89.1%, LRI_{calc} = 1501) and was initially identified as β-dihydroagarofuran by comparing its mass spectra with literature data (Adams 2007). However, its multiplicity and chemical shift at ^{13}C and ^{1}H NMR spectra are consistent with presilphiperfolane-type sesquiterpene alcohol. This structure can have five stereocenters and needs more studies for its unambiguous structure identification (Joseph-Nathan et al. 2010).

Purification of Target Bioactive Compounds with Aqueous Solvent Systems

Although non-aqueous systems may be considered a first choice for the purification of volatile and medium to low polarity compounds from essential oils due to their higher volatility and ease in recovering samples, it is possible to note from Table 1 that more than 65% of the purifications described in the literature use aqueous solvent systems for this purpose. The most cited solvent systems are *n*-hexane (or heptane)-methanol-water and *n*-hexane-ethyl acetate-methanol-water, commonly abbreviated as HEMWat. These aqueous solvent systems, however, have been used for the purification of rather more poplar compounds, such as phenylpropanoids, monoterpene alcohols, sesquiterpene lactones, some sesquiterpene ketones and coumarins (Table 1). The following selected examples show the purification of bioactive compounds from ginger, a functional food (Wang et al. 2020), as well as other phenylpropanoids from *Pimenta pseudocaryophyllus* (Gomes) Landrum, which is popularly used in Brazil as a flavoring agent in *cachaça* (a sugar cane spirit) (dos Santos et al. 2009) and the fragrant massoia lactones from the crude massoia (*Cryptocarya massoy* Kosterm) bark oil by hydrostatic CCC (Urbain et al. 2010).

The essential oil from ginger rhizomes, obtained by supercritical fluid extraction was fractionated by HSCCC with three different solvent systems, each of them optimized according to the target compound's *K* (6-gingerol, zingerone and a sesquiterpene mixture; Figure 7). The following solvent systems were tested; petroleum ether-ethyl acetate-methanol-water 5:5:3:4 and 6:4:3:7; *n*-hexane-ethyl acetate-methanol-water 7:3:5:5 and 8:3:6:5; *n*-hexane-methanol-water 3:2:1; and the non-aqueous systems *n*-hexane-chloroform-acetonitrile 6:2:5 and *n*-hexane-dichloromethane-acetonitrile 10:3:7. The two more polar phenylpropanoids 6-gingerol and zingerone were purified, respectively, with *n*-hexane-ethyl acetate-methanol-water 7:3:5:5 and *n*-hexane-methanol-water 3:2:1 as these solvent systems gave the best *K* values for these compounds in the range $0.5 < K < 2.0$. Despite petroleum ether-ethyl acetate-methanol-water 6:4:3:7 giving suitable *K* for the target phenylpropanoids, and their value was very similar and would not give a good resolution between them. The other systems would give for both compounds either too low or too high *K* values. On the other hand, the sesquiterpene mixture was obtained with the non-aqueous solvent system *n*-hexane-chloroform-acetonitrile 6:2:5. Other than optimizing solvent system composition, the authors also optimized loading capacity in respect to resolution and stationary phase retention, rotational speed and flow rate. All separations were performed in the head-to-tail (reversed-phase) elution mode, affording 35 mg of 6-gingerol, 23 mg of zingerone, and 105 mg of sesquiterpenes (mainly composed of zingiberene, β-sesquiphellandrene, and curcumene) with purities (by HPLC) of 98.6% and 99.4% for the isolated phenylpropanoids from a 350 mg sample.

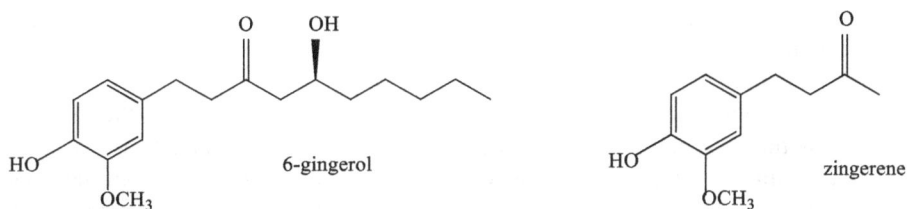

Figure 7: Chemical structures of 6-gingerol and zingerene.

Another example of the purification of bioactive phenylpropanoids with aqueous solvent systems is the obtention of chavibetol and methyleugenol from the essential oil of *Pimenta pseudocaryophyllus*. The plant is found in two Brazilian biomes (*caatinga* and *cerrado*) as well as in some regions of the Atlantic Forest (dos Santos et al. 2009). The main compounds of its EO are the above-mentioned compounds where chavibetol occurs at 34–68%, whereas methyleugenol at 11–20%. The authors reported that previous purifications of this EO by preparative TLC led to a 39% loss of the target chavibetol and so CCC was used as an efficient alternative. The K values of chavibetol and methyleugenol in several aqueous biphasic systems, such as HEMWat and its variations where methanol was replaced by ethanol or ethyl acetate by butanol, hexane-methanol-water and hexane-ethanol-water were initially calculated by GC-MS in several ratios for further optimization. Hexane-butanol-methanol-water 1:1:1:1 v/v was the one that gave partition coefficients near 1 for both chavibetol (1.48) and methyleugenol (0.73) and was selected for further optimization. Aiming to reduce the K value for both compounds, the ratio of hexane was increased and that of water was decreased, resulting in the optimized hexane-BuOH-MeOH-H_2O 3:1:1:0.75 v/v (or 12:4:4:3, v/v) where now K values were, respectively, 1.22 and 0.57. The preparative purification of 600 mg of the essential oil of leaves of *P. caryophyllus* containing 35.3% chavibetol was conducted with this solvent system, affording 200 mg of the latter and 50 mg of methyleugenol with purities of 98% and 96%, respectively. By using this procedure, the recovery rate of the target compounds was 94.4% for chavibetol and 73.7% for methyleugenol. Compared to a 39% mass loss for chavibetol when purified by prep TLC, CCC showed to be a powerful tool for its purification.

The last example shows the preparative purification of aromatic massoia lactones from 10 g of crude massoia oil by centrifugal partition chromatography with dual-mode elution (Urbain et al. 2010). When using this kind of elution mode in CCC where both the nature and direction of the mobile phase are changed during the chromatographic run (Berthod 2009), it is possible to easily elute infinitely retained compounds in a shorter time.

This EO, which has been used for centuries as massage oil is obtained by hydrodistillation from the bark of *Cryptocarya massoy* Kosterm (Lauraceae) and is comprised mainly of 5,6-dihydro α-pyrones, in which chemical structure contains an α, β-unsaturated δ-lactone and an alkyl side chain at position 6 and can have 5, 7 or 9 carbons comprising the so-called C-10, C-12, or C-14 massoia lactones (Figure 8) (Urbain et al. 2010). These target compounds have a fragrant aroma and flavor that is of commercial importance in the food industry due to their milky and butter-like aroma, especially C-10 massoia lactone.

To select an adequate solvent system for their purification from the massoia oil, the authors calculated the K values for the C-10, C-12 and C-14 lactones present as main compounds in several ratios of the solvent system hexane-methanol-water. Urbain and co-workers report that this solvent system had been used successfully in the purification of fatty acids and due to the alkyl side chain of these lactones the same biphasic system was tested. The values obtained for the C-10 derivative were too low (below 0.5) and the researchers moved to more polar *c*-hexane and toluene instead of hexane in the system, achieving good partition

(*R*)-(-)-C-10 massoia lactone

(*R*)-(-)-C-12 massoia lactone

(*R*)-(-)-C-14 massoia lactone

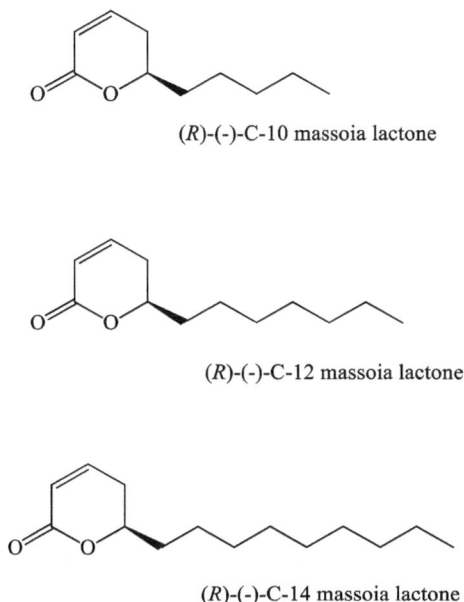

Figure 8: Chemical structures of C-10, C-12 and C-14 massoia lactones.

coefficients for C-10 massoia lactone (0.81 and 0.89 with either toluene-MeOH-H_2O or c-hexane-MeOH-H_2O 10:9:1, v/v, respectively) as well as good separation factor (α) between compounds (higher than 1.5, indicating a good resolution). The purification of 10 g of the EO was performed with the c-hex-MeOH-H_2O system affording in the ascending mode step (elution with the upper organic phase as mobile) and fractions containing pure benzyl derivatives (101 mg of benzyl salicylate and 350 mg of benzyl benzoate). The target massoia lactones, in the meantime, were obtained in the descending mode (elution with the lower aqueous phase after changing the direction of elution), affording 3.17 g of C-10 (purity over 96%), 357 mg of C-12 and 21 mg of C-14, all of them eluting in decreasing polarity order because of reversion of mobile phase. This impressive preparative separation was performed on a 1L column CPC equipment and interestingly, the authors report that despite monitoring the separation by TLC due to the high concentration of collected fractions, it was easy to follow their separation by the odor of the fractions; fractions containing the benzyl derivatives scented like flowers, whereas the fractions containing the pyrones scented like coconut.

Conclusions and Perspectives

The obtention of pure compounds from essential oils is still a challenging task due to their low to medium polarity combined with their chemical similarity, rendering it difficult to isolate and purify them. The examples discussed here show the high efficiency and versatility of countercurrent chromatography for the purification of essential oil components. In this chapter, the high loading capacity of CCC is highlighted, detaching this technique as an excellent preparative separation technique for essential oil metabolites. The best advantages are the low consumption of solvent and time, the vast array of different solvent systems which can be used according to the polarity of the target compounds and the use of either aqueous or non-aqueous solvent systems (or a combination of both); the latter being an advantage for the recovery of less polar and more volatile compounds, such as mono and sesquiterpene derivatives.

References

Adams, R.P. 2007. Identification of Essential Oil Components by Gas Chromatography/Mass Spectroscopy, 4th ed. Allured Publ. Corp, Carol Stream, IL, USA.

Andrés, G.E., G.E. Rossa, E. Cassel, R.M.F. Vargas, O. Santana, C.E. Díaz et al. 2017. Biocidal effects of *Piper hispidinervum* (Piperaceae) essential oil and synergism among its main components. Food Chem. Toxicol. 109(2): 1086–1092.

Beek, T.A.V. and D. Joulain. 2018. The essential oil of patchouli, *Pogostemon cablin*: A review. Flavour. Fragr. J. 33: 6–51.

Berthod, A. 2009. Countercurrent chromatography: From the Milligram to the Kilogram. pp. 323–352. *In*: Eli Grushka and Nelu Grinberg (eds.). Advances in Chromatography, 47. CRC Press, USA.

Berthod, A. and M. Hassoun. 2006. Using the liquid nature of the stationary phase in countercurrent chromatography: IV. The cocurrent CCC method. J. Chromatogr. A 1116: 143–148.

Berthod, A., T. Maryutina, B. Spivakov, O. Shpicun and I.A. Sutherland. 2009 Countercurrent chromatography in analytical chemistry (IUPAC technical report). Pure Appl. Chem. 81: 355–387. http://doi.org/10.1351/PAC-REP-08-06-05.

Bizzo, H.R., A.M.C. Hovell and C.M. Rezende. 2009. Óleos essenciais no Brasil: aspectos gerais, desenvolvimento e perspectivas Brazilian essential oils: General view, developments and perspectives. Quím. Nova 32(3): 588–594.

Bojczuk, M., D. Żyżelewicz and P. Hodurek. 2017. Centrifugal partition chromatography. A review of recent applications and some classic references. J. Sep. Sci. 40: 1597–609. https://doi.org/10.1002/jssc.201601221.

Burčul, F., I. Blažević, M. Radan and O. Politeo. 2020. Terpenes, phenylpropanoids, sulfur and other essential oil constituents as inhibitors of cholinesterases. Curr. Med. Chem. 27(26): 4297–4343.

Carini, J.P., G.G. Leitão, P.H. Schneider, C.C. Santos, F.N. Costa, M.H. Holzschuh et al. 2015. Isolation of achyrobichalcone from Achyrocline satureioides by high-speed countercurrent chromatography. Curr. Pharm. Biotechnol. 16: 66–71.

Chen, Q., S. Yao, X. Huang, J. Luo, J. Wang and L. Kong. 2009. Supercritical fluid extraction of *Coriandrum sativum* and subsequent separation of isocoumarins by highspeed countercurrent chromatography. Food Chem. 117(3): 504–508.

Chen, Q., X. Hu, L. Li, P. Liu, Y. Yang and Y. Ni. 2011. Preparative isolation and purification of cuminaldehyde and p-menta-1,4-dien-7-al from the essential oil of *Cuminum cyminum* L. by high-speed counter-current chromatography. Anal. Chim. Acta 689: 149–154.

Conway, W.D. 1990. Countercurrent Chromatography, Apparatus, Theory, and Applications. Weinheim: VCH Publishers, New York, USA.

Costa, F.N. and G.G. Leitão. 2010. Strategies of solvent system selection for the isolation of flavonoids by countercurrent chromatography. J. Sep. Sci. 33: 336–347. https://doi.org/10.1002/jssc.200900632.

Dang, Y.Y., X.C. Li, Q.W. Zhang, S.P. Li and Y. Wang. 2010. Preparative isolation and purification of six volatile compounds from essential oil of *Curcuma wenyujin* using high-performance centrifugal partition chromatography. J. Sep. Sci. 33: 1658–1664.

de Souza, P.A., L.P. Rangel, S.S. Oigman, M.M. Elias, A. Ferreira-Pereira, N.C. de Lucas and G.G. Leitão. 2010. Isolation of two bioactive diterpenic acids from the oleoresin of *Copaifera glycycarpa* by high-speed counter-current chromatography. Phytochem. Anal. 21: 539–543.

dos Santos, B.C.B., J.C.T. da Silva, P.G. Guerrero, G.G. Leitão and L.E.S. Barata. 2009. Isolation of chavibetol from essential oil of *Pimenta pseudocaryophyllus* leaf by high-speed counter-current chromatography. J. Chromatogr. A 1216: 4303–4306.

Gan, Z., Z. Liang, X. Chen, X. Wen, Y. Wang, M. Li et al. 2016. Separation and preparation of 6-gingerol from molecular distillation residue of Yunnan ginger rhizomes by high-speed counter-current chromatography and the antioxidant activity of ginger oils *in vitro*. J. Chromatogr. B 1011: 99–107.

Geng, Y., J. Liu, R. Lv, J. Yuan, Y. Lin and X. Wang. 2007. An efficient method for extraction, separation and purification of eugenol from *Eugenia caryophyllata* by supercritical fluid extraction and high-speed counter-current chromatography. Sep. Purif. Technol. 57: 237–241.

Hong, A.Y. and B.M. Stoltz. 2012. Enantioselective total synthesis of the reported structures of (–)-9-epi-presilphiperfolan-1-ol and (–)-presilphiperfolan-1-ol: Structural confirmation and reassignment and biosynthetic insights. Angew. Chem. 51: 9674–9678. https://doi.org/10.1002/anie.201205276.

Hong, A.Y. and B.M. Stoltz. 2014. Biosynthesis and chemical synthesis of presilphiperfolanol natural products. Angew. Chem. Int. Ed. Engl. 53: 5248–5260. https://doi.org/10.1002/anie.201309494.

Joseph-Nathan, P., S.G. Leitão, S.C. Pinto, G.G. Leitão, H.R. Bizzo, F.L.P. Costa et al. 2010. Structure reassignment and absolute configuration of 9-epi-presilphiperfolan-1-ol. Tet. Lett. 51: 1963–1965. https://doi.org/10.1016/j.tetlet.2010.02.025.

Kurniawan, C., S. Haryani, S. Kadarwati and E. Cahyono. 2019. The fractional separation of citronella, cajeput, and patchouli crude oils using spinning band distillation. J. Phys.: Conf. Ser. 1321(2019): 1–5. Doi: 10.1088/1742-6596/1321/2/022038.

Kutateladze, A.G. and D.M. Kuznetsov. 2017. Triquinanes and related sesquiterpenes revisited computationally: structure corrections of Hirsutanols B and D, Hirsutenol E, Cucumin B, Antrodins C–E, Chondroterpenes A and H, Chondrosterins C and E, Dichrocephone A, and Pethybrene. J. Org. Chem. 82: 10795–10802. https://doi.org/10.1021/acs.joc.7b02018.

Li, X.C., Q.W. Zhang, Z.Q. Yin and C.Y. Wen. 2011. Preparative separation of patchouli alcohol from patchouli oil using high performance centrifugal partition chromatography. J. Essent. Oil Res. 23: 19–24.

Liu, W., P. Wu, C. Zhuo, J. Zhang and J. Shen. 2010. One step separation and preparation of senkyunolide A and Z-ligustilide in *Ligusticum chuanxiong* Hort by high speed counter current chromatography. Zhongchengyao 32(5): 764–767.

Marques, A.M. and M.A.C. Kaplan. 2013. Preparative isolation and characterization of monoterpene isomers present in the citral-rich essential oil of *Pectis brevipedunculata*. J. Essent. Oil Res. 25(3): 210–215.

Marques, A.M., A.C. Peixoto, D.W. Provance and M.A.C. Kaplan. 2018b. Separation of volatile metabolites from the leaf-derived essential oil of *Piper mollicomum* Kunth (Piperaceae) by high-speed countercurrent chromatography. Molecules 23(12): 3064–3073.

Marques, A.M. and M.A.C. Kaplan. 2015. Active metabolites of the genus *Piper* against *Aedes aegypti*: Natural alternative sources for dengue vector control. Univ. Sci. 20(1): 61–82.

Marques, A.M., C.E. Fingolo and M.A.C. Kaplan. 2017. HSCCC separation and enantiomeric distribution of key volatile constituents of *Piper claussenianum* (Miq.) C. DC. (Piperaceae) Food Chem. Toxicol. 109: 1111–1117.

Marques, A.M., V.H.C. Aquino, V.G. Correia, A.C. Siani, M.R. Tappin, M.A.C. Kaplan et al. 2018a. Isolation of two major sesquiterpenes from the leaf essential oil of *Eugenia uniflora* by preparative-scale high-speed countercurrent chromatography Sep. Sci. Plus 1(12): 785–792.

Marston, A. and K. Hostettmann. 1994. Counter-current chromatography as a preparative tool— Applications and perspectives. J. Chromatogr. A 658: 315–341.

Marston, A. and K. Hostettmann. 2006. Developments in the application of counter-current chromatographyto plant analysis. J. Chromatogr. A 1112: 181–194.

Maurya, A., J. Prasad, S. Das and A.K. Dwivedy. 2021. Essential oils and their application in food safety. Sustain. Food Syst. 5: 653420. Doi: 10.3389/fsufs.2021.653420.

Peixoto, J.F., Y.J. Ramos, D.L. Moreira, C.R. Alves and L.F. Gonçalves-Oliveira. 2021. Potential of *Piper* spp. as a source of new compounds for the leishmaniases treatment. Parasitol. Res. 120(8): 2731–2747. Doi: 10.1007/s00436-021-07199-4. Epub 2021 Jul 10.

Pinto, S.C., G.G. Leitão, D.R. Oliveira, H.R. Bizzo, D.F. Ramos, T.S. Coelho et al. 2009a. Chemical composition and antimycobacterial activity of the essential oil from *Anemia tomentosa* var. *anthriscifolia*. Nat. Prod. Commun. 4: 1675–1678.

Pinto, S.C., G.G. Leitão, H.R. Bizzo, N. Martinez, E. Dellacassa, F.M. dos Santos et al. 2009b. (−)-epi-Presilphiperfolan-1-ol, a new triquinane sesquiterpene from the essential oil of *Anemia tomentosa* var. *anthriscifolia* (Pteridophyta). Tet. Lett. 50: 4785–4787. https://doi.org/10.1016/j.tetlet.2009.06.046.

Puertas, M.M., S. Hillebrand, E. Stashenko and P. Winterhalter. 2002. *In vitro* radical scavenging activity of essential oils from Columbian plants and fractions from oregano (*Origanum vulgare* L.) essential oil. Flavour. Fragr. J. 17: 380–384.

Queiroga, C.L., M.Q. Cavalcante, P.C. Ferraz, R.N. Coser, A. Sartoratto and P.M. de Magalhães. 2014. High-speed countercurrent chromatography as a tool to isolate nerolidol from the *Baccharis dracunculifolia* volatile oil. J. Essent. Oil Res. 26(5): 334–337.

Ribeiro, V.P., C. Arruda, J.J.M. da Silva, M.J.A. Aldana, N.A.J.C. Furtado and J.K. Bastos. 2018. Use of spinning band distillation equipment for fractionation of volatile compounds of *Copaifera oleoresins* for developing a validated Gas Chromatographic method and evaluating antimicrobial activity. Biomed. Chromatogr. 2018: e4412. Doi: 10.1002/bmc.4412.

Shi, X., X. Wang, D. Wang, Y. Geng and J. Liu. 2009. Separation and purification of a-Cyperone from *Cyperus rotundus* with supercritical fluid extraction and high-speed counter-current chromatography. Sep. Sci. Technol. 44(3): 712–721.

Silva-Santos, A., A.M.S. Antunes, H.R. Bizzo and L.A. D'Avila. 2006. Participation of the citrus oil industry in Brazilian trade balance. Plant. Med. 8(4): 8–13.

Skalicka-Woźniak, K. and I. Garrard. 2014. Counter-current chromatography for the separation of terpenoids: A comprehensive review with respect to the solvent systems employed. Phytochem. Rev. 13: 547–572. Doi: 10.1007/s11101-014-9348-2.

Syntrivanis, L.D., I. Némethová, D. Schmid, S. Levi, A. Prescimone, F. Bissegger. D.T. Major and K. Tiefenbacher. 2020. Four-step access to the sesquiterpene natural product presilphiperfolan-1β-ol and unnatural derivatives via supramolecular catalysis. J. Am. Chem. Soc. 142: 5894–5900. https://doi.org/10.1021/jacs.0c01464.

Urbain, A., P. Corbeiller, N. Aligiannis, M. Halabalaki and A.L. Skaltsounis. 2010. Hydrostatic countercurrent chromatography and ultra high pressure LC: Two fast complementary separation methods for the preparative isolation and the analysis of the fragrant massoia lactones. J. Sep. Sci. 33(9): 1198–1203.

Vetter, W., M. Müller, M. Englert and S. Hammann. 2020. Countercurrent chromatography: When liquid-liquid extraction meets chromatography. pp. 289–325. *In*: C.F. Poole (ed.). Liquid-Phase Extraction. Handbooks in Separation Science, Elsevier, Frankfurt, Germany. https://doi.org/10.1016/B978-0-12-816911-7.00010-4.

Wang, C., L. Wang, C. Li, C. Hu and S. Zhao. 2020. Anti-proliferation activities of three bioactive components purified by high-speed counter-current chromatography in essential oil from ginger. Eur. Food Res. Technol. 246(4): 795–805.

Wang, D., G. Yanling, F. Lei, S. Xikai, L. Jianhua, W. Xiao et al. 2011a. An efficient combination of supercritical fluid extraction and high-speed counter-current chromatography to extract and purify (*E*)- and (*Z*)-diastereomers of a-asarone and b-asarone from *Acorus tatarinowii* Schott. J. Sep. Sci. 34(23): 3339–3343.

Wang, D., G. Yan-ling, L. Wei, W. Xiao, L. Jian-hua, L. Jing et al. 2011b. Chemical constituents in essential oil of *Illicium verum* by supercritical CO2 extraction and high-speed counter-current chromatography. Chem. Ind. For. 31(3): 99–104.

Wang, X., C. Wang, S. Tong, G. Zuo, H. Kim, S.S. Lim et al. 2020. An off-line DPPH-GC-MS coupling countercurrent chromatography method for screening, identification, and separation of antioxidant compounds in essential oil. Antioxidants 9(8): 702–714.

Wei, Y., J. Du and Y. Lu. 2012. Preparative separation of bioactive compounds from essential oil of *Flaveria bidentis* (L.) Kuntze using steam distillation extraction and one step high-speed counter-current chromatography. J. Sep. Sci. 35: 2608–2614.

Xiao, Y., J. Xie, L. Ma and J. Ma. 2007. Separation of flavors from *Foeniculum vulgare* by high-speed counter-current chromatography. Shipin Yu Fajiao Gongye 33(7): 142–143.

Xie, J.X., S. Sun, S. Wand and Y. Ito. 2009. Isolation and purification of nootkatone from the essential oil of fruits of *Alpinia oxyphylla* Miquel by high-speed counter-current chromatography. Food Chem. 117(2): 375–380.

Xie, J.X., S. Wand, S. Sun and F. Zheng. 2008. Preparative separation and purification of b-Caryophyllene from leaf oil of *Vitex negundo* L. var. *heterophylla* (Franch.) Rehd. by high speed countercurrent chromatography. J. Liq. Chromatogr. Relat. Technol. 31(17): 2621–2631.

Xiong, Y.K., G.H. Zhu, J.Q. Zhang, Z.Y. Liu, X. Zhou, B. Nie et al. 2014. Purification of ferulic acid, senkyunolide I and senkyunolide H from the volatile oil of Rhizoma chuanxiong using high-speed countercurrent chromatography and preparative liquid chromatography. Adv. Mater. Res. 1033-1034: 259–264.

Yan, Y., G. Chen, S. Tong, Y. Feng, L. Sheng and J. Lou. 2005. Preparative isolation and purification of germacrone and curdione from the essential oil of the rhizomes of *Curcuma wenyujin* by high-speed counter-current chromatography. J. Chromatogr. A 1070(1-2): 207–210.

Zhang, D., H. Teng, L. Guisheng, G. Li, K. Liu and Z. Su. 2006. Separation and purification of Z ligustilide and senkyunolide A from *Ligusticum chuanxiong* Hort. With supercritical fluid extraction and high-speed counter current chromatography. Sep. Sci. Technol. 41(15): 3397–3408.

Zhang, Q.W., L.G. Lin and W.C. Ye. 2018. Techniques for extraction and isolation of natural products: A comprehensive review Chin. Med. 13: 1–26.

Zhang, X., J. Liang, Y. Zhang, J. Liu, W. Sun and Y. Ito. 2015. Comparative studies on performance of CCC and preparative RP-HPLC in separation and purification of steroid saponins from *Dioscorea zingiberensis* C.H. Wright. J. Steroids Horm. Sci. 6(1): 1–20.

Zhou, Y., C. Wang, R. Wang, L. Lin, Z. Yin, H. Hu et al. 2017. Preparative separation of four sesquiterpenoids from *Curcuma longa* by high-speed counter-current chromatography. Sep. Sci. Technol. 52(3): 497–503.

2

Volatile Compounds in *Piperaceae* Collected in Arauca–Colombia
Northeastern Region and Colombian–Venezuelan Plains

Geovanna Tafurt-García,[1,*] Elisa Valenzuela Vergara,[2]
Yulanderson Salguero Rodríguez,[3] Rosa Amanda
Alegría Macías[4] and Elena Stashenko[5]

Introduction

Piperaceae comes from the Latin *piper*, which means 'pepper', and from the Latin *āceae*, which means "resembling". The phylogenetic position of *Piperaceae* is among a diverse group of dicotyledons called palaeoherbs, which are plants that resemble monocots in certain vegetative traits (e.g., adaxial prophylls and scattered vascular bundles) (Jaramillo and Manos 2001).

Mainly, 11 genera are accepted for the family of *Piperaceae*, which consist of *Verhuellia, Manekia, Piperanthera, Lindeniopiper, Trianaeopiper, Ottonia, Anderssoniopiper, Zippelia, Sarcorhachis, Peperomia* and *Piper* (WFO 2021). They occur in Africa, Asia and tropical America and the Amazon region (Takeara et al. 2017). A total of 1,545 species are accepted for the genus *Piper* genus, and 1,187 species are accepted for the *Peperomia* genus which has the highest number of species among the family genera (WHO 2021).

[1] Research Group Semilla del Conocimiento del Cesar (Zajuna jwa samu). Universidad Nacional de Colombia, sede de la Paz, km 9 vía Valledupar-La Paz, Colombia..

[2] Universidad Nacional de Colombia, Sede Medellín, Facultad de Ciencias, Medellín, Colombia.

[3] Universidad Nacional de Colombia, Sede Manizales, Facultad de Ingeniería y Arquitectura, Campus Palogrande, Manizales, Colombia.

[4] Universidad Nacional de Colombia, Sede Orinoquia, km 9 vía Caño Limón, Arauca, Colombia.

[5] Universidad Industrial de Santander, Facultad de Ciencias, Escuela de Química, CENIVAM, Santander, Colombia.

* Corresponding author: gtafurg@unal.edu.co

Piper spp. are often conspicuous elements of the understory, reaching their greatest diversity in pre-montane and lowland rainforests (WFO 2021). The greatest diversity of *Piper* spp. is found in the American tropics (1000 spp.) (WFO 2021), followed by South Asia (300 spp.). Several species are restricted to a specific center of diversity (e.g., the Andes and Central America), and others are found throughout the neotropics or palaeotropics (Jaramillo and Manos 2001).

Peperomia spp. are distributed in tropical and subtropical regions of both hemispheres. Most species are found in the palaeotropics, and they are related to species distributed in neotropics. Their greatest diversification is found in Central America and northern South America (WFO 2021). The altitudinal range is from 0 to 3,800 masl. They can be found in lowland, submontane, montane, tepuis and floodplain forest areas; they are generally found in humid tropical ecosystems (Melo et al. 2016).

In Colombia, genera (species): *Manekia* (1), *Peperomia* (277) and *Piper* (415) with 693 species have been reported (Bernal 2021). These correspond to 27.27% of the genus and 25.39% of the species that are associated with the *Piperaceae* family by world report (WFO 2021).

Ethnobotanical Uses

Piper spp. are globally used as a spice and condiment as well as for their medicinal properties. Species of this genus have traditionally been used to treat stomach pain, rheumatoid arthritis, diarrhea and other general infections (Mgbeahuruike et al. 2017). In Ayurvedic medicine, *Piper* spp. have been described as having a pungent taste with uses against parasitic diseases, dyspnoea, pulmonary tuberculosis, spleen disorders, intermittent fever, piles, stiff thighs, nervous diseases, insomnia and dysentery (Kumar et al. 2020).

In Thailand, *Piper* spp. are frequently cultivated in gardens and they are used as vegetables, flavorings for cooking, decorations, medicines and for traditional ceremonies (*P. betle*, *P. longum*, *P. nigrum*, *P. pendulispicum*, *P. chaba*, *P. sarmentosum*, *P. wallichii* and *P. maculaphyllum*) (Chaveerach et al. 2012). For instance, *P. betle* is used as a mild stimulant, exhilarant, antiseptic, antiflatulent, expectorant, bronchitis, stomach tonic, kidney inflammation and thirst resulting from diabetes. *P. nigrum* is used as an aromatic, carminative, febrifuge, rubefacient, stimulant, adenitis, cancer, cholera, cold, malaria, headache, gravel, stomach colic and abortifacient. *P. longum* is used against respiratory diseases, bronchitis, stomachache, spleen diseases and tumors and can even improve appetite and expulsion of the placenta after childbirth. *P. retrofractum* is used for anti-asthmatic, antitussive and diaphoretic purposes and can even improve digestion and blood circulation. *P. wallichii* has been used for treating influenza, asthma, and flatulence and to improve blood circulation as well (Chaveerach et al. 2012).

In India, *P. longum* and *P. nigrum* are used as digestive, aperitif and laxative. They are used to treat respiratory diseases, insomnia, epilepsy, anaemia, asthma, muscular pains, inflammation and chronic fevers as well as to clear bile duct and bladder obstruction. They are also used for antiseptic, emmenagogue and abortive purposes and can improve memory (Kumar et al. 2020).

In Mexico, *Piper* spp. are used as medicinal, edible, fuel and building materials; some species are used against snakebites (*P. aduncum, P. auritum*) (Martínez-Bautista et al. 2019). In Panama, *Piper* spp. are used for the treatment of liver pain, colds, skin infections, snakebites and rheumatism. They are also used as insecticides, wound healing, antipyretic and anti-inflammatory (*P. aduncum, P. arboreum, P. auritum, P. cordulatum, P. hispidum, P. dariense, P. multiplinervium* and *P. umbellatum*) (Durant-Archibold et al. 2018).

In Brazil, *Piper* spp. are used as a tonic, carminative, stimulant, diuretic, antidiarrhoeal, antipyretic, analgesic, anaesthetic, anti-inflammatory and sudorific. They are also used for treating digestive, urinary, respiratory, gynecological and intestinal diseases as well as for dehydration, earache, headache, toothache, malaria, liver and kidney diseases, vaginitis, gall bladder pain and snake and insect bites (*P. aduncun, P. callosum, P. cavalcantei, P. cernuum, P. corcovadensis, P. gaudichaudianum, P. hispidum, P. lhostzkyanum, P. marginatum, P. regnelli* and *P. umbellatum*) (Takeara et al. 2017). In Brazil, *P. peltatum* has been used for the treatment of erysipelas, malaria, fever, leishmaniasis, hepatitis, uterine cleansing, bruises, swellings, abscesses, colds, coughs, haemorrhage, headache, hernia and arthritis pains (Fern 2021, Salehi et al. 2019).

In South and Central America, *P. hispidum* has been traditionally used to treat anemia, rheumatism, malaria, conjunctivitis, diarrhea, stomachaches, head lice, amygdalitis, mouth sores, snakebites, insect bites, wounds and symptoms of cutaneous leishmaniasis. It is also used as skin cleansing, diuretic, teeth whitening and mumps to prevent tooth decay, ease the pain of childbirth and regulate menstruation (Salehi et al. 2019).

P. marginatum is used to treat asthma, erysipelas, malaria, female disorders, problems of the urinary system, blood pressure, inflammatory diseases, gastrointestinal problems, snake bites, toothaches and liver and bile duct diseases. It is used as sedative, tonic, carminative, stimulant, diuretic, sudorific agents, and it is also used to help during childbirth (Salehi et al. 2019).

In Colombia, *P. obtusilimbum* is known as *Desvanecedora*, it is used for treating varicose veins, non-specific symptoms and general illnesses (cancer, inflammation and tumors) (Bussmann et al. 2018). Also, it is known as *Morona de monte*, and it is used as an abortifacient (Ballesteros et al. 2016). *P. obtusilimbum* has been used in the Peruvian Amazon to treat malaria, fever, body itch and as a teeth protector (Vásquez-Ocmín et al. 2021). On the other hand, in Mexico, most *Peperomia* spp. are used fresh (vegetables) or processed (flavoring/seasoning) with use as edible, ornamental or ceremonial (psychosomatic diseases with symptoms, such as headache, stomach spasms, chest tightness, teary eyes and anxiety) (Martínez-Bautista et al. 2019).

P. angustata leaves and stems are eaten as a salad (Fern 2021). Also, it has medicinal use in the treatment of rheumatism and earache (Vergara 2013). *P. quadrangularis* is used in food, as a fresh vegetable or added to beans or tamales; in addition, it has ceremonial uses (Martínez-Bautista et al. 2019). *P. serpens* is used as an ornamental in tropical South America (Centero et al. 2019). In the Jameykari indigenous community, *P. serpens* is used to potentiate the action of *Aristolochia cordiflora* (to treat unknown diseases, asthma and snakebites) (Juep 2008).

P. glabella and *P. pellucida* are used to treat skin problems such as erysipelas (Vergara 2013). *P. pellucida* is used as medicine, food and flavoring (Raghavendra and Kekuda 2018). *P. pellucida* aerial parts are used as decoctions, juices and pastes; it also used to treat conjunctivitis, fever, cold, viral diseases, abdominal and rheumatic pains, asthma, haemorrhoids, hypertension, convulsions, bone fracture, vaginal and kidney infections, acne, abscesses, white spots on the body, measles and athlete's foot (Raghavendra and Kekuda 2018, Salehi et al. 2019, WHO 2021).

In Nicaragua, *P. pellucida* is used against stings (snakes, scorpions and insects), venereal infections and female disorders (Raghavendra and Kekuda 2018). In Brazil, *P. pellucida* is used against gastrointestinal complaints and high blood pressure; it acts as a mild diuretic and treats eye diseases (Takeara et al. 2017).

Biological Activities

Piper spp. have shown demonstrated anticancer, antitumor, antimicrobial, anti-inflammatory, antioxidant, antibacterial, antifungal, antiprotozoal, insecticidal and antimalarial activities (Correa et al. 2015, da Silva et al. 2017, Mgbeahuruike et al. 2017, Takeara et al. 2017, Kumar et al. 2020). Biomedical and pharmacological findings related to the anticancer and antimicrobial properties of *Piper* spp. are described by Mgbeahuruike et al. (2017). *Piper* spp. have also shown an anti-feed effect on *Spodoptera frugiperda* larvae (Celis et al. 2008, Delgado et al. 2007, Murcia and Bermúdez 2008), and they inhibited the germination of weeds (Delgado et al. 2007).

Among the *Piper* spp. studied, *P. hispidum* has been evaluated against *Plasmodium falciparum* and *Leishmania amazonensis* and it has shown anticholinesterase activity (Takeara et al. 2017, Salehi et al. 2019). *P. eriopodon* has shown antioxidant activity (Correa et al. 2015). *P. peltatum* has shown antioxidant, anti-inflammatory and antineuralgic properties (Fern 2021, Salehi et al. 2019, Puertas-Mejía et al. 2009); its inhibitory effect on *Bothrops asper* myotoxins has been also evaluated (Núñez et al. 2005). *P. marginatum* has shown antimicrobial activity (Takeara et al. 2017). *P. obtusilimbum* has presented leishmanicidal activity (Vásquez-Ocmín et al. 2021).

Among the *Peperomias* spp. studied, *P. serpens* has been evaluated as an antifungal (Saga et al. 2006). It has shown significant anti-inflammatory and antinociceptive effects and inhibited oedema formation (Pinheiro et al. 2011). *P. pellucida* has shown activities as gastroprotective, antidiabetic, analgesic, acaricidal, anticancer, cytotoxic, antioxidant, antimicrobial, antidiarrhoeal, fibrinolytic, antiosteoporosis, anti-inflammatory, antipyretic, fracture healing, antiangiogenic, antimutagenic, immunostimulant, neuropharmacological, hypotensive and antihypercholesterolemic among others (Amarathunga and Kankanamge 2017, Raghavendra and Kekuda 2018, Salehi et al. 2019).

Phytochemistry

Piper spp. chemistry has been extensively researched, and this has allowed the isolation of biologically active compounds, which have been classified into the following categories: alkaloids/amides, propenylphenols, lignans, neolignans, terpenes, steroids, kawapyrones, piperolides, chalcones, dihydrochalcones, flavones and flavanones (Parmar et al. 1997, da Silva et al. 2017, Durant-Archibold et al. 2018, Kumar et al. 2020). Chemical composition has been reported for about 130 *Piper* spp.; however, detailed information is only available for a few species (Dyer et al. 2004, da Silva et al. 2017, Salehi et al. 2019).

Alkaloids (piperine, piperlongumine, guineensine, chabamide, methyl piperine, piperic acid, pipernonaline, piperoctadecalidine and pellitorine), which have been isolated from most *Piper* spp. exhibit anticancer properties among others (Mgbeahuruike et al. 2017, Kumar et al. 2020). Piperine is extracted from *P. nigrum*, and it is used to synthesize heliotropin, which has antiseptic and antipyretic properties. (Chaveerach et al. 2012). 4-Nerolidylcatechol and 2-(4',8'-Dimethyl-none-3',7'-dienyl)-8-hydroxy-2-methyl-2H-chromene-6-carboxylic acid methyl ester have been isolated from *P. peltatum* (Núñez et al. 2005). Phenols have also been detected by phytochemical analysis from *P. peltatum* (Puertas-Mejía et al. 2009).

It has been shown that the antioxidant capacities of *Piper* spp. are due to the presence of phenolic acids and flavonoids (quercetin, catechin, coumaric acid and protocatechuic acid) (Kumar et al. 2020). According to Correa et al. (2015), high contents of phenols, tannins, flavonoids, coumarins and anthraquinones have been found in methanolic extracts of some *Piper* spp. Anthraquinones, terpenes, steroids and moderate amounts of phenols and tannins were found in dichloromethane extracts; terpenes, triterpenes, steroidal sapogenins and steroids were detected in *n*-hexane extracts (Correa et al. 2015).

Piper spp. are rich in essential oils (EOs), which can be found in roots, stems, leaves, fruits and seeds. These EOs contain different classes of volatile organic compounds in varying amounts, which depend on polymorphism, geographical differences, environmental conditions and chemotypes among others (da Silva et al. 2017, Salehi et al., 2019). According to da Silva et al. (2017) and Salehi et al. (2019), monoterpenes, sesquiterpenes (hydrocarbons and oxygenated), and phenylpropanoids are among the most important compounds in the OEs of *Piper* spp. (da Silva et al. 2017, Thin et al. 2018, Salehi et al. 2019).

According to principal chemical constituents present in EOs, a preliminary classification of *Piper* spp. has been proposed as follows (Thin et al. 2018, Salehi et al. 2019):

- *P. demeraranum*, *P. chimonanthifolium* and *P. cubeba* have shown a high proportion of monoterpenes.

- *P. majusculum*, *P. cernuum*, *P. madeiranum*, *P. duckei*, *P. nigrum* and P. *lepturum* have shown sesquiterpenes as the majority. For instance, caryophyllene oxide, β-elemene and β-caryophyllene were the majority compounds in *P. phytolaccifoium* from Guatemala (Cruz et al. 2012).

- *P. hispidum*, *P. demeraranum* and *P. aduncum* have presented predominance of monoterpenes and sesquiterpenes.

- *P. caninum*, *P. auritum*, *P. hispidinervum*, *P. aduncum*, *P. divaricatum*, *P. betle*, *P. patulum*, *P. klotzsdhianum* and *P. marginatum* have shown phenylpropanoids as a majority.

- *P. klotzsdhianum*, *P. sarmentosum* and *P. harmandii* have presented a high proportion of benzenoids.
- *P. maclurei* and *P. caldense* have shown non-terpenoids as the majority.
- *P. aduncum*, *P. cernuum*, *P. divaricatum*, *P. marginatum*, *P. hispidum*, *P. umbellatum* and *P. guineense* are among the *Piper* spp. that have shown chemotypes (Mgbeahuruike et al. 2017, da Silva et al. 2017).

Secolignans and dinorlignans, among others, have been detected in *Peperomias* spp. (Kato and Furlan 2007). Peperomin D (secolignan) and 2-hydroxy-4,6-dimethoxyacetophenone have been isolated from *P. glabella* (Monache and Compagnone 1996, Soares et al. 2006). Sesquiterpenes (hydrocarbons oxygenated), chromenes and flavonoids have been isolated from *P. serpens* (Saga et al. 2006, Pinheiro et al. 2011). Tannins, flavonoids, xanthones, glycosides, alkaloids, saponins, inulins, phenylpropanoids, lignans, sesquiterpenes, phytosterols, steroids, resins, carbohydrates and minerals (calcium, iron and potassium) have been found in *P. pellucida* (Amarathunga and Kankanamge 2017, Kartika et al. 2020, de Moraes and Kato 2021).

Experimental Part

Piperaceae from Arauca used in this Study

According to Sua et al. (2021), the Orinoquia region of Colombia is defined as the area between the Arauca river (Arauca department) and the Meta river (Casanare department) to the north (Colombian-Venezuelan border); to the Inírida, Atabapo and Guaviare rivers to the south; and from the eastern slope of the Eastern Cordillera to the west; and to the Orinoco river, which is to the east (Colombian-Venezuelan border) (Sua et al. 2021).

Colombian Orinoco encompasses flatlands known as *llanos* and mountainous areas of the foothills of the eastern slope of the Eastern Cordillera (Sua et al. 2021). In this region, there is a botanical diversity that is represented by 167 families and 4,346 species (Rangel-Churio 2015). Specifically, for the Colombian Orinoquia Region (COR), collections of 313 and 69 *Piper* spp. and *Peperomia* spp., respectively are reported (GBIF 2021). In this region, the highest number of reports of *Piper* spp. is for *P. arboreum* (47), *P. demeraranum* (31), *P. aduncum* (30), *P. obliquum* (20), *P. peltatum* (17), *P. marginatum* (16) and *P. metanum* (12), with fewer collection reports for *P. hispidum* (7), *P. obtusilimbum* (5) and *P. phytolaccifolium* (2). For *Peperomia* spp. the reports are minors, such as *P. rotundifolia* (9), *P. macrostachyos* (9), *P. serpens* (7), *P. quadrangularis* (3), *P. glabella* (2) and *P. pellucida* (1) (GBIF 2021).

Arauca is in the Eastern Plains region, in the basin of the Orinoco River (Colombian-Venezuelan border). Departments of Arauca, Casanare, Vichada and Meta (partially, Guainía and Guaviare) are part of COR. Plants of the botanical families Rubiaceae, Asteraceae, Poaceae, Fabaceae and Orchidaceae are mostly reported in COR. However, there is a deficiency in the scientific knowledge of the regional flora.

This paper presents the results of the evaluation of the chemical composition and biological activity of *Piperaceae* collected in the foothills of the eastern cordillera of the Colombian Andes and the savannah of the department of Arauca. Some data on the record in the Herbario Nacional Colombiano (HNC) of the Instituto de Ciencias Naturales (ICN) of the Universidad Nacional de Colombia (UNAL) are *P. eriopodon* (Col 570495), *P. hispidum* (Col 570496), *P. marginatum* (Col 570497), *P. obtusilimbum* (Col 570498), and *P. phytolaccifolium* (Col 570499) were identified by A. Jara (2013). *P. peltatum* (Col 580473), *P. angustata* (Col 321 580472), *P. serpens* (Col 580471) and *P. glabella* (Col 580470) were identified by W. Trujillo-C (2013, 2014). *P. pellucida* (Col 579410), *P. serpens* (Col 321 582800) and *P. quadrangularis* (Col 582801) were identified by O. Rivera-Diaz (2015).

The information reported by Bernal (2021) and WHO (2021) about the origin and biogeographic region of the *Piperaceae* spp. used in this study are shown below (see images of *Piperaceae* spp. in Figures 1 and 2).

Piper spp. occur as a subshrub, shrub, sapling or native tree. *P. eriopodon* has been reported for Colombian Andes (500–2,800 masl: departments of Antioquia, Bolívar, Boyacá, Chocó, Magdalena, Meta and Santander), Venezuela, Ecuador and Bolivia. *P. peltatum* has been collected in Colombia (0–2,200 masl: Amazonia, Andes, Guainía, Serranía de La Macarena, Caribbean Plain, Pacific, Sierra Nevada de Santa Marta and Magdalena Valley), from México to Bolivia and Brazil. *P. marginatum* has been reported in Colombia (0–2,350 masl: Amazonia, Andes, Guainía, Serranía de La Macarena, Caribbean Plain, Orinoquia, Pacific, Sierra Nevada de Santa Marta and Magdalena Valley), from México to Perú, Brazil and Puerto Rico. *P. hispidum* has been collected in Colombia (0–1,800 masl: Amazonia, Andes, Guainía, Serranía de La Macarena, Pacific and Magdalena Valley), from México to Argentina and Brazil (Central and South America), and West Indies. *P. phytolaccifolium* has been collected in Colombia (0–1,300 masl: Amazonia, Caribbean Plain, Orinoquia and Pacific), México, Venezuela, Perú, Ecuador and Panamá. *P. obtusilimbum* is reported

Figure 1: *Piper* spp. studied: *P. eriopodon* (A), *P. peltatum* (B), *P. phytolaccifolium* (C), *P. marginatum* (D), *Piper* sp. (E), *P. obstusilimbum* (F) and *P. hispidum* (G).

Figure 2: *Peperomia* spp. studied: *P. angustata* (A), *P. glabela* (B), *P. serpens* (C), *P. serpens* (D), *P. quadrangularis* (E) and *P. pellucida* (F).

in Colombia (100–630 masl: Amazonia, Guayana and Serranía de La Macarena), Ecuador and Brazil (Bernal 2021, WHO 2021).

Peperomia spp. occur as an epiphytic herb of native origin. *P. angustata* has been collected in Colombia (50–500 masl: Amazonia, Guainía, Serranía de La Macarena, Caribbean Plain, Pacific and Magdalena Valley), Venezuela, Belize, México, Bolivia and Guyana. *P. glabella* is reported for Colombia (10–2,500 masl: Amazonia, Andes, Orinoquia, Pacific and Magdalena Valley), México, Bolivia, Brazil, Antilles (Central America and northern South America), and West Indies. *P. serpens* has been collected in Colombia (0–1,000 masl: Amazonia, Andes, Guayana, Serranía de La Macarena and Pacific), West Indies, México, Panamá, Perú, Brazil, Bolivia and the Greater Antilles. *P. quadrangularis* is reported for Colombia (400–2,000 masl: Amazonia, Andes, Guayana, Serranía de La Macarena, Caribbean Plain, Orinoquia, Sierra Nevada de Santa Marta, and Valle del Cauca), West Indies, Cuba and Central and Northern South America. *P. pellucida* has been collected in Colombia (0–3,500 masl: Amazonia, Andes, Guainía, Serranía de La Macarena, Pacific and Magdalena Valley), West Indies, USA, Spain and Tropical North and South America (Bernal 2021, WHO 2021).

Extracts and Volatiles from Piperaceae

EOs were obtained from fresh plant material (leaves) from hydrodistillation assisted by microwave radiation (HDMW) (Stashenko et al. 2004). Volatile fractions (VF) from the fresh plant (leaves) were isolated using a Likens and Nickerson microscale apparatus for high-density solvents and were modified by Godefroot et al. (1981) (Tafurt-García and Muñoz 2018). From dried, crushed and homogenized plant material were obtained ethanolic extracts (EEs). The EEs were dried by vacuum distillation (Tafurt-García et al. 2015).

Total Antioxidant Activities (TAA) and Total Phenolic Content (TPC)

The assay for ABTS radical was performed according to the method reported by Re et al. (1999) and Tafurt-Garcia et al. (2015). TAA obtained in this test was estimated regarding Trolox (mmol Trolox/kg extract). The ratio between the slopes of the curves evaluated for the EEs (dA vs kg/L) and Trolox (dA vs mM) was used to estimate TAA.

The TPC was calculated as gallic acid equivalents (g GA/g extract) according to the procedure described by Dastmalchi et al. (2007) and Tafurt-Garcia et al. (2015).

Gas Chromatography-Mass Spectrometry (GC-MS)

The separation and analysis of the components present in EOs and VFs were carried out in a gas chromatograph [Agilent Technologies (7890A)] coupled to a mass selective detector (Agilent Technologies 5975C) with split/splitless inlet (ratio split 30:1) and automatic injection system (7683 Series Injector Agilent). RTX-5MS column [60 m × 0.25 mm (d.i.) × 0.25 µm (df)] with stationary phase of 5% phenyll-poly dimethylsiloxane was used for the separation. Helium (99.999%) was the carrier gas (constant flow, 1.0 mL/min and 16.97 psi). Oven temperature programming was from 45°C (5 minutes) to 350°C (3 minutes) at a rate of 4°C/minutes. Mass spectra were obtained by electron impact (EI and 70 eV) using a quadrupole mass analyzer. Total Ion Currents (TIC) were acquired by full scan mode with a mass range of m/z 40–350. Chromatographic and spectroscopic data were processed using MSD Productivity ChemStation (NIST098.L). The secondary metabolites were identified by comparing their mass spectra with those of the databases (NIST, NIST Retention Index and Wiley) and with linear retention indices reported in the existing literature (Joulain and König 1998, Adams 2007, NIST 2015).

Results and Discussion

Chemical Composition of Piperaceae from Arauca

Chemical Composition of Piper spp.

As for the yield of *Piper* spp. EOs, *P. marginatum* showed the highest (0.574%) followed by *P. eriopodon* and *P. obstusilimbum* (0.535%). The yields of *P. hispidum* (0.169%), *Piper* sp. (0.139%) and *P. peltatum* (0.02%) were the lowest. EO or VF was not obtained for *P. phytolaccifolium*.

β-Copaene (14.5%), β-eudesmol + (5-epi-7-epi)-α-eudesmol (10.9%) and caryophyllene (8.1%) were the principal compounds detected in the *P. eriopodon* EO. Epi-13-manoyl oxide (55.1%) was identified as the major component in *Piper* sp. EO. α-Eudesmol + β-eudesmol (16.5%), caryophyllene oxide (7.8%), α-selinene (5.5%) and (10-epi)-γ-eudesmol (5.5%) were determined as the majority for *P. hispidum* EO. Safrole (99.5%) was the principal component in *P. obtusilimbum*. δ-Elemene (25.5%), β-caryophyllene (10.2%) and E-nerolidol (5.8%) were determined as the major components in *P. marginatum* EO. E-Caryophyllene (58.1%), germacrene D (29.0%) and bicyclogermacrene (12.9%) were identified as the majority for *P. peltatum* EO.

Hydrocarbon sesquiterpenes were the compound principals detected in *Piper* spp. EOs evaluated as *P. marginatum*, *P. peltatum* and *P. eriopodon*; oxygenated sesquiterpenes were the major ones for *Piper* sp. and *P. hispidum*, and phenylpropanoids (benzodioxols) were the major components of *P. obtusilimbum*.

According to Thin et al. (2018) and Salehi et al. (2019), sesquiterpenes are found in much *Piper* spp. EOs, e.g., in *P. majusculum*, *P. cernuum*, *P. madeiranum*, *P. duckei*, *P. nigrum* and *P. lepturum*. Phenylpropanoids are found as the majority compounds in the EOs of *P. caninum*, *P. auritum*, *P. hispidinervum*, *P. aduncum*, *P. divaricatum*, *P. betle*, *P. patulum*, *P. klotzsdhianum* and *P. marginatum*. Furthermore, benzenoids are found as the principal compounds in EOs of *P. klotzsdhianum*, *P. sarmentosum* and *P. harmandii* (Thin et al. 2018, Salehi et al. 2019).

Other scientific reports were found on the chemical composition of *Piper* spp. analyzed in this study, such as 1,8-cineole, β-pinene, myristicin, α-pinene, *trans*-caryophyllene and β-selinene were the principal compounds in *P. eriopodon* EO from Venezuela (Ustáriz et al. 2020). β-Caryophyllene, β-selinene and dillapiole were reported as the majority compounds in *P. eriopodon* EO from Colombia (Tangarife-Castaño et al. 2014, da Silva et al. 2017). The presence of the phenylpropanoids (benzodioxols), myristicin and dillapiole were the main differences between the *P. eriopodon* reported in the present study and they are reported by Ustáriz et al. (2020) and Tangarife-Castaño et al. (2014).

According to da Silva et al. (2017) and Salehi et al. (2019), *P. hispidum* EOs from various regions of Brazil presented the following principal compounds: (i) α-pinene, camphene, β-phellandrene, β-caryophyllene, α-guaene, γ-cadinene, γ-elemene; (ii) α-pinene, β-pinene, δ-3-carene; (iii) α-pinene, β-pinene, α-copaene; and (iv) δ-3-carene, limonene, α-copaene, β-caryophyllene, α-humulene, β-selinene, caryophyllene oxide. α-Pinene, β-pinene, δ-3-carene, β-elemene, β-caryophyllene, germacrene B, spatulenol and caryophyllene oxide were the majority compounds in *P. hispidum* collected in Venezuela (da Silva et al. 2017, Salehi et al. 2019). α-Phellandrene, eucalyptol, α-pinene, D-limonene and β-pinene were principal compounds in *P. aff hispidum* from Perú (Ruiz et al. 2015). δ-3-Carene, p-cymene, limonene, elemol, γ-elemene and β-eudesmol were the majority compounds in *P. aff hispidum* collected in Colombia (Jaramillo-Colorado et al. 2019). Eudesmol is the majority in *P. hispidum* EO from Cuba (Pino et al. 2004a); dillapiole in *P. hispidum* EO from Panamá (Santana and Gupta 2018) and trans-nerolidol (oxygenated sesquiterpenes: 46.4%) in *P. hispidum* EO from Colombia (Benitez et al. 2009).

Dos Santos et al. (2001) describe *P. hispidum* EOs mainly composed of sesquiterpenoids; however, da Silva et al. (2017) showed *P. hispidum* EOs are rich in monoterpenes. In general, monoterpenes and sesquiterpenes are the most common compounds in *P. hispidum* EOs. *P. hispidum* of this study is associated with the reports of Benitez et al. (2009) and dos Santos et al. (2001) in which sesquiterpenes (hydrocarbons and oxygenated) are the main compounds.

P. marginatum EOs are characterized by phenylpropanoids (safrole and propiopiperone); however, seven different chemotypes have been described (da Silva et al. 2017, Andrade et al. 2008). *P. marginatum* collected in Colombia presented α-phellandrene, limonene, β-elemene, *trans*-β-caryophyllene and bicyclogermacrene as major compounds (Stashenko and Martínez 2018). Elemecin, *trans*-β-caryophyllene, α-phellandrene, isoelemecine, limonene, bicyclogermacrene, β-elemene, *trans*-anetol, exalatacin, α-pinene, *cis*-methyl isoeugenol, β-phellandrene, *cis*-nerolidol, α-humulene, *trans*-methyl isoeugenol, β-pinene, *cis*-asarone and β-myrcene were reported in another *P. marginatum* EO from Colombia (Olivero-Verbel et. al. 2009). (E)-Anethole, p-anisaldehyde, anisyl methyl ketone, estragole and p-cymene were the majority in *P. marginatum* from Costa Rica (Autran et al. 2009). (i) (E)-Asarone, patchouli alcohol, (Z)-asarone, elemicin, caryophyllene, seychellene, (E)-methyleugenol and copaene; and (ii) myristicin, propiopiperone, caryophyllene and 3-carene were principal compounds in *P. marginatum* EOs from in Brazil (da Silva et al. 2017). *P. marginatum* reported in this study is associated with the report of Stashenko and Martínez (2018) in which sesquiterpenes were the main compounds.

Trans-Calamenene, caryophyllene oxide, spathulenol and α-copaene were the majority in *P. peltatum* EO from Cuba (Pino et al. 2004b). Hydrocarbon sesquiterpenes were the majority in *P. peltatum* EO reported in the present study. For *P. obtusilimbum* and *Piper* sp., no studies on chemical composition were found. Safrole, the main compound reported in *P. obtusilimbum* EO in this study has been found in other *Piper* spp.: *P. betle*, *P. hispidinervium*, *P. auritum*, *P. marginatum* and *P. umbellatum* (Mohit et al. 2014, Cremasco and Braga 2010, Rodríguez et al. 2013). Safrole has shown repellent activity, and it has been reported in an amount of 93.2% for *P. auritum* (Caballero et al. 2014). 13-epi-manoyl oxide is the main compound reported in *Piper* sp. EO in this study; it is a compound that exhibits several important properties (anti-bacterial, anticancer and anti-inflammatory), and it is a putative precursor of forskolin (which is used in traditional medicine as a supplement and as an agent for weight loss agent) (NCBI 2021).

Chemical Composition of Peperomia spp.

Z-Caryophyllene, E-caryophyllene, E-nerolidol and caryophyllene oxide were found in *P. angustata* EO, which was obtained from aerial parts (leaves and stems) with a yield of 0.0032%. α-Copaene, E-caryophyllene, α-curcumene, ω-cadinene, trans-calamenene, E-nerolidol and caryophyllene oxide were the major components in *P. glabella* leaves EO, which had a yield of 0.0033%. E-Nerolidol, germacrene-D-4-ol, viridiflorol, carotol, dillapiol and apiol were the major components in EO from *P. pellucida* aerial parts (yield of 0.06%). E-Caryophyllene, D-germacrene, γ-cadinene, ω-cadinene and caryophyllene oxide were the major components in aerial parts of *P. serpens* VF (yield of 0.02%). EO or VF was not obtained for *P. quadrangularis*.

Analysis of major compounds indicated that the *Peperomia* spp. evaluated have mostly hydrocarbon sesquiterpenes: *P. glabella*, *P. serpens* and *P. angustata*. Benzenoids (benzodioxols) and oxygenated sesquiterpenes were principal compounds for *P. pellucida*. *P. pellucida* EOs , which are typically composed of phenylpropanoids and sesquiterpenes. The most predominant chemotype is characterized by the presence of dillapiole with amounts ranging from 20.7–55.3% (Salehi et al. 2019).

Specimens collected in the Brazilian Amazon, Cameroon (Africa) and India are marked by dillapiole (Salehi et al. 2019). Trans-3-Pinanone and linalool were found in *P. pellucida* EO from Nigeria (Salehi et al. 2019). Dillapiole and carotol were found in *P. pellucida* EO from India (Verma et al. 2015). α-Humulene and β-caryophyllene were the major constituents found from a headspace analysis of *P. pellucida* fresh leaf (da Silva et al. 2006). The chemical composition of *P. pellucida* EO in this study is associated with the reports from Brazil, Africa and India in which dillapiole was found (Verma et al. 2015, Amarathunga and Kankanamge 2017, Salehi et al. 2019).

(Z)-Nerolidol acetate and (E)-nerolidol have been found in *P. serpens* EO (da Silva et al. 2006). Nevertheless, α-humulene and β-caryophyllene were the major constituents in *P. serpens* fresh-leaf analyzed by Solid Phase Micro Extraction (SPME) (da Silva et al. 2006). (E)-Nerolidol, ledol, α-humulene, (E)-caryophyllene and α-eudesmol were found in other *P. serpens* EO from Brazil (Pinheiro et al. 2011). Sesquiterpenes (hydrocarbons) have been found as the main compounds of *P. serpens* VF in the present study. α-Copaene and methyl chavicol (phenylpropanoid) were the largest components of *P. glabella* EO by Klein-Júnior et al. (2016). Sesquiterpenes have been found as the main compounds of *P. glabella* EO in the present study.

TAA and TPC of Piperaceae from Arauca

The Folin-Ciocalteau's colorimetric test allows an approximate estimation of the TPC of a substance (pure or mixed), specifically the concentration of phenolic compounds is evaluated by employing a colorimetric method involving an oxidation-reduction process, which is facilitated by the presence of hydroxyl functional groups (Tamuly et al. 2013).

As for the TPC of EE of leaves of the *Piper* spp. studied, it was found that *Piper* sp. (11.4 ± 2.2 g AG/100 g of EE) and *P. hispidum* (11.7 ± 0.6 g AG/100 g of EE), which is presented with higher TPC, while *P. phytolaccifolium* (9.8 ± 1.1 g AG/100 g of EE), *P. marginatum* (8.5 ± 0.8 g AG/100 g of EE), *P. eriopodon* (7.3 ± 0.4 g AG/100 g of EE), *P. obtusilimbum* (5.8 ± 0.6 g AG/100 g EE) and *Piper* sp. inflorescences (3.1 ± 0.7 g AG/100 g EE) showed the lowest values. EE was not obtained for *P. peltatum*.

Based on the EE yield, the TPC of *Piper* spp. leaves were obtained as follows: *P. phytolaccifolium* (0.46 ± 0.05 g AG/100 g plant), *P. obtusilimbum* (0.35 ± 0.04 g AG/100 g plant), *P. hispidum* (0.30 ± 0. 02 g AG/100 g plant), *P. marginatum* (0.30 ± 0.03 g AG/100 g plant), *Piper* sp. inflorescences (0.20 ± 0.04 g AG/100 g plant), *Piper* sp. leaves (0.19 ± 0.04 g AG/100 g plant) and *P. eriopodon* (0.096 ± 0.005 g AG/100 g plant). TPCs for *P. batle*, *P. betleoides* and *P. wallichii* [5.63, 2.51 and 1.03 (g AG/100 g extract), respectively] were shown by Tamuly et al. (2013). These were lower than the *Piper* spp. of this study (*Piper* sp., *P. hispidum*, *P. phytolaccifolium*, *P. marginatum*, *P. eriopodon* and *P. obtusilimbum*).

On the other hand, the TAA values of the *Piper* spp. of this study were very similar to each other. The highest value was found for *P. obtusilimbum*

[0.009 ± 0.001 (mmol of Trolox/kg substance)] and *Piper* sp. inflorescences [0.006 ± 0.001 (mmol of Trolox/kg substance)] followed by *P. eriopodon* [0.0043 ± 0. 0003 (mmol of Trolox/kg substance)], *P. hispidum* [0.00419 ± 0.00005 (mmol of Trolox/kg substance)] and *P. marginatum* [0.004 ± 0.001 (mmol of Trolox/ kg substance)]. *P. phytolaccifolium* [0.0025 ± 0.0003 (mmol Trolox/kg substance)] and *Piper* sp. [0.0023 ± 0.0003 (mmol Trolox/kg substance)] were the species with the lowest TAA values.

Piper spp. with high TAA have been reported in the literature; some of these are *P. betle*, *P. aleyreanum*, *P. lanceaefolium*, *P. aduncum*, *P. acutifolium*, *P. glabratum*, *P. heterophyllum* and *P. pilliraneum* (Annegowda et al. 2013, Ruddock et al. 2011, Facundo et al. 2012). Also, 0.000516 ± 0.000004 (mmol Trolox/kg substance) was estimated for *P. bredemeyeri* and 0.000585 ± 0.000002 (mmol Trolox/kg substance) was found for *P. eriopodon*, which are values lower than those obtained in this research (Castañeda, 2007).

The antioxidant activity measured by TAA is related to the total reducing capacity of a substance or a mixture in a reaction with the ABTS[+] (radical-cation). TAA is related to substances with a high reduction capacity (e.g., phenols, carotenoids and other anti-radical compounds), which facilitates reactivity with ABTS[+] (Re et al. 1999).

In general, the higher the content of reducing compounds, the higher the TAA values are expected to be (Castañeda 2007). It should be noted that for a substance to be a good antioxidant, it must have high solubility in the medium and a suitable structure to interact with free radicals [e.g., ABTS[+] or Reactive Oxygen Species (ROS)]; it must be a highly reactive reductant (Mesa et al. 2010), with reactive functional groups or the possibility of structural conjugation (Miller and Rice-Evans 1997). For instance, sesquiterpenes with a high degree of unsaturation have a good reducing effect (Beyer and Walter 1987).

Conclusions

Among the objectives of this work were to: (i) determine the chemical composition of *Piperaceae* spp. collected in Colombian-Venezuelan plains; (ii) establish relationships with *Piperaceae* spp. reported in the scientific literature; (iii) propose or estimate substances with potential for commercial or industrial application; and (iv) establish the possible relationship between antioxidant activity (TAA), phenol content (TPC) and chemical composition for *Piper* spp.

Hydrocarbon sesquiterpenes were the compound principals detected in *Piperaceae* spp. EOs evaluated *P. marginatum*, *P. peltatum*, *P. eriopodon*, *P. glabella*, *P. serpens* and *P. angustata*; oxygenated sesquiterpenes were the major ones for *Piper* sp. and *P. hispidum*. Phenylpropanoids (benzodioxols) were the major components of *P. obtusilimbum* and *P. pellucida*.

P. hispidum, *P. marginatum* and *P. peltatum* showed some similarities with species previously reported in the literature. On the other hand, *P. eriopodon* showed notable differences from previous studies, which indicated phenylpropanoid (benzodioxol) compounds present in this species. Safrole and 13-epi-manoyl oxide, which were found in *P. obtusilimbum* and *Piper* sp., respectively, have commercial or industrial potential, which is according to previous reports on biological activities (repellent, anti-bacterial, anticancer and anti-inflammatory).

Piper spp. had similar or somewhat higher TPCs than those reported for other *Piper* spp. However, they are lower than other botanical families or species. *Piper* sp. and *P. hispidum*, with oxygenated sesquiterpenes as principal compounds, showed the highest TPC values. Also, *Piper* spp. had a very lower TAA than other botanical families or species. However, they were similar to other *Piper* spp. previously reported. *P. obtusilimbum* and *Piper* sp. inflorescences EEs with phenylpropanoids and oxygenated sesquiterpenes as principal compounds, respectively, showed the highest TAA values.

Acknowledgment

The authors are grateful for the financial support offered by the Patrimonio Autónomo Fondo Nacional de Financiamiento para la Ciencia, la Tecnología y la Innovación, Francisco José de Caldas, Contract RC-0572-2012-Bio-Red-Co-CENIVAM, and to the Instituto de Estudios de la Orinoquia (IEO), of the UNAL.

References

Adams, R.P. 2007. Identification of Essential Oil Components by Gas Chromatography/Mass Spectrometry. Allured Publishing Co., Illinois, USA.

Amarathunga, A.A.M.D.D.N. and S.U. Kankanamge. 2017. A review on pharmacognostic, phytochemical and ethnopharmacological findings of *Peperomia pellucida* (l.) kunth: pepper Elder. International Research Journal of Pharmacy 8: 16–23.

Andrade, E.H., L.M. Carreira, M.H. da Silva, J.D. da Silva, C.N. Bastos, P.J. Sousa et al. 2008. Variability in essential oil composition of *Piper marginatum sensu lato*. Chemistry & Biodiversity 5: 197–208.

Annegowda, H., P. Tan, M. Mordi, S. Ramanathan, M.R. Hamdan, M.H. Sulaiman et al. 2013. TLC-bioautography-guided isolation, HPTLC and GC-MS-assisted analysis of bioactives of *Piper betle* leaf extract obtained from various extraction techniques: *In vitro* evaluation of phenolic content, antioxidant and antimicrobial activities. Food Analytical Methods 6: 715–726.

Autran, E.S., I.A. Neves, C.S. da Silva, G.K. Santos, C.A. da Câmara and D.M. Navarro. 2009. Chemical composition, oviposition deterrent and larvicidal activities against *Aedes aegypti* of essential oils from *Piper marginatum* Jacq. (Piperaceae). Bioresource Technology 100: 2284–2288.

Ballesteros, J.L., F. Bracco, M. Cerna, P. Vita Finzi and G. Vidari. 2016. Ethnobotanical research at the Kutukú scientific station, Morona-Santiago, Ecuador. BioMed Research International, 9105746.

Benitez, N.P., E.M. Melendez Leon and E.E. Stashenko. 2009. Essential oil composition from two species of *Piperaceae* family grown in Colombia. Journal of Chromatographic Science 4: 804–807.

Bernal, R. In Catálogo de plantas y líquenes de Colombia. 2015. UNAL, Bogotá. http://catalogoplantasdecolombia.unal.edu.co. Accessed on: 27 Jul 2021.

Beyer, H. and W. Walte. 1987. Manual de Química Orgánica. Reverté S.A., Barcelona (España).

Bussmann, R.W., N.Y. Paniagua Zambrana, C. Romero and R.E. Hart. 2018. Astonishing diversity-the medicinal plant markets of Bogotá, Colombia. Journal of Ethnobiology and Ethnomedicine 14: 43.

Caballero, K., J. Olivero, N. Pino and E. Stashenko. 2014. Chemical composition and bioactivity of *Piper auritum* and *P. multiplinervium* essential oils against the red flour beetle, *Tribolium castaneum* (Herbst). Boletín Latinoamericano y del Caribe de Plantas Medicinales y Aromáticas 13: 10–19.

Castañeda, M. 2007. Estudio de la composición química y la actividad biológica de los aceites esenciales de diez plantas aromáticas colombianas. Degree work presented as a partial requirement to obtain the degree in chemistry. Universidad Industrial de Santander, Facultad de Ciencias, Escuela de Química, Bucaramanga, Colombia.

Celis, A., C. Mendoza, M. Pachón, J. Cardona, W. Delgado and L.E. Cuca. 2008. Extractos vegetales utilizados como biocontroladores con énfasis en la familia *Piperaceae*. Una revisión. Agronomía Colombiana 26: 97–106.

Centero, J.J., C.O. Núñez, G. Bernardello, A. Amuchastegui, J. Mulko, P. Brandolin et al. 2019. Las Plantas de Importancia económica en Argentina. UniRio, Argentina.

Chaveerach, A., P. Mokkamul, R. Sudmoon and T. Tanee. 2012. Ethnobotany of the genus *Piper* (*Piperaceae*) in Thailand. Ethnobotany Research and Applications 4: 223–231.

Correa Navarro, Y.M., R.P. García and O. Mosquera. 2015. Actividad antioxidante y antifúngica de *Piperaceae*s de la flora colombiana. Revista Cubana de Plantas Medicinales 19: 167–181.

Cremasco, M. and M. Brag. 2010. Isomerização do óleo essencial de pimenta-longa (*Piper hispidinervium* C. DC) para a obtenção de isosafrol. Acta Amazónica 40: 737–740.

Cruz, S.M., A. Cáceres, L.E. Álvarez, M.A. Ape and A.T. Henriques. 2012. Chemical diversity of essential oils of 15 *Piper* species from Guatemala. Acta Horticulturae 964: 39–46.

Da Silva, A.C.M., E.H.A. Andrade, L.M.M. Carreira, E.F. Guimarães and S.M.J. Guilherme. 2006. Essential oil composition of *Peperomia serpens* (Sw.) Loud. Journal of Essential Oil Research 18: 269–271.

Da Silva, J.K., R. da Trindade, N.S. Alves, P.L. Figueiredo, J.G.S. Maia and W.N. Setzer. 2017. Essential oils from Neotropical *Piper* species and their biological activities. International Journal of Molecular Sciences 18: 2571.

Dastmalchi, K., D. Dorman, M. Kosarb and R. Hiltunen. 2007. Chemical composition and *in vitro* antioxidant evaluation of a water soluble Moldavian balm (*Dracocephalum moldavica* L.) extract. LWT - Food Science and Technology 40: 239–248.

Delgado, W., M.E. Pachón, A. Celis, C. Mendoza, J.O. Cardona, M. Bustamante, M. Daza and L.E. Cuca. 2007. Informe técnico de avance proyecto Bioprospección participativa de comunidades vegetales asociados a la familia *Piperaceae* en la región del Sumapaz medio bajo occidental. Colciencias-Universidad Nacional de Colombia-Universidad de Cundinamarca, Bogotá, Colombia.

De Moraes, M.M. and M.J. Kato. 2021. Biosynthesis of pellucidin A in *Peperomia pellucida* (L.) HBK. Frontiers in Plant Science 12: 413.

Dos Santos, P.R., D. de Limas Moreira, E.F. Guimarães and M.A. Kaplan. 2001. Essential oil analysis of 10 *Piperaceae* species from the Brazilian Atlantic forest. Phytochemistry 58: 547–551.

Durant-Archibold, A.A., A.I. Santana and M.P. Gupta. 2018. Ethnomedical uses and pharmacological activities of most prevalent species of genus *Piper* in Panama: A review. Journal of Ethnopharmacology 217: 63–82.

Dyer, L.A., J. Richards and C.D. Dodson. 2004. Isolation, synthesis, and evolutionary ecology of *Piper* amides. In: *Piper*: A Model Genus for Studies of Phytochemistry, Ecology, and Evolution. Kluwer Academic/Plenum Publishers, New York, NY, USA.

Facundo, V.A., L.J.J. Bálico, D.K.S. Lima, A.R.S. Santos, S.M. Morais, G.V.J. da Silva et al. 2012. Non-substituted B-ring flavonoids and an indole alkaloid from *Piper aleyreanum* (*Piperaceae*). Biochemical Systematics and Ecology 45: 206–208.

Fern K. Tropical Plants Database. tropical.theferns.info. <tropical.theferns.info/viewtropical.php?id=Piper+peltatum>. Accessed on: 11 Nov 2021.

GBIF Occurrence Download https://www.gbif.org/dataset/5f78d639-26ec-473b-90c9-4efe59a0632d. Accessed on: 29 Jul 2021.

Godefroot, M., P. Sandra and M. Verzele. 1981. New method for quantitative essential oil analysis. Journal of Chromatography A 203: 325–335.

Jaramillo, M.A. and P.S. Manos. 2001. Phylogeny and patterns of floral diversity in the genus *Piper* (*Piperaceae*). American Journal of Botany 88: 706–716.

Jaramillo-Colorado, B.E., N. Pino-Benitez and A. González-Coloma. 2019. Volatile composition and biocidal (antifeedant and phytotoxic) activity of the essential oils of four *Piperaceae* species from Choco-Colombia. Industrial Crops and Products 138: 111463.

Joulain, D. and W.A. Koenig. 1998. The Atlas of Spectral Data of Sesquiterpene Hydrocarbons. E.B.-Verlag, Hamburg, Alemania.

Juep Bakuants, A. 2008. Rescate del conocimiento tradicional y biológico para el manejo de productos forestales no maderables en la comunidad indígena Jameykari. Programa de educación para el desarrollo y la conservación escuela de posgrado. Tesis sometida a consideración de la Escuela de

Posgrado, Programa de Educación para el Desarrollo y la Conservación del Centro Agronómico Tropical de Investigación y Enseñanza- Magister Scientiae en Manejo y Conservación de Bosque Tropical y Biodiversidad. Costa Rica.

Kato, M. and M. Furlan. 2007. Chemistry and evolution of the *Piperaceae*. Pure and Applied Chemistry 79: 529–538.

Kartika, I.G.A.A., I.J. Bang, C. Riani, M. Insanu, J.H. Kwak, K.H. Chung et al. 2020. Isolation and characterization of phenylpropanoid and lignan compounds from *Peperomia pellucida* [L.] Kunth with estrogenic activities. Molecules 25: 4914.

Klein-Júnior, L.C., M.A. Recalde-Gil, C. dos Santos Passos, F. Gobbi de Bitencourt, J. Salton, L. Jacobi Danielli et al. 2016. The monoamine oxidase inhibitory activity of essential oils obtained from *Peperomia* Ruiz. & Pav. (*Piperaceae*) species and their chemical composition. Journal of Essential Oil Bearing Plants 19: 1762–1768.

Kumar, B., S. Tiwari, V. Bajpai and B. Singh. 2020. Phytochemistry of plants of genus *Piper*. First edition. CRC Press, Taylor and Francis Group, Boca Ratón, USA.

Martínez-Bautista, B.G., L.A. Bernal-Ramírez, D. Bravo-Avilez, M.S. Samain, J.M. Ramírez-Amezcua and B. Rendón-Aguilar. 2019. Traditional uses of the family *Piperaceae* in Oaxaca, Mexico. Tropical Conservation Science 12: 1–22.

Melo, A., E.F. Guimarães and M. Alves. 2016. Synopsis of the genus *Peperomia* Ruiz & Pav. (*Piperaceae*) in Roraima State, Brazil. Hoehnea 43: 119–134.

Mesa, A., C. Gaviria, F. Cardona, J. Sáez, S. Blair and B. Rojano. 2010. Actividad antioxidante y contenido de fenoles totales de algunas especies del género *Calophyllum*. Revista Cubana de Plantas Medicinales 15: 13–26.

Mgbeahuruike, E.E., V.H. Yrjönen Teijo, H. Vuorela and Y. Holm. 2017. Bioactive compounds from medicinal plants: Focus on *Piper* species. South African Journal of Botany 112: 54–69.

Miller, N. and C. Rice-Evans. 1997. The relative contributions of ascorbic acid and phenolic antioxidants to the total antioxidant activity of orange and apple fruit juices and blackcurrant drink. Food Chemistry 60: 331–337.

Mohit, S., K. Naveen, S. Priyanka, V. Kodakandla and K. Santosh. 2014. Antimicrobial activity and chemical composition of leaf oil in two varieties of *Piper betle* from northern plains of India. Journal of Scientific & Industrial Research 73: 95–99.

Monache, F.D. and R.S. Compagnone. 1996. A secolignan from *Peperomia glabella*. Phytochemistry 43: 1097–1098.

Murcia, A.M. and H. Bermúdez. 2008. Evaluación de la actividad insecticida de extractos vegetales de la familia *Piperaceae*, sobre *Spodoptera frugiperda* Smith, en condiciones semicontroladas. Trabajo de grado. Facultad de Ciencias Agropecuarias, Universidad de Cundinamarca, Fusagasugá, Cundinamarca, Colombia.

National Center for Biotechnology Information (NCBI). PubChem Compound Summary for CID 6432025, Labd-14-ene, 8,13-epoxy-, (13S)-. https://pubchem.ncbi.nlm.nih.gov/compound/Labd-14-ene_-8_13-epoxy-_-_13S. Accessed on: 01 Nov 2021.

Núñez, V., V. Castro, R. Murillo, L.A. Ponce-Soto, I. Merfort and B. Lomonte. 2005. Inhibitory effects of *Piper* umbellatum and *Piper peltatum* extracts towards myotoxic phospholipases A2 from Bothrops snake venoms: Isolation of 4-nerolidylcatechol as active principle. Phytochemistry 66: 1017–1025.

Olivero-Verbel, J., J. Güette-Fernandez and E. Stashenko. 2009. Acute toxicity against *Artemia franciscana* of essential oils isolated from plants of the genus *Lippia* and *Piper* collected in Colombia. Boletín Latinoamericano y del Caribe de Plantas Medicinales y Aromáticas 8: 419–427.

Online Archive of National Institute of Standards and Technology (NIST) [Available from: http:/ / webbook.nist.gov/chemistry]. Department of Commerce. U.S. Accessed on: 27 Jul 2021.

Parmar, V.S., S.C. Jain, K.S. Bisht, R. Jain, P. Taneja, A. Jha et al. 1997. Phytochemistry of the genus *Piper*. Phytochemistry 46: 597–673.

Pinheiro, B.G., A.S. Silva, G.E. Souza, J.G. Figueiredo, F.Q. Cunha, S. Lahlou et al. 2011. Chemical composition, antinociceptive and anti-inflammatory effects in rodents of the essential oil of *Peperomia serpens* (Sw.) Loud. Journal of Ethnopharmacology 138: 479–486.

Pino, J.A., R. Marbot, A. Bello and A. Urquiola. 2004a. Composition of the essential oil of *Piper hispidum* Sw. from Cuba. Journal of Essential Oil Research 16: 459–460.

Pino, J.A., R. Marbot, A. Bello and A. Urquiola. 2004b. Essential oils of *Piper peltata* (L.) Miq. and *Piper aduncum* L. from Cuba. Journal of Essential Oil Research 16: 124–126.

Puertas-Mejía, M.A., L. Gómez-Chabala, B. Rojano and J.A. Sáez-Vega. 2009. Capacidad antioxidante *in vitro* de fracciones de hojas de *Piper peltatum* L. Revista Cubana de Plantas Medicinales 14: 1–11.

Raghavendra, H.L. and T.R. Prashith Kekuda. 2018. Ethnobotanical uses, phytochemistry and pharmacological activities of *Peperomia pellucida* (l.) kunth (*Piperaceae*)—A review. International Journal of Pharmacy and Pharmaceutical Sciences 10: 1–8.

Rangel-Churio, J.O. 2015. The richness of flowering plants in Colombia. Caldasia 37: 279–307.

Re, R., N. Pellegrini, A. Proteggente, A. Pannala, M. Yang and C. Rice-Evans. 1999. Antioxidant activity applying an improved ABTS radical cation decolorization assay. Free Radical Biology & Medicine 26: 1231–1237.

Rodríguez, E.J., Y. Saucedo-Hernández, Y.V. Heyden, E.F. Simó-Alfonso, G. Ramis-Ramos, M.J. Lerma-García et al. 2013. Chemical analysis and antioxidant activity of the essential oils of three *Piperaceae* species growing in the central region of Cuba. Natural Product Communications 8: 1325–1328.

Ruddock, P., M. Charland, S. Ramírez, A. López, G.H. Neil Towers, J.T. Arnason et al. 2011. Antimicrobial activity of flavonoids from *Piper lanceaefolium* and other Colombian medicinal plants against antibiotic susceptible and resistant strains of *Neisseria gonorrhoeae*. Sexually Transmitted Diseases 38: 82–88.

Ruiz, C., C. Díaz and R. Rojas. 2015. Composición química de aceites esenciales de 10 plantas aromáticas peruanas. Revista de la Sociedad Química del Perú 81: 81–94.

Saga Kitamura, R.O., P. Romoff, M.C.M. Young, M.J. Kato Massuo and J.H.G. Lago. 2006. Chromenes from *Peperomia serpens* (Sw.) Loudon (*Piperaceae*). Phytochemistry 67: 2398–2402.

Salehi, B., Z.A. Zakaria, R. Gyawali, S.A. Ibrahim, J. Rajkovic, Z.K. Shinwari et al. 2019. *Piper* species: A comprehensive review on their phytochemistry, biological activities and applications. Molecules 24: 1364.

Santana, A.I. and M.P. Gupta. 2018. Potential of Panamanian aromatic flora as a source of novel essential oils. Biodiversity International Journal 2: 405–413.

Soares, M.G., A.P.V. de Felippe, E.F. Guimarães, M.J. Kato, J. Ellena and A.C. Doriguetto. 2006. 2-Hydroxy-4,6-dimethoxyacetophenone from Leaves of *Peperomia glabella*. Journal of the Brazilian Chemical Society 17: 1205–1210.

Stashenko, E. and J.R. Martínez. 2018. The expression of biodiversity in the secondary metabolites of aromatic plants and flowers growing in Colombia, potential of essential oils. *In*: H.A. El-Shemy (ed.). Potential of Essential Oils. Cairo University. Egypt.

Stashenko, E., B. Jaramillo and J. Martínez. 2004. Comparison of different extraction methods for the analysis of volatile secondary metabolites of *Lippia alba* (Mill.) N.E. Brown, grown in Colombia, and evaluation of its *in vitro* antioxidant activity. Journal of Chromatography A 1025: 93–103.

Sua Tunjano, S.M., M.F. González Giraldo and S. Sua. 2021. Base de Datos de Flora de la Cuenca del Río Orinoco en Colombia. Version 2.8. (Instituto Amazónico de Investigaciones Científicas – SINCHI, 2021). Occurrence dataset https://doi.org/10.15472/2dbndj. Accessed on: 29 Jul 2021.

Tafurt-García, G. and A. Muñoz. 2018. Volatile secondary metabolites in Cascarillo (*Ocotea caparrapi* (Sandino-Groot ex Nates) Dugand - Lauraceae). Journal of Essential Oil Bearing Plants 21: 374–387.

Tafurt García, G., L. Jiménez Vidal and A. Calvo Salamanca. 2015. Antioxidant capacity and total phenol content of *Hyptis* spp., *P. heptaphyllum*, *T. panamensis*, *T. rhoifolia*, and *Ocotea* sp. Revista Colombiana de Química 44: 28–33.

Takeara, R., R. Goncalves, A.V.F. dos Santos and A. Cavalcante Guimaraes. 2017. Biological properties of essential oils from the *Piper* species of Brazil: A review. *In*: H.A. El-Shemy (ed.). Aromatic and Medicinal Plants, Cairo University. Egypt.

Tamuly, C., M. Hazarika, J. Bora, J. Boroloi, M. Boruah and P. Gajurel. 2013. *In vitro* study on antioxidant activity and phenolic content of three *Piper* species from North East India. Journal of Food Science and Technology 52: 1–12.

Tangarife-Castaño, V., J.B. Correa-Royero, V.C. Roa-Linares, N. Pino-Benitez, L.A. Betancur-Galvis, D.C. Durán et al. 2014. Anti-dermatophyte, anti-fusarium and cytotoxic activity of essential oils and plant extracts of *Piper* genus. Journal of Essential Oil Research 26: 221–227.

Thin, D.B., H.V. Chinh, N.X. Luong, T.M. Hoi, D.N. Dai and I.A. Ogunwande. 2018. Chemical analysis of essential oils of *Piper laosanum* and *Piper acre* (*Piperaceae*) from Vietnam. Journal of Essential Oil Bearing Plants 21: 181–188.

Ustáriz Fajardo, F.J., M.E. Lucena de Ustáriz, F.G. Urbina Carmona, D.M. Villamizar Sánchez, L.B. Rojas Fermín, Y.E. Cordero de Rojas et al. 2020. Composition and antibacterial activity of the *Piper eriopodon* (miq.) C.DC. Essential oil from the venezuelan Andes. Pharmacologyonline 2: 13–22.

Vásquez-Ocmín, P.G., A. Gadea, S. Cojean, G. Marti, S. Pomel, A.C. Van Baelen et al. 2021. Metabolomic approach of the antiprotozoal activity of medicinal *Piper* species used in peruvian amazon. Journal of Ethnopharmacology 264: 113262.

Vergara Rodríguez, D. 2013. Diversidad y distribución de las especies del género *Peperomia* (*Piperaceae*) en el estado de Veracruz. Centro de Investigaciones Tropicales. Tesis para obtener el grado de maestra en ecología tropical. Universidad Veracruzana. México.

Verma, R.S., R.C. Padalia, P. Goswam and A. Chauhan. 2015. Essential oil composition of *Peperomia pellucida* (L.) Kunth from India. Journal of Essential Oil Research 27: 89–95.

World Flora Online (WHO). Published on the Internet; http://www.worldfloraonline.org. Accessed on: 27 Jul 2021.

3

Chemical Diversity of Essential Oils and Volatile Fractions of Some Plants Found in the Tropical Forest/Savannah from Colombia

Amner Muñoz-Acevedo,[1,*] María C. González,[1]
Osnaider J. Castillo,[1] Cindy P. Guzmán,[1]
Nubellys M. Peralta[1] and Martha Cervantes-Díaz[2]

Introduction

Colombia is the second richest country in the world in plant biodiversity (20299 species - angiosperms), distributed between natural forests and savannahs among others; nonetheless, these biomes are deteriorating due to uncontrolled anthropogenic activity (illegal mining, extensive agriculture/animal breeding and illicit crops related to drug trafficking/armed conflict) that has put them at risk. Despite this, these ecosystems have important chemical, biological and genetic resources, whose applications in medicine, food, pharmaceutical, cosmetics and perfumery would be sustainable alternatives that would favor/help to conserve them; some of the plants have shown promising results due to the chemical compositions of their essential oils or volatile fractions and/or their biological activities.

These vegetable species of different botanical families are fragrant trees or shrubs/herbs with flowers and/or fruits and resins/exudates; the most of these plants have some ethnobotanical use or application for indigenous/afro-descendant/ peasant communities in the urban and/or rural areas. Some of them have restricted geographical distribution as well. The promising plants are *Annona purpurea*

[1] Departamento de Química y Biología, Universidad del Norte, Barranquilla, Colombia.
[2] Grupo Investigaciones Ambientales para el Desarrollo Sostenible, Facultad de Química Ambiental, Universidad Santo Tomás, Bucaramanga, Colombia.
* Corresponding author: amnerm@uninorte.edu.co

(guanábana matimbá), *Astronium graveolens* (quebracho/Santa Cruz), *Bursera glabra* (caraño de monte), *B. graveolens* (caraña/palo santo), *B. simaruba* (indio desnudo/almácigo), *Chromolaena barranquillensis* (rositavieja), *Cordia curassavica* (maíz tostado), *Croton fragilis* (llora sangre), *Cr. fragrans* (cascarillero), *Cr. malambo* (malambo/palo matías), *Cr. niveus* (plateado/colpachí), *Cyanthillium cinereum* (venadillo), *Dalea carthagenensis* (añís/escobilla), *Eugenia procera* (arrayán), *Lantana cámara* (venturosa/cinco negritos), *Lippia alba* (prontoalivio/ poleo), *Miconia* sp. (clavo de monte), *Piper eriopodon* (cordoncillo), *P. marginatum* (Santa María/oloroso), *Plumeria rubra* (azuceno rojo/frangipani), *Pl. alba* (azuceno/ White frangipani), *Protium heptaphyllum* (anime/elemí) and *Xilopia emarginata* (escobillo/yaya).

This chapter contains information about science dynamics related to these plants, geographical distribution, scientific and vernacular names, botanical and chemical aspects (including some structures) and biological properties.

Text Mining

The analysis of text mining related to bibliometric exploration on the topic of interest in this chapter was based on the following search equation, i.e., (TITLE-ABS-KEY) (*"Annona purpurea"* OR *"Astronium graveolens"* OR *"Bursera glabra"* OR *"Bursera graveolens"* OR *"Bursera simaruba"* OR *"Chromolaena barranquillensis"* OR *"Cordia curassavica"* OR *"Croton fragrans"* OR *"Croton fragilis"* OR *"Croton malambo"* OR *"Croton niveus"* OR *"Cyanthillium cinereum"* OR *"Dalea carthagenensis"* OR *"Eugenia procera"* OR *"Lantana camara"* OR *"Lippia alba"* OR *"Miconia* sp." OR *"Piper eriopodon"* OR *"Piper marginatum"* OR *"Plumeria rubra"* OR *"Plumeria alba"* OR *"Protium heptaphyllum"* OR *"Xylopia emarginata"*) and pub-year > 1999 and [LIMIT-TO (DOCTYPE, "ar")], which was applied in the Scopus database (Elsevier and BV 2021). The data were analyzed by the VantagePoint software (academic version 12, Search Technology). The most important result after cleaning the data was that 3,217 indexed records were found, which were related to the names of the plants and essential oils.

Figure 1 exhibits the scientific dynamics from 2000 to 2021, which tends to an exponential growth in the observation window considered. Accordingly, 2021, 2020, 2012 and 2018 have been the years of greatest science activity with 229, 227, 217 and 212 articles for each year; surprisingly, in the two years of a pandemic due to the SARS-CoV-2 virus (2020/2021), the highest numbers of records were found. If De Solla Price´s Law (De Solla Price 1976) is taken into consideration, the growth rate of the number of publications since 2000 was ca. 14% per year (R^2 0.958); despite the fact that, in the years 2007, 2008, 2016, 2017 and 2019, the number of records decreased significantly.

In other subtopic of text mining, the six main areas of knowledge related to essential oils and plants of interest were: (i) agricultural and biological sciences (32% of the records), (ii) pharmacology/toxicology and pharmaceutics (14% of the records), (iii) biochemistry/genetics and molecular biology (12% of the records), (iv) environmental science (10% of the records), (v) medicine (9% of the records) and (vi) chemistry (7% of the records). Alike, the principal science journals where the articles were published are *Journal of Ethnopharmacology* (119 records), *Journal of Essential Oil Research* (50 records), *Revista Brasileira de Plantas Medicinais* (40 records), *Industrial Crops and Products* (34 records), *Natural Products Communications* (31 records), *Revista Brasileira de Farmacognosia* (28 records), *Boletín Latinoamericano y del Caribe de Plantas Medicinales y Aromáticas* (21 records), *Flavour and Fragrance Journal/Journal of Essential Oil-Bearing Plants* (26 records). Furthermore, the top five countries in the world with the highest records were India (847 records), Brazil (784 records), the United States of America (255 records), Mexico (179) and Colombia (159 records), whereas in Latin America there were, in addition to the three countries mentioned above (Brazil/Mexico/Colombia), Venezuela (38 records), Argentina (31 records) and Cuba (27 records).

It is very interesting to note that when the names of plants with biological activities were crossed in the bibliometric analysis, the most important bioproperties were antimicrobial/antibacterial (22/13% of the records), antioxidant (15% of the records), cytotoxic (10% of the records), insecticide (7% of the records) and anti-inflammatory (6% of the records), while the top five of plants that contained these activities were *Lantana camara* (34% of the records), *Lippia alba* (14% of the records) and *Plumeria rubra/Cyanthillium cinereum* (5% of the records, each).

Lastly, when the vegetable species under study were correlated to Colombia during the cleaning phase of the global bibliometric data, *Lippia alba* (42), *Piper marginatum* (6), *Croton malambo/Lantana camara* (5) presented the highest number

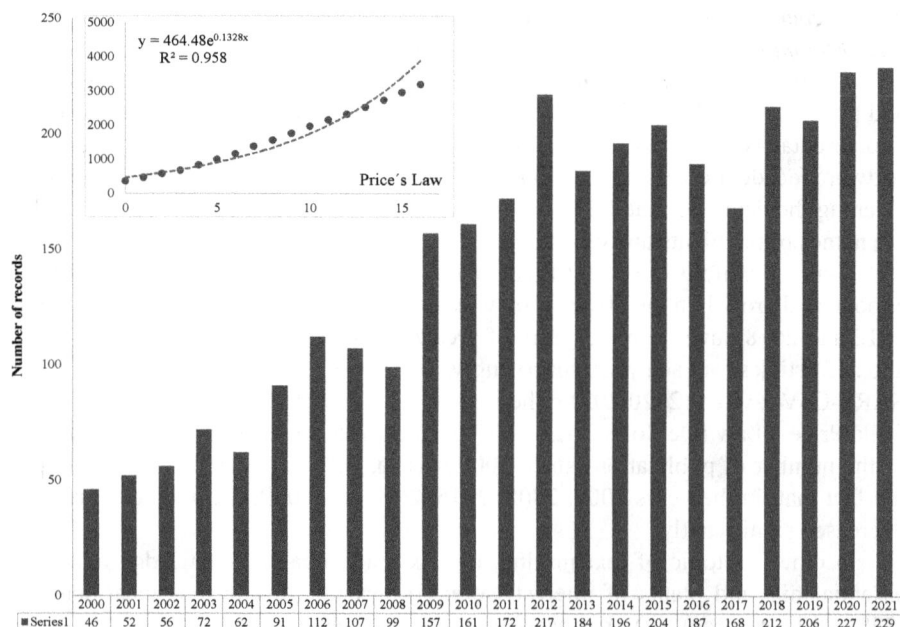

	2000	2001	2002	2003	2004	2005	2006	2007	2008	2009	2010	2011	2012	2013	2014	2015	2016	2017	2018	2019	2020	2021
Series1	46	52	56	72	62	91	112	107	99	157	161	172	217	184	196	204	187	168	212	206	227	229

Inset equation: $y = 464.48e^{0.1328x}$, $R^2 = 0.958$

Figure 1: Source: Bibliometry unit-CRAI library, Universidad Santo Tomás (Bucaramanga). Calculations based on the scopus data (Elsevier, B.V., 2021) analyzed with VantagePoint software (version 12.0, search technology).

of records, while *Annona purpurea*, *Chromolaena barranquillensis*, *Cyanthillium cinereum* and *Dalea carthagenensis* each had one record that related to the essential oils.

Some Plants Found in the Tropical Forest and Savannah from Colombia

Approximately 23 native and/or introduced species belonging to 12 families of angiosperms were selected; e.g., *Annona purpurea*, *Astronium graveolens*, *Bursera glabra*, *B. graveolens*, *B. simaruba*, *Chromolaena barranquillensis*, *Cordia curassavica*, *Croton fragrans*, *Cr. fragilis*, *Cr. malambo*, *Cr. niveus*, *Cyanthillium cinereum*, *Dalea carthagenensis*, *Eugenia procera*, *Lantana camara*, *Lippia alba*, *Miconia* sp., *Piper eriopodon*, *P. marginatum*, *Plumeria rubra*, *Pl. alba*, *Protium heptaphyllum* and *Xilopia emarginata*, which will be briefly described in terms of the botanical generalities, distribution, vernacular names, traditional uses and biological (properties) and chemical (volatile composition) aspects.

Anacardiaceae

Astronium graveolens Jacq. (synonyms - *A. conzattii* S.F. Blake, *A. planchonianum* Engl., *A. zongolica* Reko) was the only species considered in this family. The deciduous tree (up to 35 m of high) contains a sticky resin with a smell (strong) reminiscent of turpentine as well as on the leaves/branches. It is distributed from southern North America to South America (tropic and subtropic zones). It is commonly named as *Santacruz/quebracho/diomate*, *gateado(-Galán)*, *palo de cera/culebra/fierro* (Mexico), *gusanero* (Colombia), *gateado* (Venezuela), *jejuíra/guarita/gonçalo alves* (Brazil), *guasango* (Ecuador), *tigrillo* (Panama), *ronron, palo obrero, jobillo, zorrowod/tigerwood/zebrawood*. Between the ethnobotanical uses of the plant can be mentioned as the treatment of skin allergy and rash (in baths), inflammation, antiseptic, antibiotic, diarrhea and ulcers. The principal non-medicinal use is timber for construction/furniture. Some bioproperties reported are antioxidant, antimicrobial, antiangiogenic and repellent against ants (Bernal et al. 2011, Chen et al. 1984, Fundación BioColombia 2013, Grandtner 2005, Grandtner and Chevrette 2014, HCN-ICN 2021, Hernández et al. 2012, 2013, 2014, Loureiro 2014, Marin and Flores 2002, Quattrocchi 2012, Rodríguez et al. 2012, Sarmiento Bernal et al. 2017, The Plant List 2013).

Table 1 includes the constituents identified in the volatile fractions and essential oils of *As. graveolens* from the northern Colombian region. According to the Table, myrcene (**1**) (57–95%) was the most representative volatile metabolite of the leaves/bladder of the trees in the northern region; though, other terpenoids [(E)-β-ocimene (**2**), β-bisabolol, germacrene B (**3**)] were also present as main components. The essential oil (1.5%) of the aerial parts from Venezuelan trees was constituted by (E)-β-ocimene (24%), α-pinene (19%) and δ-3-carene (16%) as reported by Hernández et al. (2013). The same authors described that the EO was active against *Staphylococcus aureus* (MIC 0.2 μg/mL, MBC 0.2–0.5 μg/mL) and *Pseudomonas aeruginosa* (MIC 0.5 μg/mL).

Table 1: Principal secondary metabolites identified in the volatile fractions and essential oils of different parts from *A. graveolens*.

Type of VSM	Parts of the Plant	Main Chemical Constituents (Relative Amounts, %)	References
Astronium graveolens			
VF by SDE	Leaves	Myrcene (57–95).	Muñoz-Acevedo et al. 2017
	Bladder	Myrcene (73) and α-pinene (18).	
	Branches	Germacrene B (29), germacrene D (22), β-elemene (16) and γ-elemene (10).	
	Bark	β-Bisabolol (31), β-curcumene (11) and α-acorenol (10).	
EO (0.15%)	Leaves	Myrcene (93).	
EO (0.1%)	Leaves	Myrcene (75) and δ-3-carene (21).	
EO (0.1%)	Leaves	(E)-β-Ocimene (40) and myrcene (17).	Rodríguez-Burbano et al. 2010

VSM: Volatile Secondary Metabolites; EO: Essential Oil; VF: Volatile Fraction; SDE: Simultaneous-Distillation Extraction.

Annonaceae

Annona purpurea and *Xilopia emarginata* were the two selected species from the Annonaceae family. Thus, *Annona purpurea* Moc. and Sessé ex Dunal (synonyms - *A. involucrata* Baill., *A. manirote* Kunth and *A. prestoei* Hemsl.) is a flowering tree (up to 20 m), medicinal and has edible fruit, wild-grown or cultivated in home gardens; it is a native species of Mesoamerica and is also distributed in Colombia, Venezuela and Ecuador. In the departamentos de Atlántico, Bolívar, Cordoba and Sucre of the Colombian Caribbean region, the plant has been commonly found. The current names by which it is known are cabeza negro/de indio, catiguire, catuche, cowsap, gallina gorda, guanábana cimarrona/matimbá/ñeca/de monte, guanabarjo, guanacona, mal olor (bad smell due to the "unpleasant" smell of its leaves), manire, matimbá, nejo, soncoya annona, suprecaya, tiragua, toreta, torete, tucuria, etc. Some uses of it are reported in ethnomedicine such as for the treatment of diarrhea and dysentery, inflammation, body pain, fever, pulmonary infection, antimalarial and *mal de ojo* (the evil eye), which is prepared as decoctions/baths. The tree has been used for non-medicinal purposes as wood (agricultural implements and construction). Regarding the most bioactive metabolites of higher molecular-weight, *A. purpurea* leaves/branches/seeds contained aporphinic alkaloids and acetogenins; and some of its bioactivities were cytotoxicity on tumor cell-line, antiplatelet, antibacterial and leishmanicidal (Barriga 1992, Bernal et al. 2011, 2019, Cárdenas et al. 2005, Cepleanu et al. 1993, Chang et al. 1998a, b, 2000, Chávez and Mata 1999, Fundación BioColombia 2013, Gómez-Estrada et al. 2011, Grandtner 2005, Grandtner and Chevrette 2014, HCN-ICN 2021, Jiménez-Escobar 2012, Luna-Cazáres and González-Esquinca 2008, Murillo 2001, Quattrocchi 2012, Rodríguez et al. 2012, Sarmiento Bernal et al. 2017, The Plant List 2013, Vásquez-Londoño 2012).

The chemical composition of the essential oils present in the leaves of young and old trees from Colombia is reported in Table 2. The most important constituents

Table 2: Principal secondary metabolites identified in the essential oils of different parts from *A. purpurea* and *X. emarginata*.

Type of VSM	Parts of the Plant	Main Chemical Constituents (Relative Amounts in %)	Biological Activities	References
Annona purpurea				
EO (0.06%)	Leaves (young plant)	β-Eudesmol (69) and α-eudesmol (17).	TAA: 165 ± 8 mmol Trolox/kg; HC$_{50}$: 490 ± 48 μg/mL; LC$_{50}$: 145.5 ± 0.7 μg/mL	Muñoz-Acevedo et al. 2016
EO (0.1%)	Leaves (old plant)	Germacrene D (56) and bicyclogermacrene (20).	TAA: 602 ± 38 mmol Trolox/kg; HC$_{50}$: 4.3 ± 0.6% (1000 μg/mL); LC$_{50}$: 346 ± 8 μg/mL	
Xylopia emarginata				
EO (0.4%)	Leaves	Germacrene D (18), bicyclogermacrene (15) and δ-elemene (8).	Antifungal against *Colletotrichum gloeosporioides*	Páez Aranzalez and Sánchez-Corredor 2018
EO (1.6%)	Fruits	β-Pinene (19), eucalyptol (15), α-pinene (14) and α-phellandrene (11).	NI	
EO (0.2%)	Bark	*m*-Mentha-2,8-diene (17), α-phellandrene (12) and δ-2-carene (7).	Antifungal against *C. gloeosporioides*	

VSM: Volatile Secondary Metabolites; EO: Essential Oil; NI: Not Inhibited.

were eudesmane [β-/α-eudesmols (**4,5**)] and germacrane (germacrene D (**6**)/ bicyclogermacrene) sesquiterpenoids; and although both EO showed biological effects (antioxidant and cytotoxic), the most active EO as an antioxidant and a cytotoxic (on human erythrocytes and lymphocytes) was that from the old and young tree, in that order. An interesting feature of the young tree EO is that it was a solid (Muñoz-Acevedo et al. 2016).

The second species of Annonaceae family is *Xylopia emarginata* Mart. (synonyms - *X. emarginata* var. *duckei* R.E. Fr., *Xylopicrum emarginatum* (Mart.) Kuntze). This evergreen tree (up to 15 m tall) has fragrant leaves and drupe-shaped fruits that are reddish when ripe, and the birds use them as feed. It is native to South America (particularly Colombia, Venezuela, Peru, Brazil and Guyana). Some of its common names are: escubiyu/escobilla, carguero (de hormiga), palo astilloso, yaya (Colombia), pintana amarilla zancuda (Peru), pindaíba/pindaíba-d'água, pindaúba and pindaubuna (Brazil). The main ethnobotanical applications of tree leaves/fruits/ bark have been to treat skin disorders, bronchitis and malaria; the demonstrated bioactivities are antiplasmodial and acaricidal. A number of hydrocarbons, sesquiterpenoids, alkaloids, flavonoids, condensed tannins and diterpene adducts have been isolated/identified as secondary metabolites. Between the non-medicinal uses can be mentioned as timber and fiber (Cruz et al. 2021, de Mesquita et al. 2007, Fischer et al. 2004, Grandtner 2005, Grandtner and Chevrette 2014, HCN-ICN 2021, Medeiros et al. 2021, Moreira et al. 2003, 2006, 2007, Quattrocchi 2012).

Páez Aranzalez and Sánchez-Corredor (2018) studied the chemical compositions and the antifungal capacities (against *Colletotrichum gloeosporioides*) of the EO from different parts of *escobillo* (Table 2). These authors found that EO leaves, fruits and bark were constituted by sesquiterpenes (e.g., 6), monoterpenoids (e.g., β-pinene (7)/eucalyptol) and monoterpenes [e.g., *m*-mentha-2,8-diene (8)], correspondingly. On the other hand, the leaf and fruit EO at 50% and 100%, one-to-one, inhibited the growth of the fungus completely. Other authors (Lago et al. 2005, Maia et al. 2005) have reported the compositions of the leaf EO from the Brazilian tree. Based on the research by Lago et al., the EO contained spathulenol (34%) and caryophyllene oxide (25%), while according to Maia et al., spathulenol (73%) was the major component of the EO, which were different from what was reported by Páez Aranzalez and Sánchez-Corredor (2018).

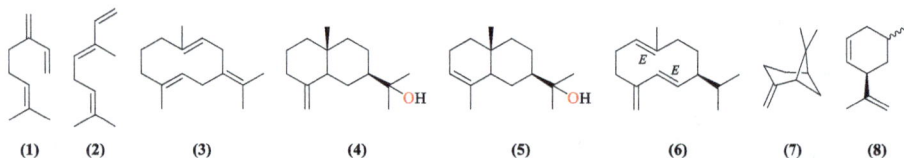

(1)   (2)   (3)   (4)   (5)   (6)   (7)   (8)

Apocynaceae

Two species and some of their ecotypes, belonging to Apocynaceae family, are included in this part. Accordingly, *Plumeria rubra* (frangipani) and *Pl. alba* (white frangipani) are small trees/shrubs up to 5–10 m of high, erects and succulents with small flowers in terminal cymes, which are fragrant. Fruits are glabrous pods containing seeds. These plants are native to Central and South America and in all tropical regions are distributed (≤ 1,500 m.a.s.l.); also, the entire plant contains white latex. Their flowers (orange/yellow, red/pink, tricolor and white) are widely used in the perfume/cosmetic industries of the world by their exquisite fragrances; and, the volatile compounds emitted by plants could play an ecological role as attractants for pollinators (Bernal et al. 2019, Duke et al. 2009, Grandtner 2005, IFRA 2015, Lim 2014, Quattrocchi 2012, Raguso and Pichersky 1999, The Plant List 2013).

(i.) *Plumeria alba* L. (synonyms: *Pl. alba* var. *fragrans* Kunth, *Pl. alba* var. *fragrantissima* G. Don). Common names include alelí blanca, amancayo, bois de lait, caracucha blanca, caterpillar tree, flor de mayo/pán/ensarta/cruz, florón blanco, franchipanier blanc, franjipanye blan, jasmim de cayenne/anta brava/leita, graveyard, milktree, milky bush, plumeria, sabak nikté, suche, tamaiba, West Indian jasmine, white champa and white/wild frangepan/frangipani.

(ii.) *Plumeria rubra* L. Synonyms are *Pl. acuminata* W.T. Aiton, *Pl. acutifolia* Poir, *Pl. angustifolia* A. DC., *Pl. aurantia* Endl., *Pl. bicolor (tricolor)* Ruiz and Pav., *Pl. purpurea* Ruiz and Pav. Common names are alelí de la tierra, atabaiba rosada, bellaco caspi, cacaloxochitl, campechana, florón, franchipanier rose/rouge, frangijapone, frangipán, frangipani, Honh Ji Dan Hua, jasmin caiana, juche, lirio tricolor, mayflower, nicte de monte, pagoda tree, q´arakuchu, red fragepan/jasmine/paucipan/plumeria, sacuanjoche, suche rojo/amarillo/blanco/turumbaco, súchil, turumbaco, etc.

Ethnomedicinal uses for both plants include the different parts of trees (flowers, seed, latex, bark and roots), which were used as a treatment for abscesses, blennorhagia, constipation, cholera/dysentery, erysipelas/dropsy/dermatosis, fungus/bacteria/herpes, hemorrhoids, fever/pain/vermifuge/purgative, rheumatism, syphilis, or toothache/headache as well as sedative, sudorific, vermifuge, stimulant, laxative, hypoglycemic, abortifacient, etc. Some biological properties of them are anti-bacterial/fungal, anti-tumoral/cancer, toxic, antioxidant/anti-inflammatory, antiparasitic, anesthetic, antidiabetic, etc. The flowers are mainly prepared as a tea; also, fruits and flowers are edible. Latex is the most widely used part of the tree as medicine, but they are also the bark and leaves which are beneficial. The main secondary metabolites responsible for biological activities are iridoids (pyranofurofuran skeleton such as plumericin, isoplumericin, β-dihydroplumericin, protoplumericine A, fulvoplumierin, allamcin/allamandin, plumieride, plumieride and β-dihydroplumericinic acids and plumeridoids A-C) and alkaloids (plumericidine, plumerianine, plumerinine and indol)/triterpenoids (lupeol, rubrinol, ursolic/ oleanolic/betulinic acids and α-/β-amyrins) (Afifi et al. 2006, Alhozaimy et al. 2017, Anggoro et al. 2020, Baghel et al. 2010, Bawa et al. 2019, Bernal et al. 2011, 2019, Bihani 2021, Comisión Técnica Subregional para la Política de Acceso a Medicamentos 2014, Dabhadkar and Zade 2012, Devprakash et al. 2012, Dey and Mukherjee 2015, Dey et al. 2011, Duke et al. 2009, González-Rocha and Cerros-Tlatilpa 2015, Grandtner 2005, Gupta et al. 2016, Khan 2017, Lim 2014, Pasaribu et al. 2020, Quattrocchi 2012, Rodríguez et al. 2012, Shinde et al. 2014, Sibi et al. 2014, Sirisha et al. 2013, Srivastava et al. 2017, Sura et al. 2018, Syakira and Brenda 2010, The Plant List 2013, Xia et al. 2018, Zaheer et al. 2010, Zhao et al. 2015).

The uses as teas (among other exploitation) and enfleurage/absolutes (concretes) of the flowers of some *Plumeria* spp., can be related to good organoleptic properties associated with its aromas. Thus, the chemical compositions determined for the secondary metabolites identified in the volatile fractions of the *P. rubra* and *P. alba* flowers from the northern region of Colombia are listed in Table 3.

Based on the Table, *Pl. rubra* had the highest chemical diversity; however, one variety of *Pl. alba* was enriched (> 50%) with linalool (one of the most important ingredients in the perfume industry) (Aprotosoaie et al. 2014, Fahlbusch et al. 2005, Kamatou and Viljoen 2008, Uter et al. 2013). The comparison between these compositions with those reported in the consulted scientific literature was different. That is,

*white flowers
   - benzyl salicylate (34%) and benzyl benzoate (12%)/germacrene B (10%) (Sahoo et al. 2021);
   - benzyl salicylate (27–34%) and geraniol (17%) or benzyl benzoate (22%) (Goswami et al. 2016);
   - benzyl salicylate (39%) and benzyl benzoate (17%) (Tohar et al. 2006b);
   - butyl oleate (14%) and butyl palmitate (12%) (Pino et al. 1994)
   - (E)-beta-ocimene (64%) and alpha-farnesene (11%)/nerolidol (9%) (Knudsen and Tollsten 1993);

Table 3: Main secondary metabolites identified in the volatile fractions of flowers from *Pl. rubra* varieties and *Pl. alba.*

Type of VSM	Parts of the Plant	Main Chemical Constituents (Relative Amounts in %)	References
Plumeria rubra			
VF by SDE	Yellow flowers (*P. rubra* var. *Lutea*)	Linalool (9 and 18), *trans*-nerolidol (**12**, 16), *cis*-linalool oxide (14, furanoid), phenethyl alcohol (11) and *trans*-linalool oxide (10, pyranoid).	Gutiérrez et al. 2010
	White flowers	α-Tolualdehyde (10 and 35), linalool (17) and phenethyl alcohol (13).	
	Pink flowers	α-Tolualdehyde (29), phenethyl alcohol (11 and 22), linalool (20).	
	Tricolor flowers	Linalool (21), α-tolualdehyde (19) and (E)-β-ocimene (12).	
Plumeria alba			
VF by SDE	White flowers	Linalool (53), α-tolualdehyde (10), (E)-β-ocimene (12), pelargonaldehyde/*cis*-linalool oxide (13 and furanoid) (8 each).	Gutiérrez et al. 2010
		Lauric acid (64), (E)-3-hexen-1-ol/myristic acid (6, each).	

VSM - Volatile Secondary Metabolites, VF - Volatile Fraction, SDE - Simultaneous-Distillation Extraction.

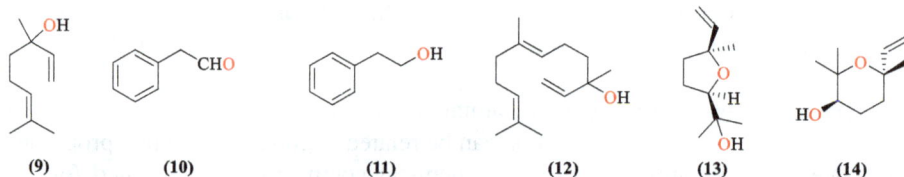

| (9) | (10) | (11) | (12) | (13) | (14) |

*flowers
  - (E)-non-2-en-1-ol (16%) and limonene (11%) (Lawal et al. 2015);
  - limonene (9%) and (E)-alpha-bergamotene/linalool/caryophyllene oxide (8%, each) (Lawal et al. 2014);
  - palmitic (36%)/myristic (11%) acids, flowers: 9-hexacosene (15%)/octadecanal (12%) (Liu et al. 2012);

*yellow flowers
  - palmitic (36%)/linoleic (17%) acids, pink flowers: lauric (31%)/ myristic (17%) acids (Tohar et al. 2006a);
  - phenylacetaldehyde (16%) and linalool (14%) (Omata et al. 1991);

*orange flowers
  - benzyl salicylate (21%)/(E)-nerolidol (14%) (Tohar et al. 2006a);

*reddish-orange flowers
  - phenethyl benzoate (12%)/lauric acid (12%) (Tohar et al. 2006b);

*red flowers
  - palmitic acid (27%) and linoleic (21%)/myristic (19%) acids (Tohar et al. 2006b);

*tricolor flowers
- phenethyl alcohol (32%) and phenylacetaldehyde (12%)/2-methylbutan-1-ol (10%) (Omata et al. 1992).

To close, the essential oils from two varieties of *Pl. rubra* were active against five strains of bacteria as reported by Liu et al. (2012).

Asteraceae

Two scented shrubs were selected from one of the most abundant/important and widely distributed plant families of Compositae: *Chromolaena barranquillensis* and *Cyanthillium cinereum*. In the case of *Chromolaena barranquillensis* (Hieron.) R.M. King and H. Rob. (synonym – *Eupatorium barranquillense* Hieron), it is a branched bush (1–3 m) with triangulated leaves and glandular points on its underside as well as velvety indumentum; in addition, it contains tubular inflorescences (white-lilac) arranged in corymbs. The plant is native to Colombia (considered weed), and in the northern region of the country it is distributed. The plant is named *rositavieja/rosavieja, saragüay, falso guaco, calentura vieja, Santa Cruz/María, jackin bush*, and it is commonly confused with another closely related species (*Ch. odorata* L.) with which it shares most of the morphological characteristics. The plant (leaves) is used in botanical medicine to treat rashes, insect bites and colics/kidney stones prepared as a poultice, decoction or baths; it has "promising" value because of the pharmacological properties shown, such as antifungal, antibacterial, antioxidant and cytotoxic (Álvarez et al. 2011, HNC-ICN 2021, Méndez et al. 2014b, Molina et al. 2014, Muñoz-Acevedo et al. 2011b, Rodríguez et al. 2014).

The main volatile secondary metabolites (volatile fractions/essential oils) of the fresh/dried leaves/flowers as well as some biological activities (toxicity and antioxidant capacities) of the EO are disclosed in Table 4. Thus, sesquiterpenoids [e.g., β-caryophyllene (15) and β-elemene (16)] were the predominant constituents of both volatile fractions and EO; nonetheless, some monoterpenes (e.g., myrcene, α-pinene and limonene) were prominent in the volatiles of fresh leaves and flowers. On the other hand, EO was toxic against *A. salina* (LC_{50} 8.9 ± 0.6–19 ± 1 µg/mL), the most toxic being that from dry leaves. In the same way, the EO from dry parts (leaves/flowers) showed the greatest protective effect against oxidation.

The other shrubby species is *Cyanthillium cinereum* (L.) H. Rob. (synonyms – *Blumea esquirolii* H. Lév. and Vaniot, *Cacalia cinerea* (L.) Kuntze, *Conyza chinensis* L., *Co. cinerea* L., *Co. ivifolia* Burm., *Eupatorium myosotifolium* Jacq., *Senecioides cinereum* (L.) Kuntze, *Vernonia abbreviata* DC., *V. arguta* Baker and *V. cinerea* Less.). The plant is an annual erect herb (15–70 cm tall) with alternate/spatulate leaves (1–6 cm long) and with a pleasant characteristic odor and scattered/fine trichomes; highly branched terminal inflorescences and purple/pink or white flowers. This cosmopolitan species is considered a weed. In addition to being present in Colombia, it is distributed in the USA, Central/South America and Asia. Some common names by which it is known are blue fleabane, ironweed, little/small ironweed, strongman bush, vernonia and *altamisa o venadillo* (Colombia). In the ethnobotanical practice of the Mokaná Indigenous Reservation (Tubará/Galapa, Atlántico, Colombia), the species has been used as an infusion/decoction for the

Table 4: Major volatile metabolites determined in the volatile fractions and essential oils of different parts from *Ch. barranquillensis* and *Cy. cinereum*.

Type of VSM	Parts of the Plant	Main Chemical Constituents (Relative Amounts in %)	Biological Activities	References
Chromolaena barranquillensis				
VF by SDE	Flowers	*F* – β-Elemene (24), α-pinene (20) and limonene (16); *D* – (E)-β-caryophyllene (21), germacrene D (17) and caryophyllene oxide (14).	N/A	Álvarez et al. 2010, 2011a, b, Álvarez-Carrillo 2011, Muñoz-Acevedo et al. 2011b
	Leaves	*F* – Myrcene (39), γ-curcumene (18) and limonene (10); *D* – (E)-β-caryophyllene (14), γ-curcumene (10) and β-elemene (8).		
EO	Flowers (0.2–0.4%)	*F/D* – (E)-β-Caryophyllene (23–24), β-elemene (14–21) and germacrene D (14–16).	Toxicity on *Artemia salina* - LC_{50} (µg/mL): FF: 12 ± 1, DF: 19 ± 1, FL: 14 ± 2, DL: 8.9 ± 0.6; protector effect (oxidation inh., %): DF (42)/DL (48) > BHT (29)	
	Leaves (0.06–0.1%)	*F* – (E)-β-Caryophyllene (22), limonene (12) and δ-cadinene (7); *D* – β-caryophyllene (29), germacrene D (13) and caryophyllene oxide (12).		
Cyanthillium cinereum				
VF by SDE	Flowers	*JA* – α-Cadinol (14), δ-cadinene (11) and 2,5-dimethoxy-*p*-cymene (7).	N/A	Méndez Beltrán and Ortega Morales 2012, Muñoz-Acevedo et al. 2012
	Leaves	*JA* – α-Cadinol (20), δ-cadinene (12), germacrene D-4-ol (9); *P* – α-cadinol (24), δ-cadinene (11), τ-muurolol (9)[†]; caryophyllene oxide (25), α-tolualdehyde (17) and α-humulene (13)[*].		
EO	Flowers (0.09–0.1%)	*JA* – δ-Cadinene/α-cadinol (16, each one), α-humulene (10), α-muurolol/2,5-dimethoxy-*p*-cymene (6 each); *P* – α-cadinol (34), τ-muurolol (17), δ-cadinene (11).	Toxicity on *A. salina* - LC_{50} (µg/mL): JAF: 17 ± 2, JAL: 16 ± 1, PF: 13 ± 2, PL: 11 ± 2; %inh. by ABTS[+]: JAF: 13 (13600 ppm), JAL: 10 (14200 ppm), PF: 10 (14300 ppm), PL: 6 (15000 ppm); protector effect (oxidation inh., %): JAF/JAL ≥ BHT; PF ≥ BHT	
	Leaves (0.02–0.1%)	*JA* – α-Cadinol (23), elemol (11) and δ-cadinene (10); *P* – α-cadinol (37), τ-muurolol (14), δ-cadinol (6)[†]; humulene epoxide II (36), δ-cadinol (11) and 2,5-dimethoxy-*p*-cymene (7)[*].		

F/D: fresh/dry; F/L: flowers/leaves; VSM: volatile secondary metabolites; EO: essential oil; VF: volatile fraction; SDE: simultaneous-distillation extraction; JA: Juan de Acosta (town); P: Piojó (town); [†]in reproductive state; [*]in vegetative state; N/A – not applicable.

treatment of colic/fevers as a vermifuge, sedative, galactagogue, against infections and general pains, according to the knowledge of the healer Urbano González (RIP) and the elderly José Félix González (HCN-ICN 2021, Muñoz-Acevedo et al. 2012, Quattrocchi 2012, The Plant List 2013).

The volatile metabolites identified in the aerial parts of *C. cinereum* from Colombia (in two municipalities) were represented mainly by cadinane-type bicyclic sesquiterpenoids [Table 4, α-cadinol (17) and δ-cadinene (18)]. Furthermore, there were some differences both in relative amounts and minority constituents based on the report by Mendez Beltran and Ortega Morales (2012). Moreover, when the plant was studied at a different time of its biological cycle (vegetative), the chemistry of its volatile fraction and EO varied [the main constituents were caryophyllene oxide (19) and humulene epoxide II (20)]. Regarding the biological properties, the EO were effective on brine shrimp nauplii with LC_{50} values between 11 ± 2–17 ± 2 µg/mL [the most active EO was from leaves (Piojó)], which could be considered toxic. Considering the protective effect against oxidation and the radical-scavenging capacity (ABTS$^{+\cdot}$), the EO of leaves/flowers from the Juan de Acosta plant together with the flowers from the Piojó plant acted as oxidation protectors with effects greater than or equal to BHT, while the capacities of the EO by scavenging the radical-cation ABTS$^{+\cdot}$ were very poor.

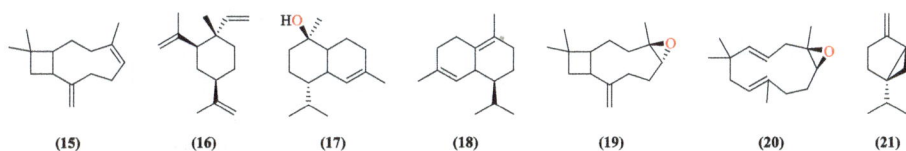

(15)   (16)   (17)   (18)   (19)   (20)   (21)

Boraginaceae

Cordia curassavica (Jacq.) Roem. and Schult. (Synonyms – *Co. brevispicata* M. Martens and Galeotti, *Co. chepensis* Pittier, *Co. graveolens* Kunth, *Co. linearis* A. DC., *Co. rugosa* Willd., *Lantana bullata* L., *Lithocardium cuneiforme* (DC.) Kuntze, *L. lanatum* Kuntze, *Montjolya bullata* (L.) Friesen, *M. guianensis* (Desv.) Friesen, *Varronia curassavica* Jacq., *V. guianensis* Desv., *V. lanata* (Kunth) Borhidi and *V. macrostachya* Jacq.) is an erect/large/branched shrub (up to 3 m) or small tree (up to 4 m) with fragrant foliage, spicate inflorescence, small white-yellow flowers and small red drupe-shaped fruits, which is distributed in the Caribbean and Latin America as well as in Madagascar. Its vernacular names are barredor, pava canelilla, black/wild sage, basura pretu, basura/bolita prieta, cariaqyuito, cariaquito negro, chocobaupo, cuero negro serrano, erva-baleeira, Juan prieto, liane mouton, maíz tostado, majao negro, mulato, negrito, pata de gallina(judío), saguinay de costa, ak-sheb and blacka oema. The most relevant ethnobotanical uses of the plant aerial parts prepared in infusions/teas/decoctions are as a treatment for respiratory (cold/cough/influenza), gastrointestinal such as diarrhea, dermatological disorders and as a remedy for headache/stomachache, inflammation, fever, muscle/menstrual pains and parasites. As well, the macerated leaves have been externally used in veterinary medicine to rid mites in chicken pens. In addition, the plant has some non-medicinal applications as wood for household materials. Referring to certain extracts/isolated constituents/essential oils of the plant, they have demonstrated their capacities as anti-bacterial/fungal, anti-inflammatory, antinociceptive and insecticidal (larvicidal) agents. And, among the secondary metabolites, the meroterpenoid naphthoquinones (cordiaquinones A, B, J, K) stand out as well as their essential oils (Bayeux et al. 2002, Bernal et al. 2011, Bristot et al. 2020, Grandtner and Chevrette 2014, HCN-ICN 2021, Hernández et al. 2003, Ioset et al. 2000, Oza and Kulkarni 2017, Quattrocchi 2012, Roth and Lindorf 2002).

The fragrant smell of the bush from the Caribbean region of Colombia led to the study of the volatile fractions and the essential oil by Rodriguez et al. (2014) (Table 5). As stated by these authors, the most abundant constituent present in volatile fractions/EO was sabinene [(21), 29–77%] and the volatile chemistry between the different parts varied.

Table 5: Leading secondary metabolites identified in the volatile fractions and essential oil of different parts from *C. curassavica.*

Type of VSM	Parts of the Plant	Main Chemical Constituents (Relative Amounts in %)	References
Cordia curassavica			
VF by SDE	Flowers	Sabinene (77).	Rodríguez et al. 2014
	Leaves	Sabinene (32), (E)-β-caryophyllene (13) and β-elemene (10).	
	Branches	Sabinene (51), terpinen-4-ol (11) and γ-terpinene (9).	
EO (0.2%)	Leaves	Sabinene (29), (E)-β-caryophyllene (14) and β-elemene (10).	

VSM: Volatile Secondary Metabolites; EO: Essential Oil; VF: Volatile Fraction; SDE: Simultaneous-Distillation Extraction.

The comparison of this composition with the other reports of scientific literature evidenced significant differences between the leaf/aerial part EO; e.g., Mexican plants contained as principal components to (i) 4-methyl,4-ethenyl-3-(1-methylethenyl)-1-(1-methyl methanol)cyclohexane (37%), β-eudesmol (19%) and spathulenol (11%) (Hernández et al. 2007); (ii) rainy season – germacrene (24%), tricyclene (19%) and α-pinene (10%); dry season – tricyclene (20%), isocaryophyllene (12%) and α-pinene (10%) (Hernández et al. 2014). Venezuelan plant EO (leaves) was composed by tricyclene (24%) and bicyclogermacrene (12%) (Meccia et al. 2009); Brazilian tree EO (leaves) presented α-pinene (20%), bicyclogermacrene (14%), β-pinene (13%) and β-caryophyllene (12%) (Santos et al. 2006).

Burseraceae

Four deciduous trees distributed in Latin America from the Burseraceae family were chosen. The first of them is *Bursera glabra* (Jacq.) Triana and Planch (synonyms – *B. howellii* Standl., *Amyris elaphrium* Spreng. and *Elaphrium glabrum* Jacq.) is a very branchy tree (up to 15 m) with fragrant leaves/stems and resin [pleasant terebinthaceous odor (like camphor) characteristic of the genus], coppery outer rind, non-exfoliating. Additionally, it presents immature yellowish-green flowers and globous fruits from yellowish-green (immature) to reddish (mature). The tree has been found in Mexico, Costa Rica, Colombia and Venezuela; in Colombia, it is mainly distributed in the northern region (departamentos Atlántico, Bolívar, Guajira and Magdalena). The vernacular names are aria, aliia, bálsamo real (de incienso), bija, caraña(o) de monte (silvestre), glabrous bursera, gommier glabre, maluua, maruwa and palo santo. The resin, trunk bark, leaves or branches are prepared in infusion/decoction, maceration, poltice or in baths to alleviate respiratory symptoms associated with colds and flu, pain, infections, skin allergies or rashes and as an insect repellent (Benítez et al. 2014a, Bernal et al. 2019, Castro-Laportte 2013, Grandtner and Chevrette 2014, HNC-ICN 2021, Muñoz-Acevedo et al. 2019b, PNN 2005, The Plant List 2013, Zamora-Villalobos et al. 2000).

Table 6 lists the secondary metabolites with a volatile nature identified in *B. glabra*. Thus, the volatile fractions (by SDE/HS-SPME) of the different parts of *B. glabra* were predominantly represented by monoterpene hydrocarbons [myrcene (18–40%) and α-pinene (22) (25–36%)], but the flowers (by SPME) was characterized by sesquiterpene hydrocarbons (germacrene D). Likewise, the EO was typified by sesquiterpenoids such as germacrene D and cubebol (24). The EO yields were 0.01–0.25%. Only one article (Cáceres-Ferreira et al. 2019) that mentions the chemical composition of *B. glabra* resin was available in the science literature consulted, i.e., limonene (78%) and *cis*-ocimene (8%) were the most abundant constituents of the Venezuelan species that differed from the composition of any of the parts of the Colombian species.

The second plant is *Bursera graveolens* (Kunth) Triana and Planch (synonyms – *Amyris caranifera* Willd. ex Engl., *B. penicillata* (DC.) Engl., *B. tatamaco* (Tul.) Triana and Planch., *Elaphrium graveolens* Kunth, *E. penicillatum* DC., *E. pubescens* Schltdl., *Spondias edmonstonei* Hook. f., *Terebinthus graveolens* Rose and *T. pubescens* (Schltdl.)), which is a species widely studied/exploited (considered a promising species) in Latin America where it is distributed (Brazil, Colombia, Costa Rica, Cuba, Ecuador, Honduras, Mexico, Nicaragua, Perú and Venezuela). This tree (up to 15 m and the whole plant is fragrant) is known as bijá, caraño(a), crespín, oloroso, palo santo, Santa-María, sasafrás, tamajaco/tatamaco (Colombia), caragana (Bolivia), caraño (Nicaragua), copalillo (Honduras), sasafrás (Cuba), Crispín, Huacoe (Perú), strongmelling bursera and gommier caraña. The leaves/stems/bark/resin has been used in botanical medicine (Latin America) in the form of fumes/infusions/poultices/compresses/baths to treat wounds, diarrhea, anemia, rheumatism, dermatitis, asthma, colic, kidney stones, hernia, "cracks" (feet) and also as healing, abortifacient, anti-inflammatory, antitumor, analgesic, depurative, diaphoretic, expectorant, insecticide and mosquito repellent. In addition, it is used to extract epidermal foreign bodies. Among the main isolated constituents, the mono-/sesqui-terpenoids [benzofuran compounds (mint lactone and their derivatives) and daucane/juneol/eudesmane/agarofurane derivatives] and triterpenoids (α-/β-amyrins) can be highlighted. Lastly, some biological effects evaluated/determined/demonstrated by the extracts/EO/isolated molecules are anti-inflammatory, antiacaricidal, antimicrobial, antioxidant and cytotoxic (Alonso-Castro et al. 2011, Álvarez 1991, Bernal et al. 2019, Castro 2012, Castro-Laportte 2013, Correa-Barrera 2006, Correa

Table 6: Main Chemical constituents determined in the volatile fractions and essential oils of different parts from *B. glabra*, *B. graveolens* and *B. simaruba*.

Type of VSM	Parts of the Plant	Main Chemical Constituents (Relative Amounts in %)	Biological Activities	References
Bursera glabra				
VF by SDE	Leaves	Myrcene (37) and germacrene D (25).	N/A	Benítez et al. 2014a, Muñoz-Acevedo et al. 2019b
	Branches	Myrcene (40), α-pinene (20) and β-pinene (9).		
	Bark	α-Pinene (36), β-pinene (13) and α-copaene (10).		
VF by HS-SPME	Flowers	Germacrene D (19), myrcene (15), α-pinene (10) and α-copaene (10).		
	Leaves	α-Pinene (25), α-copaene (20) and myrcene (12).		
	Branches	Myrcene (32), α-pinene (26), β-pinene (13) and α-copaene (9).		
	Branch bark	α-Pinene (28), β-pinene (17) and α-copaene (10).		
	Trunk bark	Myrcene (18), α-pinene (16), β-pinene (8) and α-copaene (7).		
EO	Leaves (0.01%)	Germacrene D (22) and myrcene (20).		
	Branches (0.25%)	Cubebol (17), geranyl α-terpinene (9), δ-cadinol (9), epi-cubebol (6) and α-copaene (5).		
	Bark (0.25%)	Cubebol (25), geranyl α-terpinene (14), α-copaene/δ-cadinol (8, each), δ-cadinene (7) and epi-cubebol/α-muurolene (6 each).		
Bursera graveolens				
VF by SDE	Immature fruits	3-Hydroxy-mint furanone (7), mint furanone (6), carvone (5) and limonene/*trans*-carveol/limonene-1,2-diol (4 each).	N/A	Muñoz-Acevedo et al. 2013
	Leaves	Germacrene D (21), (E)-β-caryophyllene (18), viridiflorol (8), limonene (7) and linalool (6).		
	Branches	Mint furanone (44), iso-mint furanone (7) and 3-hydroxy mint furanone (6).		
	Stem bark	Mint furanone (45), 3-hydroxy-mint furanone (16) and iso-mint furanone (6).		
	Resin	Limonene (23), mint furanone (16), mint furanone derivative (15), pulegone (12), 3-hydroxy-mint furanone (9) and menthofuran (6).		

Table 6 contd. ...

...Table 6 contd.

Type of VSM	Parts of the Plant	Main Chemical Constituents (Relative Amounts in %)	Biological Activities	References
EO	Leaves/ stems (0.3%)	Limonene (42), pulegone (21) and carvone (8).	*T. castaneum* – (i) fumigant effect: $LC_{50/95}$ 108–119 μg/mL; (ii) repellency (1% EO): 88–89%–2–4 h.	Jaramillo-Colorado et al. 2019
Bursera simaruba				
VF by SDE	Fruits	Sabinene (18), terpinen-4-ol (17), α-pinene (12) and γ-terpinene (9).	N/A	Benítez et al. 2014b, Muñoz-Acevedo et al. 2019b
	Leaves	Sabinene (31), α-pinene (21), β-pinene (7) and terpinen-4-ol (6).		
	Branches	Sabinene (24), α-pinene (15), terpinen-4-ol (12) and γ-terpinene (8).		
VF by HS-SPME	Fruits	Terpinen-4-ol (45), sabinene (13), α-pinene (13) and bicyclogermacrene (10).		
	Leaves	*p*-Cymene (35), (E)-β-caryophyllene (16), terpinen-4-ol (9) and germacrene (7).		
	Branches	*p*-Cymene (19), terpinen-4-ol (13), sabinene (12), bicyclogermacrene (11) and γ-terpinene (8).		
EO	Leaves (0.2%)	Sabinene (38), α-pinene (17) and β-pinene/terpinen-4-ol/myrcene (6 each).		
	Branches (0.1%)	Sabinene (39), α-pinene (16), terpinen-4-ol (7) and myrcene/β-pinene (6 each).		
	Bark (0.1%)	Sabinene (23), α-pinene (17), terpinen-4-ol (8) and β-pinene/γ-terpinene (6 each).		

VSM: Volatile Secondary Metabolites; VF: Volatile Fraction; EO: Essential Oil; SDE: Simultaneous-Distillation Extraction; HS-SPME: Headspace Solid Phase Micro-Extraction; N/A: Not Applicable.

and Bernal 1990a, Duke et al. 2009, Grandtner and Chevrette 2014, Gupta 1995, HNC-ICN 2021, Muñoz-Acevedo et al. 2013, Nakanishi et al. 2005, Noel-Martínez et al. 2021, Quattrocchi 2012, Rey-Valeirón et al. 2017, Robles et al. 2005, Sánchez et al. 2006, Sotelo-Méndez et al. 2017, Soukup 1970, Tene et al. 2007, The Plant List 2013, Yukawa et al. 2004a, b, 2005, Zuñiga et al. 2005).

The chemical composition of the different parts of *caraña* from the northern region of Colombia had some similarities/differences with the reports of scientific literature consulted, i.e., Muñoz-Acevedo et al. (2013) reported the chemical compositions of the volatile fractions isolated by SDE from the immature fruits, leaves, branches, stem bark and resin of the tree (Table 6) that were represented by (i) terpene lactones [3-hydroxy-mint furanone (25)]/ketones – fruits; (ii) sesquiterpenes (germacrene D) – leaves; (iii) terpene lactones [mint furanone (26)/(25)] – branches/

stem bark; (iv) monoterpenes (limonene) and terpene lactones [mint furanone **(26)**/ **(25)**]/ketones – resin.

Jaramillo-Colorado et al. (2019) identified limonene (42%)/pulegone (21%) as the main constituents of the leaf/stem EO (Colombia) as well as Leyva et al. (2007) stated that leaf and branch EO (Colombia) contained (i) limonene (48%) and caryophyllene oxide (14%) in leaves and (ii) limonene (42%), myrcene (20%), menthofuran (15%) and *cis*-β-ocimene (14%) in branches as the most abundant constituents. The principal chemical components from the branch, fruit and stem EO of the Ecuadorian species were reported by Fon-Fay et al. (2019), Rey-Valeirón et al. (2017), Manzano-Santana et al. (2009) and Young et al. (2007), respectively, which were (i) viridiflorol (71%) in the branch, (ii) limonene (50%)/α-phellandrene (38%) in mature fruits and (iii) limonene (35–59%)/α-terpineol (11–13%) in the stem. The EO from Peruvian species was characterized by (i) mint furanone (44–45%) in branch and bark (Yukawa and Iwabuchi 2003), (ii) α-terpinene (32%) in wood (Sotelo-Méndez et al. 2017), (iii) limonene (22%)/β-elemene (12%) in leaves (Leyva et al. 2020) and (iv) limonene (77%) in the trunk (Noel-Martínez et al. 2021). Other EO from the aerial parts/leaves of the Cuban species were riched with limonene (26–31%), β-elemene (11–14%) and (E)-β-ocimene (13–21%) (Carmona et al. 2007, Monzote et al. 2012).

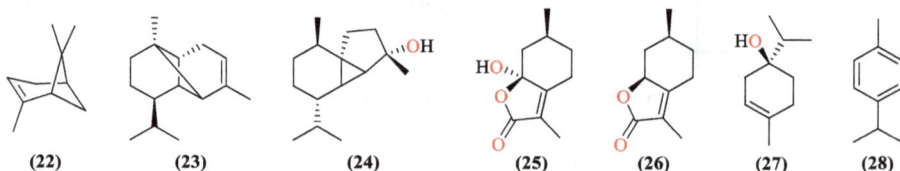

(22)   **(23)**   **(24)**   **(25)**   **(26)**   **(27)**   **(28)**

Referring to the biological properties of these EO, the Colombian EO showed powerful fumigant and repellent effects on *Tribolium castaneum* (Jaramillo-Colorado et al. 2019); while antibacterial/antifungal effects of the Peruvian EO were between moderate and high (Sotelo-Méndez et al. 2017) on six bacterial/one fungi strains. The Ecuadorian EO were larvicidal agents against *Rhipicephalus* (*Boophilus*) *microplus* larvae (55% of death, at 1.25% concentration) (Rey-Valeirón et al. 2017) and on *Aedes aegypti* larvae (LC_{50} 10–32 µg/mL; LC_{90} 27–64 µg/mL) (Leyva et al. 2020). The Cuban EO was antitumoral (on MCF-7 line and IC_{50} 49 ± 4 µg/mL) and antileishmanial (IC_{50} 37 ± 5 µg/mL) agent (Monzote et al. 2012).

B. simaruba (Kunth) Triana and Planch (synonyms – *Bursera bonairensis* Bold., *B. gummifera* L., *B. integerrima* (Tul.) Triana and Planch., *B. ovalifolia* (Schltdl.) Engl., *B. subpubescens* (Rose) Engl., *Elaphrium integerrimum* Tul., *E. ovalifolium* Schltdl., *E. simaruba* (L.) Rose, *E. subpubescens* Rose, *Icicariba simaruba* M. Gómez, *Pistacia simaruba* L., *Tapirira macrophylla* Lundell, *Terebinthus simaruba* (L.) W. Wight) is the third plant selected from the Burseraceae family. The tree (up to 20 m) is semi-deciduous with very fast-growing, shiny reddish-brown scaly bark, odd-pinnate leaves, inconspicuous creamy-white flowers and greenish-reddish drupe fruits. As well as *B. graveolens*, *B. simaruba* has been a plant widely studied, and it is distributed in Central America and the Caribbean and northern South America (Bahamas, Barbados, Belize, Brazil, Colombia, Costa Rica, Cuba, Dominican Republic, Guatemala, Guianas, Haiti, Honduras, Jamaica, Mexico,

Netherlands Antilles, Nicaragua, Panamá, Puerto Rico, Trinidad and Tobago, Venezuela and West Indies). The species is named as *almácigo(a)*, *indio encuero* (naked indian), *aceitero*, *balsam-tree*, *gum elemi*, *gommier*, *gumbo-limbo*, *chacajiota*, *palo mulato*, *copal* or *palo de jiote*, *chicchica*, *cholo pelao*, *caratero*, *mararo*, *palo mulato*, *resbalamono*, *turpentine*, *Faca-naca* or *Aría*. The different parts (e.g., leaves, branches, trunk, bark, resin, seeds, flowers, young fruits and buds) from the plant are used in the herbal medicine prepared in form of infusions/decoctions/macerates as a stomach tonic to treat infections, colds, flu, fevers, intestinal pains, diarrhea, gastric ulcers and as a carminative and slimming agent and as poultice or baths to relieve insect bites, rash, itching, bruises, wounds, abscesses, as healing and even for snake bites. It is used as insecticides and repellents and as a treatment for rheumatism, syphilis, measles, swelling, hypertension, nephrolithiasis, urethritis, nephrosis, obesity, anemia, etc. Moreover, a non-medicinal application worth mentioning is that the tree is an important forest/food resource in its distribution area for birds and primates. Among the pharmacological effects of the plant can be mentioned: anti-inflammatory, anti-leishmanicidal (on *Leishmania amazonensis*), cytotoxic/ toxic (*A. salina*)/antineoplastic (MK-1, HeLa, B16F10 cell lines), antiviral (herpes simplex viruses, HSV-1/HSV-2) and antimicrobial (*St. aureus*, *Candida albicans* and *Cladosporium cucumerinum*). The constituents responsible for these effects are sterols and lupene-related pentacyclic triterpenoids (Álvarez et al. 2015, Balick and O'Brian 2004, Benítez et al. 2014a, Bernal et al. 2019, Bussman et al. 2018, Carretero et al. 2008, Castro-Laportte 2013, Correa and Bernal 1990b, Duke et al. 2009, García et al. 2010, García-Barriga 1992, Gil-Otaiza et al. 2006, Grandtner and Chevrette 2014, Gupta 1995, HNC-ICN 2021, Kinjo et al. 2016, Muñoz-Acevedo et al. 2019b, Peraza-Sánchez et al. 1995, Quattrocchi 2012, Rahalison et al. 1993, Scott and Martin 1984, The Plant List 2013, Volpato et al. 2009, Wedler 2013, Yasunaka et al. 2005).

The EO and/or volatile fractions (by SDE) of the leaves, bark, branches and fruits from Colombian "naked Indian" were mainly represented by monoterpenoids as sabinene (18–39%), α-pinene (12–17%) and terpinen-4-ol (6–17% – **27**). Nonetheless, the chemical compositions of the volatile fractions (by SPME) of the same parts of the plant were characterized by *p*-cymene (19–35% – **28** for leaves/branches) and terpinen-4-ol (45%, for fruits). These compositions differed when compared to the scientific literature related to the EO from other Latin American/Caribbean countries. That is, Costa Rica contained α-terpinene (26%), γ-terpinene (20%) and α-pinene (18%) (fruits); *o*-cymene (65%) (leaves); α-phellandrene (29%) and β-caryophyllene (19%) (bark). Guadeloupe had limonene (47%) and β-caryophyllene (15%) (leaves). Venezuela presented caryophyllene oxide (18%) and caryophyllene (10%) (branches), sabinene (59%) and terpinen-4-ol (12%) (fruits) and α-pinene (52–67%), β-phellandrene (25–31%), germacrene D (12–18%) and β-caryophyllene (15%) (resins). Jamaica had α-pinene (28%), β-pinene (24%) and terpinen-4-ol (13%) (fruits), α-pinene (10%), (E)-cadina-1(6),4-diene (10%) and β-caryophyllene (9%) (leaves), α-pinene (32%) and β-pinene (14%) (bark). The yields for this EO were 0.007–2.0% (Cáceres-Ferreira et al. 2019, Junor et al. 2007, 2008, Marcano et al. 2013, Rosales-Ovares and Cicció 2002, Setzer 2014, Sylvestre et al. 2007). Finally, the fruit and stem EO from the Jamaican plant showed some antibacterial effects (moderate to high) against six bacterial strains [*E. coli, St. aureus, Proteus mirabilis, P. aeruginosa*, MRSA and β-haemolytic *Streptococcus* group A (BHSA)].

One species belonging to the genus *Protium* was also selected. This was *Pr. heptaphyllum*, which is an arboreal species (5–10 m) characterized by having glabrous and imparipinnate leaves with oblong-lanceolate/elliptic leaflets, axillary inflorescences with yellow-greenish to green-reddish flowers and resinous drupaceous (oblique-ovoid/globose) fruits. In addition, it contains a hard translucent whitish resin. Tree with neotropical distribution predominantly in Colombia, Brazil, Panama and Venezuela, even though in Bolivia, Peru, Paraguay and Argentina, the plant has also been found. Synonyms of the scientific name are *Amyris ambrosiaca* Willd., *A. brasiliensis* Willd. ex Engl., *Icica ambrosiaca* Mart., *I. surinamensis* Miq., *Pr. angustifolium* Swart, *Pr. multiflorum* Engl., *Pr. octandrum* Swart, *Tingulonga heptaphylla* (Aubl.) Kuntze, *T. multiflora* Kuntze., etc. Their vernacular names are al(r)mésca, ánime (blanco), Brazilian elemí, brea(u), breu-branco-do-campo, cánime, caraño, catamajaca, curruc(s)ay, guate de galina, hitamaká, incienzo, isiga(o), matupa, pepa de loro, pergamín, sevenleaflet resintree, taca-mahaca gum, tacamaj(h)aca(o), tingimoni, tobonae, ysy(i), etc. The main part of the tree applied in ethnomedicine is its resin (exudate), although the leaves/stem/bark prepared as a decoction/infusion/poultice have also been used. They are used to treat stomach disorders, colds/catarrh, ulcers, tumors, syphilis, as astringent, cicatrizing, abortive, etc.; some of their biological properties are anti-inflammatory, antiplatelet, acaricidal/cercaricidal/molluscicidal/larvicidal, antiparasitic, antitumoral/cytotoxic, anticolinergic, analgesic, antinociceptive, antimicrobial, hepato-/gastroprotective, etc. Also, another non-medicinal use for the resin is for incense/varnishes. The main types of constituents isolated are essential oils, terpenoids (*p*-menth-3-ene-1,2,8-triol), triterpenoids (α-/β-amyrins, α-/β-amyrones, lupenone and friedelin), flavonoids (quercetin, quercetin-3-O-rhamnosyl and catechin), coumarins (scopoletin, fraxetin, propacin and cleomiscosin A) (Almeida et al. 2002, Aragão 2004, Bandeira et al. 2002, Bernal et al. 2011, Cabral et al. 2021, Case et al. 2003, Correa and Bernal 1990, DeCarlo et al. 2009, Duke et al. 2009, Frischkorn et al. 1948, García-Barriga 1992b, Grandtner and Chevrette 2014, Gupta 1995, HNC-ICN 2021, Lima-Junior et al. 2006, Maia et al. 2000, Mobin et al. 2016, Oliveira et al. 2004a, b, 2005a, b, Pontes et al. 2007, Quattrocchi 2012, Roth and Lindorf 2002, Rüdiger et al. 2007, Siani et al. 1999, Susunaga et al. 2001, Swart 1942, The Plant List 2013, Vieira-Junior et al. 2005, Zoghbi et al. 1995).

The most representative chemical constituents of the volatile fractions from the Arauca region of Colombia were the sesquiterpenoids [germacrene D (14–28%) and guaiol (14%–29)] for the flowers/bark and leaves, respectively. The monoterpenes [*p*-cymene (30%) and α-pinene (22%)] were for resin (Table 7). Whilst, the chemical compositions of the different volatile fractions (by SDE/HS-SPME) of all the studied parts from the Sucre region were predominantly characterized by limonene (31–77% – 30), although the second constituents depended on part of the plant, i.e., α-pinene (15–23%) for flowers/trunk, *p*-cymene (16–24%) for bark and eucalyptol (17%–31)/α-terpineol (15% – 32) for resin and α-phellandrene (22–24% – 33) for fruits. Furthermore, the trunk and bark EO had limonene (64–84%) as the main

Table 7: Principal constituents identified as volatile secondary metabolites from different parts of colombian *Pr. heptaphyllum.*

Type of VSM	Parts of the Plant	Main Chemical Constituents (Relative Amounts in %)	References
Arauca Region			
VF by SDE	Flowers	Germacrene D (14), germacrene B (13), bicyclogermacrene (12) and limonene (8).	Tafurt-Garcia and Muñoz-Acevedo 2012
	Leaves	Guaiol (14), α-copaene (9), 1,10-di-epi-cubenol (8) and (E)-β-cariophyllene (6).	
	Bark	Germacrene D (28), 1,10-di-epi-cubenol (8), guaiol (7) and γ-cadinene (7).	
	Resin	*p*-Cymene (30), α-pinene (22) and limonene (14).	
Sucre Region			
VF by SDE	Flowers	Limonene (58), α-pinene (23) and β-pinene (15).	Muñoz-Acevedo and Castillo-Contreras 2023
	Fruits	Limonene (57), α-phellandrene (24) and β-pinene (6).	
	Leaves	Limonene (46), (E)-β-caryophyllene (15) and α-pinene (6).	
	Trunk	Limonene (51), α-pinene (22) and β-pinene (17).	
	Bark	Limonene (77) and *p*-cymene (16).	
	Resin	Limonene (31), α-terpineol (15), *p*-cymene (14), eucalyptol (13) and α-phellandrene (10).	
VF by HS-SPME	Flowers	Limonene (63), α-pinene (15) and β-pinene (13).	
	Fruits	Limonene (63) and α-phellandrene (22).	
	Bark	Limonene (64) and *p*-cymene (24).	
	Resin	Limonene (39), eucalyptol (17), *p*-cymene (16) and α-phellandrene (13).	
EO	Leaves	(E)-β-Caryophyllene (24), α-copaene (14), limonene (12) and γ-cadinene/α-humulene (7, each).	
	Trunk	Limonene (82-84), α-pinene (4-6) and β-pinene (3-4).	
	Bark	Limonene (64) and *p*-cymene (8).	

VSM: Volatile Secondary Metabolites; VF: Volatile Fraction; EO: Essential Oil; SDE: Simultaneous-Distillation Extraction; HS-SPME: Headspace Solid Phase Micro-Extraction; N/A: Not Applicable.

component, whereas, the leaf EO was composed of β-caryophyllene (24%). The comparison between the volatile metabolites of the same plant from the other Latin American countries showed certain similarities/differences, i.e., the essential oils from the Venezuelan tree presented germacrene D (24%)/α-phellandrene (22%)/(E)-β-caryophyllene (11%) in the leaves and α-phellandrene (57%)/sabinene (12%) in the stems. On the other hand, EO from Brazilian species contained (i) limonene (93%), α-pinene (71%), δ-3-carene (64%)/α-pinene (15%) and α-terpinene (48%) for fruits; (ii) (E)-β-caryophyllene (32%)/germacrene B (17%)/(E)-β-ocimene (16%), (E)-β-caryophyllene (29%)/(E)-β-ocimene (14%)/bicyclogermacrene (13%), β-elemene (22%)/terpinolene (15%), 9-epi-(E)-caryophyllene (21%)/14-hydroxy-9-epi-(E)-caryophyllene (17%), myrcene (19%)/β-caryophyllene (18%) and β-caryophyllene (15%)/α-amorphene (9%) for leaves; (iii) terpinolene (40%) and α-pinene

(40%)/*p*-mentha-1,4(8)-diene (12%)/α-phellandrene/*p*-cymene (10% for each) for stem and aerial parts, respectively; and, (iv) δ-3-carene/iso-sylvestrene (69-80%), eucalyptol (59%)/α-terpinene (14%), δ-3-carene/iso-sylvestrene (56%)/*p*-cymene (14%), limonene (50%)/(E)-β-ocimene (12%)/eucalyptol/*p*-cymene (11% for each), *p*-cymene/terpinolene (38% for each), terpinolene (36%)/*p*-cymene (27%), *p*-cymene (36%)/α-terpinene (18%)/γ-terpinene (12%), myrcene (35%)/α-pinene (27%), limonene (34%)/eucalyptol (21%)/*p*-cymene (17%), *p*-cymene (33%)/δ-3-carene/iso-sylvestrene (15%), terpinolene (33%)/limonene (22%)/3-carene (15%), limonene (29%)/*p*-cymene (27%)/α-terpineol (18%), terpinolene (28%)/limonene/α-phellandrene (17% for each), terpinolene (28%)/*p*-cymene (16%), *p*-cymene (26%)/terpinolene (20%)/apiole (16%) and dillapiole (16%)/terpinolene (15%)/*p*-cymene/*p*-cymen-8-ol (11% for each) for oleoresins. This EO from the leaves/resins evidenced different biological properties: anti-inflammatory, antimicrobial (*C. albicans, C. parasilopsis, C. krusei, C. guilliermondii, St. aureus, Serratia marcescens, K. pneumoniae, Proteus mirabilis, E. coli, Prevotella nigrescens, Str. mutans* and *Str. mitis*), antimutagenicity, antinociceptive, antioxidant, cytotoxicity (SP2/0, Neuro-2A, J774 and MCF-7 cell lines), gastroprotective, wound-healing, leishmanicidal (*L. amazonensis*), toxicity (against *A. salina*) and vasorelaxant (Amaral et al. 2009, Araujo et al. 2011, Bandeira et al. 2001, 2006, Bernadi et al. 2015, Cabral et al. 2021, 2020, Citó et al. 2006, da Silva et al. 2016, de Lima et al. 2016, Jiménez-Medina 2002, Mobin et al. 2016, 2017, Pinto et al. 2015, Pontes et al. 2007, Rao et al. 2007, Siani et al. 1999a, b, 2011, Silva et al. 2009, Zoghbi et al. 1995).

Euphorbiaceae

One of the most prominent genera (more numerous and diverse) of the Euphorbiaceae family is *Croton*, which includes herbs and trees with colored exudates. Some species grow in dry/open vegetation and are used in traditional medicine. Four representative species of this genus are *Croton fragilis, Cr. fragrans, Cr. malambo* and *Cr. niveus*. The first of these species is *Croton fragilis* Kunth (synonyms – *Cr. fragilis* var. *sericeus* Müll.Arg., *Cr. cienagensis* Rusby, *Oxydectes fragilis* (Kunth) Kuntze and *O. schlechtendaliana* Kuntze) which is a highly branched shrub (up to 3 m) with fragrant stipulate/tomentose leaves, racemose inflorescence (white/yellow flowers with glabrous petals that have simple hairs/trichomes) and drupe fruits (reddish-brown when ripe) containing a fragrant exudate (pale yellowish-brown) as well as its bark and branches. It is native to Colombia/Venezuela and is distributed in Mexico, Guatemala, Honduras, Guayana and Peru. The common names are llora sangre, vara blanca, aceitillo de hoja ancha, Taan-ché and Tanché. In Colombia, the bark and leaves of the plant are used in ethnomedicine as an infusion to treat colds and in baths to treat skin allergies and rashes (HCN-ICN 2021, Luján et al. 2015, Martínez Gordillo 1988, Murillo Aldana 1991, RBG Kew 2022, The Plant List 2013). The chemical constituents of the volatile fractions (VF) and essential oil of the leaves from the Colombian shrub were mainly β-caryophyllene (19–23%)/germacrene D (11–13%), whilst the most abundant components of the VF for fruits and branches were α-pinene (18%)/γ-terpinene (13%) and *p*-cymene (19%), in that order, based on Table 8 (Méndez et al. 2014a).

Table 8: Major chemical components determined in the volatile fractions and essential oils of different parts from *Cr. fragilis*, *Cr. fragrans*, *Cr. malambo* and *Cr. niveus*.

Type of VSM	Parts of the Plant	Main Chemical Constituents (Relative Amounts in %)	Biological Activities	References
Croton fragilis				
VF by SDE	Fruits	α-Pinene (18), γ-terpinene (13), germacrene D (12) and (E)-β-caryophyllene (10).	ND	Méndez et al. 2014a
	Leaves	(E)-β-Caryophyllene (19), germacrene D (11), myrcene (10) and γ-terpinene (9).		
	Branches	*p*-Cymene (19) and caryophyllene oxide (9).		
EO (0.07%)	Leaves	(E)-β-Caryophyllene (23), germacrene D (13) and β-elemene (8).		
Croton fragrans				
VF by SDE	Flowers	α-Pinene (68), linalool (7) and methyleugenol (6).	ND	Aristizabal-Córdoba and Pacheco-Castro 2014, Aristizabal-Córdoba et al. 2014
	Leaves	α-Pinene (70) and geraniol (8).		
EO (0.2%)	Leaves	α-Pinene (26), *trans*-nerolidol (11) and eudesm-7(11)-en-4-ol/ (E)-α-atlantone (6, each).	TAA: 50 ± 2 mmol Trol/kg; LC_{50}: 52 ± 3 μg/mL	
Croton malambo				
EO (1.6%)	Leaves	Methyleugenol (68), (E)-γ-bisabolene (7) and γ-curcumene (6).	%Inh. $ABTS^+$: $50 \pm 2\%$ (2045 μg/mL); LC_{50}: 310 ± 17 μg/mL (*in vitro*)	Altamar et al. 2016, Muñoz-Acevedo et al. 2014a,b
EO (2.7%)	Branches	Methyleugenol (85) and elemicin (2).	%Inh. $ABTS^+$: $28 \pm 1\%$ (2045 μg/mL); LC_{50}: 311 ± 5 μg/mL (*in vitro*); LC_{50}: 28 ± 6 μg/mL (*in vivo*);	
Croton niveus				
VF by SDE	Leaves (F/D)	F – Linalool (13) and germacrene D/β-pinene (12 each). D – Methylpyrrole (22), germacrene D (10) and methyl (E)-farnesoate (7).	ND	Abadía-Bejarano and Chamorro-Crespo 2011, Chamorro et al. 2011, Muñoz-Acevedo et al. 2011c
	Root bark (F/D)	F – Eucalyptol (12), borneol/α-pinene/bornyl acetate (8 each). D – α-Pinene (13) and eucalyptol/bornyl acetate (11 each).		

Table 8 contd. ...

...Table 8 contd.

Type of VSM	Parts of the Plant	Main Chemical Constituents (Relative Amounts in %)	Biological Activities	References
EO (0.08– 0.2%)	Leaves (F/D)	F – β-Pinene (25), germacrene D (11), linalool (10) and (E)-caryophyllene (8). D – β-Eudesmol (11), methyl (E)-farnesoate/spathulenol (9 each) and α-eudesmol (8).	LC$_{50}$ (μg/mL) on *A. salina* FL: 61 ± 2, DL: 18 ± 5; %inh. ABTS$^+$: FL: 39 ± 1 (13.9 mg/mL), DL: 39.8 ± 0.1 (14.2 mg/mL); inh. ox. (linoleic acid - 3 h): FL: 90%, DL: 30%	
EO (0.4– 0.6%)	Root bark (F/D)	F – Eucalyptol (20), α-pinene (18), bornyl acetate (10) and camphene (8). D – α-Pinene (23), eucalyptol (16), bornyl acetate (12) and camphene (9).	LC$_{50}$ (μg/mL) on *A. salina* FRB: 76 ± 8, DRB: 40 ± 5; %inh. ABTS$^+$: DRB: 6.7 ± 0.4 (13.5 mg/mL); inh. ox. (linoleic acid - 3 h): FRB: 60%, DRB: 60%	

F/D: Fresh/dried; VSM: Volatile Secondary Metabolites; EO: Essential Oil; VF: Volatile Fraction; SDE: Simultaneous-Distillation Extraction; ND: Not Determined.

The second species is *Croton fragrans* Kunth (synonym – *Oxydectes fragrans* (Kunth) Kuntze), which is a fragrant shrub-treelet (up to 10 m) whose branches/twigs and petioles secrete a colorless/yellowish, sticky/crystalline exudate with wrinkled leaves, green/spherical inflorescence and small yellow/white/cream flowers. It has persistent sepals in the fruit, covered by fine villi and trilobed subglobular/capsule fruits. The vernacular names of the species are algodón, algodoncillo, cascarillero, colpachi, cordoncillo, croton odorant, fragrant croton and mosquero, and it is native from Panama to Venezuela. It is used for animal food (wildlife birds), fuel (wood) and as a medicine (antirheumatic) (David Higuita et al. 2014, Grandtner and Chevrette 2014, HCN-ICN 2021, RBG Kew 2022, The Plant List 2013). The compositions of the VF from the Colombian tree were mostly characterized by α-pinene (68–70%), but the EO of the leaves contained α-pinene (26%), *trans*-nerolidol (11%) and atlantones [Z/E-α- (9% – 34/35); Z/E-γ- (3% – 36/37)]. The latter are sesquiterpene ketones which are rare constituents for the *Croton* genus. Into the bargain, the EO showed a strong cytotoxic effect on human lymphocytes with an LC$_{50}$ value of 52 ± 3 μg/mL.

The other species from genus is *Croton malambo* H. Karsten (synonym – *Oxydectes malambo* (H. Karst.) Kuntze), which is a deciduous shrub/tree (up to 10 m) of alternate leaves with a yellowish fragrant bark (translucent exudate), white/ pink-purple flowers and fruitful, and oil glands on all sides. Among its common names can be mentioned alouka, car(ca)naripe, cáscara de lombrices, malambo, malambo/Winter's bark, palo matias (palomatias), palomitas and torco. It is a species native to Colombia and Venezuela. In the Colombian/Venezuelan traditional medicine, the bark infusion is used for the treatment of diabetes, diarrhea/stomach disorders, rheumatism, gastric ulcer and cough and as an anti-inflammatory, antiparasitic, antiflu, antipyretic, hypoglycemic, sedative, analgesic and healing. Other pharmacological effects have been demonstrated as cytotoxic, anticoagulant and antinociceptive. From the leaves have been isolated phenylpropanoids (methyleugenol and elemicin), diterpenoids (*neo*-clerodane-*ent*-15,16-dihydroxy-cleroda-3,14-diene, *ent*-kaur-15-en-17-ol, 16β,17α-diolenkaurane, 5-hydroxy-*cis*-dehydrocrotonine and cajucarinolide/isocajucarinolide), triterpenoids (lupeol, botulin and stigmasterol), alkaloids (julocrotol and julocrotone). The different parts of the tree (leaves, branches/bark) contain essential oils constituted by methyleugenol (38) or (Z)-methyl isoeugenol. (Ávila Ayala 2014, Bernal et al. 2011, HNC-ICN 2021, Jaramillo et al. 2007, 2010, Jaramillo-Colorado et al. 2014, Mendoza et al. 2012, Morales et al. 2005, Pérez Arbeláez 1996, Quattrocchi 2012, RBG Kew 2022, Ruíz-Baquero et al. 2021, Suárez et al. 2003, 2005, 2008, The Plant List 2013).

According to Table 8, methyleugenol (68–85%) was the principal component of the EO from Colombian "malambo" leaves and branches with a difference only in the relative amounts; the EO yields were high (1.6–2.7%). Though, these chemical compositions presented some similarities/differences based on what was reported by Jaramillo et al. (2007, 2010), Jaramillo-Colorado et al. (2014), Mendoza-Meza et al. (2014) and Suárez et al. (2005, 2008), (i) bark EO contained methyleugenol (64–75%) and Z-methyl isoeugenol (54%), (ii) leaf EO contained methyleugenol (94%), (iii) SDE/SFE extracts were composed by methyleugenol (64–65%) and (iv) the EO yields were 0.4–1.2%, which were low values. Lastly, the essential oils were acaricidal (on *Dermatophagoides farinae*; Hughes, 1961), antibacterial (on *B. cereus* and *P. aeruginosa*), antioxidant (by ABTS^{+}) and toxic (on *A. salina*, *A. franciscana* and *Dario rerio*)/cytotoxic (on human lymphocytes and MCF-7 cell line).

The last species is *Croton niveus* Jacq. (synonyms – *Berhamia hispida* Klotzsch; *B. macrostachya* Klotzsch; *Cr. mollii* Müll. Arg.; *Cr. populifolius* Mill.; *Cr. pseudochina* Schltdl.; *Cr. septemnervius* McVaugh; *Cr. syringifolius* Kunth; *Kurkas populifolium* (Mill.) Raf.; *Oxydectes nivea* (Jacq.) Kuntze; *O. populifolia* (Mill.) Kuntze), which is a fragrant shrub (up to 5 m) with silvery-white leaves on the underside, brown flower buds and copious white latex. It is named as *algodoncillo*, *cascarillo*, *chulché*, *colpachi(í/e)*, *croton blanc-de-neige*, *guayabito*, *huilote*, *juilocuahuitl*, *olith*, *oroquemas*, *plate(i)ado*, *platero*, *platino*, *quisarrá colpachi*, *snowwhite croton*, uvitas, *vara/palo blanca*(o) and *vidrioso*. The shrub is native to tropical America and it is distributed from Mexico (all of Central America, including the Great Antilles) to Guyana. The parts of the plant (e.g., bark and leaves) have been applied as infusions/poultice/baths to treat insect bites, heal sores, intermittent fevers, malaria, rheumatism, burns, dyspepsia, hypoglycemic, etc. From the bark/roots extract has been isolated a diterpene lactone (nivenolide) (Abadía-Bejarano and Chamorro-Crespo 2011, Barros et al. 2015, Bernal et al. 2011, Grandtner and Chevrette 2014, HNC-ICN 2021, Pérez Arbeláez 1996, Quattrocchi 2012, RBG Kew 2022, Rojas and Rodriguez-Hahn 1978, The Plant List 2013).

The volatile secondary metabolites (VF and EO) from Colombian shrub leaves and root bark (fresh and dried) were represented by β-pinene (12–25%), α-pinene (13–23%), methylpyrrole (22%), eucalyptol (11–20%), linalool (10–13%), germacrene D (10–12%), bornyl acetate (8–12%) and β-eudesmol (11%) (Table 8).

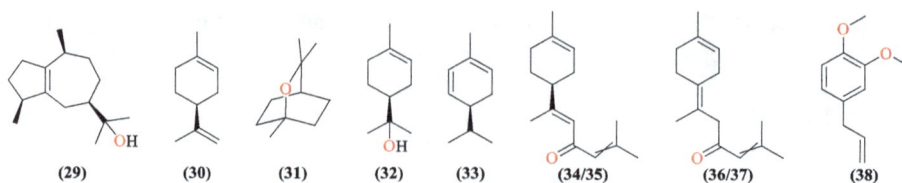

(29)   (30)   (31)   (32)   (33)   (34/35)   (36/37)   (38)

These compositions showed some similarity to those reported by Setzer (2006), which mentioned that the bark OE from Costa Rican species contained α-pinene (14%), eucalyptol (12%) and borneol (8%); this author only studied the bark of the plant. In other reports by Schmidt Werka et al. (2007) on the biological properties of the same EO, they reported that the EO was active against MCF-7 cell line (57 ± 2%, 100 μg/mL), *A. salina* (LC$_{50}$ 18 μg/mL), *B. cereus* (MIC 625 μg/mL) and *St. aureus* (MIC 78 μg/mL). Moreover, the bioactivities demonstrated by the Colombian EO were toxicity against *A. salina* (LC$_{50}$ 18 ± 5–76 ± 8 μg/mL, with the dried leaves and root bark being the most active) and the inhibition of the linoleic acid oxidation (30–90%, being the most active the fresh leaves). Additionally, leaf EO presented *in vitro* cyto-/genotoxicity in human lymphocytes and sperm (Hernández et al. 2015).

Fabaceae

The chosen shrubby plant from the Fabaceae family is *Dalea carthagenensis* (Jacq.) J.F. Macbr. (synonyms – *D. domingensis* L., *D. emphysodes* L., *D. enneaphylla* (L.) Willd., *D. phymatodes* L., *Parosela carthagenensis* (Jacq)., *Psoralea carthagenensis* Jacq. and *P. enneaphylla* L.,) which is distributed from Mexico to Bolivia and the northern Colombian region has been found on the bridle path. This species is an evergreen and highly branched sub-shrub (up to 2 m) with petiolate/glabrous leaves and intra-petiole glands, and villose branches (leaves/branches give off a pleasant smell), yellow flowers and fruits. Some of its vernacular names are Cartagena prairie-clover, prairieclover, escobilla, añil or añis. The leaves are applied to crafts as dyes and in Mexico, they are used as a treatment for rashes and digestive infections and as an anti-inflammatory. If the bioproperties of certain extracts [(CH$_3$)$_2$CO/ CH$_2$Cl$_2$/MeOH/hexane]/fractions (BuOH) from this herb (whole plant/flowers/stem) are considered, these result in cytotoxic (KB cell line), anti-inflammatory (on factor NF-κB) and antimicrobial (*St. aureus*, *C. albicans*, *C. tropicalis* and *Trychophyton mentagrophytes*) and the presumed constituents responsible for these bioactivities would be some terpenic and phenolic compounds (Ankli et al. 2002, HNC-ICN 2021, Montes-de-Oca et al. 2017, Muñoz-Acevedo et al. 2019, Piñeros and González 2020, The Plant List 2013).

The main constituents of volatile nature isolated by three extraction methods from the different parts of *añis* are mentioned in Table 9; they were (E)-β-ocimene and β-pinene by SDE for both parts and β-pinene and β-caryophyllene by HS-SPME for leaves/branches. These components varied in their percentage relative areas (10–50%). The greatest chemical variability (caryophyllene oxide, β-pinene and (E)-β-ocimene) was observed in the composition of the EO from leaves, which could be explained by the effects of the dry and rainy seasons in the collection area of the

Table 9: The most abundant volatile components found in the volatile fractions and essential oils of different parts from *D. carthagenensis*.

Type of VSM	Parts of the Plant	Main Chemical Constituents (Relative Amounts in %)	Biological Activities	References
Dalea carthagenensis				
VF by SDE	Leaves	(E)-β-Ocimene/β-caryophyllene (21, each) and β-pinene (19).	N/A	Álvarez et al. 2015, Muñoz-Acevedo et al. 2019
	Branches	(E)-β-Ocimene (36), β-pinene (20), *trans*-pinocarvyl acetate (14) and myrcene (10).		
VF by HS-SPME	Leaves	β-Pinene (50), β-caryophyllene (23) and α-pinene (11).		
	Branches	β-Pinene (20), β-caryophyllene (10) and caryophyllene oxide (9).		
EO (0.08–0.4%)	Leaves	L_1 – Caryophyllene oxide (32), β-pinene (13) and β-caryophyllene (9). L_2 – β-Pinene (16), (E)-β-ocimene (13) and myrcene/caryophyllene oxide/β-caryophyllene (10 each). L_3 – (E)-β-Ocimene (36), β-caryophyllene (21) and β-pinene (14).	%Rep. on *Sitophyilus zeamais* (2–4 h, 1 µg/cm³) L_1: 45 ± 6–58 ± 11. L_2: 47 ± 6–82 ± 13; %inh. by ABTS⁺: L_1: 24 (1964 ppm), L_2: 31 (2109 ppm); LC_{50} on lymphocytes: L_2: 115 ± 1 µg/mL.	
	Branches	B_3 – *trans*-Pinocarvyl acetate (29) and caryophyllene oxide (22).	N/A	

VSM: Volatile Secondary Metabolites; VF: Volatile Fraction; EO: Essential Oil; SDE: Simultaneous-Distillation Extraction; HS-SPME: Headspace Solid Phase Micro-Extraction; L_1: Leaves of Dry Season (2013); L_2: Leaves in Rainy Season (2014); L_3/B_3: Leaves/Branches Collected in Rainy Season (2017); N/A: Not Applicable.

plant samples. It is noteworthy that the branch EO was a yellowish solid constituted by *trans*-pinocarvyl acetate (41) (reported for the first time for this genus), which is a flavoring agent for non-/alcoholic beverages/chewing gums/hard/soft candies (GRAS flavoring substance), as well as a fragrance for perfume/cosmetic products. From the point of view of bioproperties, two EO (L_1/L_2) showed good repellent effects ($\geq$ the positive control) against *S. zeamais* and one EO (L_2) was moderately toxic (100 µg/mL < LC_{50} < 1,000 µg/mL) and two EO (L_1/L_2) showed poor capability for scavenging the ABTS⁺ radical-cation (Álvarez et al. 2015, Api et al. 2021, Cohen et al. 2015, IFRA 2021, Muñoz-Acevedo et al. 2019a).

Melastomataceae

Miconia sp. Ruiz and Pav. is a very branched and small tree (up to 12 m) with large, opposite, and symmetrical/elliptic/basiauriculate leaves and with woody and scaly stems and drupe-shaped fruits. It is native to the tropical areas of America and the *Miconia* genus is the largest of the family Melastomataceae. Its common name is *clavo de monte* because of its smell reminiscent of "clove". In herbal medicine, the

Table 10: Main chemical constituents determined in the hydrolate of bark from *Miconia* sp.

Type of VSM	Parts of the Plant	Main Chemical Constituents (Relative Amounts in %)	Reference
Miconia sp.			
Hydrolate	Bark	Eugenol (56) and 1-octen-3-ol (44)	Páez Aranzalez and Sánchez Corredor 2018

tree has been used as a vermifuge to cure throat and neck pain, diarrhea, tuberculosis, toothaches, oral infections, fungal infections, scabies and as an antidote against "Conga" ant bite. The principal non-medicinal uses are wood, ornamental and ecological restoration. EtOH extracts from the plant have demonstrated antibacterial (*St. aureus*, *K. pneumoniae*) and antifungal (*C. albicans*) properties (Celotto et al 2003, Gómez Chimá and Vargas Campos 2018, Paéz Aranzalez and Sánchez Corredor 2018). This plant did not produce essential oil, although its wood was fragrant. Nevertheless, the result of the hydrodistillation process made it possible to obtain hydrolate-rich eugenol (42) and 1-octen-3-ol (43) (Table 10).

Myrtaceae

One species (*Eugenia procera* (Sw.) Poir.) from Myrtaceae family was included in this chapter. This small tree (up to 3 m) has dark purple fruits, and it is distributed from West Indies (SE North America) to Colombia/Trinidad and Tobago (N South America). The synonyms are *Eugenia parkeriana* DC., *Myrtus brachystemon* DC., *M. cerasina* Eggers non Vahl and *M. procera* Sw. Its vernacular names are rock myrtle, eugénie grandissime, merisier, arraiján/arrayán, guayabito arrayán and hoj(y) a menuda. Its bark is medicinal and is used in the treatment of diabetes (Grandtner and Chevrette 2014, HCN-ICN 2021, RBG Kew 2022).

The chemical composition of the volatile fraction and essential oil from Colombian plant leaves is reported in Table 11. The VF and EO from leaves were characterized by α-pinene (30–33%); LC_{50} value on human lymphocites was 99.9 ± 0.6 µg/mL. If the radical-scavenging capacity (by ABTS⁺ method) of the EO is considered, this showed poor scavenging capacity (inh. 7.8 ± 0.5%, 2091 µg/mL).

Table 11: Main chemical components determined in the volatile fraction and essential oil of leaves from *E. procera*.

Type of VSM	Parts of the Plant	Main Chemical Constituents (Relative Amounts, %)	Biological Activities	References
Eugenia procera				
VF by SDE	Leaves	α-Pinene (33), eucalyptol (8) and viridiflorol (7).	ND	Bru Hernández and Fernández-Aparicio 2014, Bru et al. 2014
EO (0.07%)		α-Pinene (30), viridiflorol (6) and eucalyptol/aromadendrene (5 each).	%Inh. ABTS⁺: 7.8 ± 0.5% (2091 µg/mL); LC_{50}: 99.9 ± 0.6 µg/mL	

VSM: Volatile Secondary Metabolites; EO: Essential Oil; VF: Volatile Fraction; SDE: Simultaneous-Distillation Extraction; ND: Not Determined.

Piperaceae

One of the 604 species of Piperaceae found in Colombia is *Piper eriopodon* (Miq) C. DC. (synonyms – *Artanthe eriopoda* Miq., *P. leptophyllum* C. DC. *P. tomasi* Trel.; common name: *cordoncillo*) which is a shrub (up to 4 m) with erect inflorescences and leaves with a characteristic odor. It is native to Colombia, Venezuela and Ecuador. Some traditional uses of the plant are as an analgesic, diuretic and antirheumatic as well as a treatment for kidney stones, bronchial conditions and antidote against snake bites. Some extracts (e.g., hexane, MeOH, EtOH, BuOH and CH_2Cl_2), essential oils (inflorescences/leaves/stems) and isolated compounds (e.g., gibbilimbol B – 44) from the species have showed antibacterial (e.g., *Mycobacterium bovis*, *M. tuberculosis* and *St. aureus*), antifungal (e.g., *Aspergillus fumigatus*, *Botrytis cinerea*, *Fusarium solani*, *F. oxysporum*, *Trichophyton mentagrophytes* and *T. rubrum*), herbicide, anticancer (e.g., A549, HeLa, HepG-2, MDAMB-231, MCF-7 and PC-3 cell lines)/ cyto-toxic (e.g., *A. franciscana*, human dermal fibroblast, RAW264.7 macrophages and Vero cell line) and antioxidant (e.g., $ABTS^{+}$, $DPPH^{•}$, ORAC) properties as well as a chemical variability (Amaya and Acevedo 2011, Arbeláez-Cortés 2013, Castañeda Muñoz 2007, Correa Navarro et al. 2001, Cuervo Cuervo 2017, Guzmán et al. 2010, HCN-ICN 2021, Muñoz 2008, Muñoz et al. 2018, Olivero-Verbel et al. 2009, RBG Kew 2022, Saavedra Barrera 2015, Tangarife-Castaño et al. 2014, The Plant List 2013, Ustáriz Fajardo et al. 2020, Velandia et al. 2018).

Table 12 contains the chemical compositions of the VF and EO from cordoncillo fruits, inflorescences and leaves, which were represented by gibbilimbol B (10–72%), (E)-β-caryophyllene (7–43%), myrcene (9–31%) and β-pinene (9–23%). Nonetheless, simple phenols predominated in SDE extracts of the fruits and leaves and monoterpenes in the inflorescences; sesquiterpenes prevailed in the inflorescences/ leaves by HS-SPME. The main constituent of the leaf EO from the northern region of Colombia was gibbilimbol B, which was identified by Muñoz (2008) in the fruit EO (Bogotá). Nevertheless, the comparison of the chemical compositions with other literature reports showed some differences; Tangarife-Castaño et al. (2014)/ Castañeda et al. (2007), Uztáris-Fajardo et al. (2020), Velandia et al. (2018) and Valenzuela-Vergara et al. (2014), whose EO were respectively constituted by dillapiole (~39%)/β-caryophyllene (~8%), 1,8-cineole (~37%)/β-pinene (~9%), α-pinene (~19%)/β-pinene (~16%)/β-caryophyllene (~12%)/caryophyllene oxide (~11%) and β-copaene (14%)/β-eudesmol/5-epi-7-epi-α-eudesmol (11%).

Referring to the bioproperties of the EO, the cytotoxicity on erythrocytes based on the CH_{50} value allowed the determination that EO was moderately hemolytic (100 μg/mL < HC_{50} < 1,000 μg/mL). The EO, meanwhile, was active (LC_{50} < 100 μg/mL) against lymphocytes and Hep-2 cell line, but more on the cell line than normal cells showing a selectivity index of 2.2 on Hep-2 line compared to human lymphocytes. Data on cytotoxicity of EO from *P. eriopodon* have shown a notable cytotoxic potential according to described by Velandia et al. on HEK293 (IC_{50}: 153 ± 10 μg/mL), MCF-7/HeLa (IC_{50}: 50 μg/mL for each) and HepG-2 cells (IC_{50}: 140 ± 24 μg/mL) by EO composed by α-pinene/β-pinene/β-caryophyllene/caryophyllene oxide; the gibbilimbol B isolated and evaluated on the cell lines produced IC_{50} values of 11 μg/mL (on MDAMB-231 line), 12 μg/mL

Table 12: Major compounds present in the volatile fractions and essential oils of different parts from *P. eriopodon* and *P. marginatum.*

Type of VSM	Parts of the Plant	Main Chemical Constituents (Relative Amounts in %)	Biological Activities	References
Piper eriopodon				
VF by SMP	Inflorescence	Gibbilimbol B (70) and (E)-β-caryophyllene (7).	N/A	González et al. 2016b, Muñoz-Acevedo et al. 2016, 2022
	Leaves	(E)-β-Caryophyllene (19), phytol (11) and gibbilimbol B/β-selinene (10 each).		
VF by SDE	Inflorescence	Myrcene (31), β-pinene (23) and gibbilimbol B/α-pinene (14 each).		
	Fruits	Gibbilimbol B (60), β-pinene (10) and myrcene (9).		
	Leaves	Gibbilimbol B (46), (E)-β-caryophyllene (11) and β-pinene (9).		
VF by HS-SPME	Inflorescence	(E)-β-Caryophyllene (23), myrcene (20), β-pinene (19) and α-pinene (10).		
	Leaves	(E)-β-Caryophyllene (43) and β-selinene (20).		
EO (%)	Leaves	Gibbilimbol B (72) and (E)-β-caryophyllene (9).	Cytotoxicity (μg/mL) on eryth.: HC_{50}: 115 ± 2, lymph.: LC_{50}: 71 ± 4, Hep-2 line: LC_{50}: 33 ± 2; antibacterial inh. (φ inh. zone, mm; 4 μg EO): 23 ± 2 (*L. monocytogenes*), 22.5 ± 0.4 (*St. aureus*), 23 ± 2 (*E. coli*); TAA value: 2249 ± 130 mmol Trolox®/kg; AChE inh.: IC_{50}: 13 ± 1 μg/mL.	
Piper marginatum				
VF by SDE	Fresh leaves	(E)-β-Caryophyllene (12), linalool/α-pinene (9, each), β-pinene (8) and β-elemene (7).	N/A	Borrero Meza and Hurtado Palacio 2013, Muñoz-Acevedo et al. 2011b
EO (0.3–0.4%)	Fresh leaves	α-Pinene (17), β-pinene (13), linalool/(E)-β-caryophyllene (7, each) and β-elemene (6).		
	Dry leaves	Linalool (14), (E)-β-caryophyllene (13) and β-elemene (8).		

VSM: Volatile Secondary Metabolites; VF: Volatile Fraction; EO: Essential Oil; SMP: Simple Maceration Process; SDE: Simultaneous-Distillation Extraction; HS-SPME: Headspace Solid Phase Micro-Extraction; Eryth.: Erythrocytes; Lymph.: Lymphocytes; AChE: Acetylcholinesterase Enzyme; HC: Hemolytic Concentration; LC: Lethal Concentration; IC: Inhibitory Concentration; N/A: Not Applicable.

(on MCF-7 cells), 32 µg/mL (on PC-3 cells) and 40 µg/mL (on A549 line). Lastly, Tangerife-Castaño et al., determined that the EO (rich in alkenylbenzodioxole and sesquiterpene derivatives) showed an IC_{50} value of 16 ± 1 µg/mL on the Vero cell.

Remarkable bioactivity demonstrated by *P. eriopodon* EO (4–16 µg) was its high efficacy for inhibiting the growth of *St. aureus*, *E. coli* and *L. monocytogenes*; the descending order of bacterial susceptibility (from highest to lowest) was *St. aureus* > *E. coli* > *L. monocytogenes*. According to the report by Ustáriz-Fajardo et al., the Venezuelan EO was not active against *E. coli/K. pneumoniae*, and moderate on *St. aureus* (MIC 2500 µg/mL). And, finally, the TAA value of the EO (2249 ± 130 mmol Trolox®/kg) was slightly higher (ratio 1.04) than that of the BHA.

A second *Piper* species of interest is *Piper marginatum* Jacq. [synonyms – *Artanthe alaris* (Ham.) Miq., *A. catalpaefolia* (Kunth) Miq., *A. caudata* (Vahl) Miq., *A. marginata* (Jacq.) Miq., *P. alare* Desv. ex Ham., *P. anisatum* Kunth, *P. catalpaefolium* Kunth, *P. caudatum* Vahl., *P. patulum* Bertol., *P. quiriguanum* Trel., *P. sanjoseanum* C. DC., *P. uncatum* Trel., *Schilleria catalpaefolia* (Kunth) Kunth, *Sch. caudata* (Vahl) Kunth and *Sch. marginata* (Jacq.) Kunth.], a small perennial shrub (up to 3 m) with fragrant leaves (reminiscent of anise), white flowers in spikes and drupaceous fruits. It is distributed from SE North America to Brazil/Peru (including the Caribbean and Antilles), and its common names are anis, aniseto, anisillo, bachar, buttonwood, caapeba cheirosa, cake bush, canilla de grulla, corazón de la Virgen, cordon(s)illo (negro), deshinchadora, guayuyo, higuillo (oloroso), katio, margined/ marigold pepper, oloroso, Santa María (de anís), spanish alder and zorillo. It is used in ethnomedicine for the treatment of headaches, fevers, nosebleeds, flu, inflammations, muscle pains, wound healing, amenorrhea/blennorrhea, emmenagogue, carminative, digestive, diuretic, amebiasis, malaria, mosquitoes repellent and prepared in infusion/ decoction/poultice/bath. The essential oils/extracts from the plant have shown cytotoxicity, antimicrobial, antimalarial/antiprotozoal, deterrent and antilarval (against *A. aegypti*), antileishmanial and repellent activities. Some molecules type terpenoids, phenylpropa(e)noids, flavonoids, alkaloids, amides have been isolated and could be responsible for the therapeutic effects (Barros et al. 2015, Bernal et al. 2011, Borrero Mendoza and Hurtado Palacio 2013, Bru and Guzmán 2016, García-Barriga 1992b, Gonçalves et al. 2019, Grandtner and Chevrette 2014, HNC-ICN 2021, Jaramillo-Colorado et al. 2015, Leal et al. 2013, Olivero-Verbel et al. 2009, Paz et al. 2017, Quattrocchi 2012, RBG Kew 2022, Reigada et al. 2007, Tangarife-Castaño et al. 2014, The Plant List 2013).

The main chemical constituents identified in VF and EO of "Santa María" from the northern region of Colombia (Table 12) were (E)-β-caryophyllene (12–13%), α-pinene (9–17%) and linalool (7–14%). These compositions differed from those described by Olivero-Verbel et al. (2009), Jaramillo-Colorado et al. (2015) and Valenzuela-Vergara et al. (2014) for Colombia, which reported that their EO contained as major components elemecin (18%)/(E)-β-caryophyllene/α-phellandrene (11%, each); *cis-p*-anethole (46%)/estragol (29%), germacrene D (37%)/β-elemene (13%); and δ-elemene (20%), respectively. As well as the plant EO from Costa Rica contained (E)-anethole (46%)/*p*-anisaldehyde (22%)/anisyl ketone (14%) (Vogler et al. 2006), while the Panamanian plant EO was constituted by Z-isosafrol (34%)/

croweeacin (11%) (Zachrisson et al. 2019) and in the Ecuadorian EO, curzerene (22%) was the most abundant constituent (Moncayo et al. 2021).

The greatest chemical diversity in the composition of the *P. marginatum* EO is found in Brazil. There have been reported as major components myristicin (9%)/3,4-methylenedioxypropiophenone (8–9%) (Ramos et al. 1986); safrole (64%), safrole (41–52%)/3,4-(methylenedioxy)propiophenone (12–30%), 3,4-(methylenedioxy)propiophenone (33%)/safrole (24%)/(E)-β-ocimene (14%), *p*-mentha-1(7),8-diene (23–39%)/3,4-(methylenedioxy)propiophenone (19–41%), 3,4-(methylenedioxy)propiophenone (17%)/myristicin (16%)/safrol (10%), (E)-β-ocimene (15%)/3,4-(methylenedioxy)propiophenone (14%)/germacrene D (10%), 3,4-(methylenedioxy)propiophenone (23–25%)/(E)-β-caryophyllene (10–11%), 3,4-(methylenedioxy)propiophenone (40%)/(E)-β-ocimene (14%), 3,4-(methylenedioxy)propiophenone (18–30%)/γ-terpinene (9–14%), (E)-β-caryophyllene (13%)/germacrene D (9%)/3,4-(methylenedioxy)propiophenone (8%), α-copaene (11%)/(E)-β-caryophyllene/3,4-(methylenedioxy)propiophenone (10%, each), (E)-isoosmorhizole (29–47%)/ isoosmorhizole (11–24%)/(E)-anethole (14–26%), 2-methoxy-4,5-(methylenedioxy)propiophenone (26%)/ methoxy-4,5-(methylenedioxy)propiophenone (22%)/(E)-isoosmorhizole (16%), (E)-β-caryophyllene (14%)/bicyclogermacrene (12%)/(E)-asarone (11%) (Andrade et al. 2008); (Z)-asarone(30%)/patchouli alcohol (16%), (E)-asarone (22–33%)/patchouli alcohol (23–26%) (Autran et al. 2009); *p*-mentha-1(7),8-diene (39%)/3,4-methylenedioxypropiophenone (19%), (E)-isoosmorhizole (32%)/ (E)-anethole (26%) (Souto et al. 2012), 3,4-methylenedioxypropiophenone (22%) (da Silva et al. 2016); 3,4-(methylenedioxy) propiophenone/bicyclogermacrene/ germacrene D (10%, each) (Gonçalves et al. 2019); δ-3-carene (13%) (Majolo et al 2019); kakuol (13%)/myristicin (13%)/sarisan (12%) (de Souza et al. 2020); and, exalatacin (9%)/α-pinene (8%) (Dutra et al. 2020).

Verbenaceae

Lantana camara is the first species considered in this chapter from the Verbenaceae family. It is an evergreen, deciduous and highly branched fragrant shrub (up to 3 m) with a rough surface and simple/opposite leaves with toothed margins, inflorescences in capitula with small flowers (white, yellow, orange, pink or mauve) which change color as they mature; shiny bluish-black drupe-shaped fruits. It is native to Latin America, but it is widely distributed in the neo-tropical regions and is a troublesome weed, forming impenetrable thickets. The best known vernacular names are alantana/lantana/lampana/lauraimana, albahaca de caballo, aya manchana, bandera española, bubita negra, bwa wa tau, cámara, cariaquit(ll)o(a), carnica, carraquillo, cinco negritos (coloraditos), confite, cuasquito, cura verrugas, erva sagrada, filigrana, flor de duende/San Cayetano/sangre, gurupacha, hierba de Cristo, ingarosa, jaral, mabizou, maestrante del Brasil, mamizou, matizadilla, mille fleurs, mora, peonia negra, petelkin, quita pesar, San Rafaelito, red sage, sanguinaria, siete negritos (colores), socorrete, sonora, supirosa, té de Bahamas, tucnai, tupirosa, venturosa,venturosa colorado, verbena morado, yerba de la maestranza, zapotillo, zarzamora and zorrito. Synonyms of the scientific name are *Camara aculeata*

Kuntze, *C. vulgaris* Benth., *L. aculeata* L., *L. annua* C.B. Clarke, *L. antillana* Raf., *L. armata* Schauer, *L. crenulata* Otto and A. Dietr., *L. crocea* Jacq., *L. cummingiana* Hayek, *L. glandulosissima* Hayek, *L. horrida* Kunth, *L. melissifolia* Sol., *L. mixta* Medik., *L. moritziana* Otto and A. Dietr., *L. mutabilis* Weigel, *L. polyacantha* Schauer, *L. sanguínea* Medik., *L. scabrida* Sol., *L. scandens* Moldenke, *L. scorta* Moldenke, *L. spinosa* Le Cointe, *L. tiliifolia* Cham. and *L. urticifolia* Mill. This plant is toxic, hepatotoxic and nephrotoxic in Ayurveda medicine; nevertheless, in other types of traditional medicine, prepared in decoction/paste/chewed/inhalation/ extract forms, it has been effective against tetanus, rheumatism, malaria, chickenpox, boils, swellings, wounds/cuts, snakebite, toothache, headache, fever, colds, coughs, sore throat, conjunctivitis, etc. The extracts/EO/isolated compounds (triterpenoids) from the plant have demonstrated anti-inflammatory, antifeedant, antileishmanicidal, antimicrobial, antimycobacterial, antiplasmodial, antioxidant, cytotoxic/antitumor, enzyme inhibitory (thrombin and protein kinase), insecticidal, larvicidal, mosquito repellent, nematicidal, virucidal and sedative properties (Ali et al. 2019, Ayalew 2020, Ayub et al. 2019a, b, Begum et al. 1995, 2000, 2002, 2008, 2013, 2014, Bernal et al. 2011, Costa et al. 2010, Delgado-Altamirano et al. 2021, Dougnon and Ito 2020, Fatope et al. 2002, García et al. 2010, Gupta et al. 2017, Herbert et al. 1991, Hiremath et al. 2021, HNC-ICN 2021, Litaudon et al. 2009, Melanie et al. 2020, O'Neill et al. 1998, Ono et al. 2020, Pour and Sasidharan 2011, Qamar et al. 2005, Quattrocchi 2012, Rajakrishnan et al. 2022, RBG Kew 2022, Saleh et al. 1999, Sharma et al. 2021, Sousa et al. 2015, The Plant List 2013, Wangrawa et al. 2021, Wu et al. 2020, Zoubiri and Baaliouamer 2012).

The main constituent of the leaf EO (from the Colombian savannah) was *cis*-cadin-4-en-7-ol (28%, 45) which differed from the constituents identified in the other species of the *L. camara* from Colombia and other Latin American/ African/Asian countries. Thus, other Colombian EO presented (E)-β-farnesene (39%)/bicyclosesquiphellandrene (18%)/(E)-β-caryophyllene (14%), bicyclosesquiphellandrene (24–31%)/γ-muurolene+γ-curcumene (10–14%)/(E)-β-caryophyllene/α-zingiberene (9–13%, each) (Caroprese Araque et al. 2011). The Brazilian EO were constituted by bicyclogermacrene (19%)/isocaryophyllene (17%)/valencene (13%)/germacrene D (12%) (De Sousa et al. 2020); β-elemene (7–23%)/β-caryophyllene (7–53%)/α-humulene (7–39%)/α-curcumene (34%)/ germacrene D (9–48%)/bicyclogermacrene (8–19%)/spathulenol (11%)/ caryophyllene oxide (8–18%)/ledol (8–13%) (Pereira et al. 2020); germacrene D (26–36%)/germacrene B (11–13%)/β-caryophyllene (9–13%) (Dos Santos et al. 2019); β-caryophyllene (11–33%)/bicyclogermacrene (11–14%)/spathulenol (7–16%)/caryophyllene oxide (6–17%)/isocaryophyllene (10%) (Medeiros et al. 2012); germacrene D/(E)-β-caryophyllene (20%, each) (Passos et al. 2012); bicyclogermacrene (26%)/β-caryophyllene (24%)/germacrene D (19%)/valecene (12%) (Sousa et al. 2012); isocaryophyllene (17%)/germacrene D (12%) (Sousa et al. 2010); limonene (16%)/α-phellandrene (16%)/germacrene D (13%)/β-caryophyllene (11%), germacrene D (28%)/germacrene B (9%), γ-curcumene+ar-curcumene (28–32%)/α-zingiberene (16–19%)/α-humulene (10–11%) (da Silva et al. 1999). For its part, the Argentinian EO contained spathulenol (35%) (García et al. 2010); the

Table 13: Principal compositions of the volatile fractions and essential oils of the different parts from *La. camara* and *L. alba.*

Type of VSM	Parts of the Plant	Main Chemical Constituents (Relative Amounts in %)	Biological Activities	References
Lantana camara				
EO (0.11%)	Leaves	*Cis*-Cadin-4-en-7-ol (28), β-bisabolene (7) and β-caryophyllene/germacrene-D-4-ol (5 in each).	Antifungal against *C. gloeosporioides*	Páez Aranzalez and Sánchez Corredor 2018
Lippia alba				
VF by SMP	Flowers	*Cis*-Piperitone oxide (44–56), germacrene D (16–32) and limonene (8–15).	N/A	González et al. 2016a, González 2019, Muñoz-Acevedo et al. 2016, 2019
	Leaves			
VF by SDE	Flowers	*Cis*-Piperitone oxide (17–24), limonene (10–16) and germacrene D (11–14).		
	Leaves			
VF by HS-SPME	Flowers	PDMS: Germacrene D (28), limonene/*cis*-piperitone oxide (13 each). PA: *cis*-piperitone oxide (38), germacrene D (25) and limonene (15).		
	Leaves	PDMS: Germacrene D (29), (E)-β-caryophyllene (7)/*cis*-piperitone oxide (7 each). PA: *cis*-piperitone oxide (34), germacrene D (30) and limonene (11)		
EO (0.5%)	Leaves	*Cis*-Piperitone oxide (44 ± 2), limonene (18 ± 5) and germacrene D (16 ± 3).	Cytotoxicity (µg/mL) on: eryth.: HC_{50} 580 ± 1, lymph.: LC_{50} 127 ± 3, Hep2 cell: LC_{50} 38 ± 2; AChE inh.: IC_{50} 28 ± 2 µg/mL	

VSM: Volatile Secondary Metabolites; VF: Volatile Fraction; EO: Essential Oil; SMP: Simple Maceration Process; SDE: Simultaneous-Distillation Extraction; HS-SPME: Headspace Solid Phase Micro-Extraction; Eryth.: Erythrocytes; Lymph.: Lymphocytes; AChE: Acetylcholinesterase Enzyme; LC: Lethal Concentration; IC: Inhibitory Concentration; N/A: Not Applicable.

Cuban EO was composed of (E)-nerolidol (43%) (Pino et al. 2004); the Peruvian EO presented carvone (76%)/limonene (17%) (Benites et al. 2009).

Considering the African countries, the Benin EO was constituted by sabinene (39%)/eucalyptol (29%) (Dougnon and Ito 2020); the Burkina Faso EO contained sabinene (32%)/eucalyptol (21%)/β-caryophyllene (13%) (Wangrawa et al. 2018), the Cameroon EO presented *ar*-curcumene (25%)/β-caryophyllene (13%) (Ngassoum et al. 1999); in the Madagascar EO was identified davanone (18–29%)/β-caryophyllene (8–17%)/sabinene (5–14%)/*ar*-curcumene (13%), β-caryophyllene (18–44%)/β-bisabolene (10–20%)/sabinene (3–19%)/*epi*-cubenol (12%)/eucalyptol/α-humulene (11%, each) (Randrianalijaona et al. 2006); and, davanone (15%)/β-caryophyllene (12%) (Ngassoum et al. 1999). β-Caryophyllene (25%)/α-humulene (20%) were the main constituents in the Nigerian EO (Oyedeji et al. 2003), whilst sabinene (17%)/β-caryophyllene (14%) were for the Yemen EO (Satyal et al. 2016).

On the other hand, the Chinese EO was characterized by germacrene D (16%)/β-caryophyllene (12%) (Sundufu and Shoushan 2004). The Indian EO showed a wide chemical diversity: β-caryophyllene (14–22%)/bicyclogermacrene (8–18%)/sabinene (7–16%)/eucalyptol (6–10%)/α-humulene (6–11%) (Gotyal et al. 2016, Sefidkon 2002); 10-dodecatriene (29%)/β-caryophyllene (12%) (Kurade et al. 2010); β-caryophyllene (25%)/δ-selinene (18%) (Sarma et al. 2020); (E)-nerolidol (41%)/β-caryophyllene (13%), germacrene D (28%)/β-caryophyllene (10%) (Padalia et al. 2015); β-caryophyllene (23%)/α-humulene (12)/germacrene D (11%) (Rana et al. 2005); *cis*-davanone (37–52%)/β-caryophyllene (10–13%), *cis*-davanone (7–48%)/β-caryophyllene (10–27%)/bicyclogermacrene (5–12%) (Misra and Saikia 2011); germacrene D (20%)/γ-elemene (10%), β-elemene (14%)/germacrene D/α-copaene (11%, each one) (Khan et al. 2002); palmitic acid (23–33%)/stearic acid (13–24%) (Khan et al. 2003).

The Iranian EO contained α-humulene (23%)/*cis*-caryophyllene (16%) (Zandi-Sohani et al. 2012); the Nepali EO was represented by davanone (44%)/ (E)-nerolidol (13%) (Satyal et al. 2016); in the Saudi Arabia EO was found *cis*-3-hexen-1-ol (11%)/1-octen-3-ol/spathulenol (9% each), caryophyllene oxide (11%)/β-caryophyllene (10%) (Khan et al. 2015); the Vietnamese EO had β-caryophyllene (20%)/sabinene (16%), germacrene D (19–25%)/β-caryophyllene (11–14%), sabinene (18%)/ β-caryophyllene (16%)/eucalyptol (11%) (Hung et al. 2021). The Italian EO was represented by *ar*-curcumene (39%)/α-humulene (10%) (Marongiu et al. 2007).

The second shrub from the Verbenaceae family is *Lippia alba* (Mill.) N.E. Br. ex Britton and P. Wilson (synonyms – *Camara alba* Kuntze, *Lantana alba* Mill., *L. geminata* (Kunth) Spreng., *L. malabárica* Hayek, *Lippia capensis* (Thunb.) Spreng., *L. carterae* (Moldenke) G.L. Nesom, *L. citrata* Willd. ex Cham., *L. geminata* Kunth, *L. globiflora* Kuntze, *L. havanensis* Turcz., *L. lantanoides* (Lam.) Herter, *L. panamensis* Turcz., *L. unica* Ramakr., *Verbena capensis* Thunb., *V. globiflora* L'Hér., *V. globulifera* Spreng., *V. lantanoides* Willd. Ex Spreng., *Zappania geminata* Gibert, *Z. globiflora* Juss., *Z. lantanoid* Lam. and *Z. odorata* Pers.), which is widely distributed in Central and South America. The species is a perennial, highly branched and fragrant plant (up to 3 m), and it is used in botanical medicine as infusions/ tinctures to treat colds and digestive problems as well as febrifuge/sudorific, sedative/ analgesic, digestive/stomachic, emmenagogue/expectorant but also as a condiment. The plant is commonly named achueriala, alecrim do campo, bushy matgrass, cidrela, curalotodo, erva cidreira, falsa melissa, hierbaluisa, juani(s)lama, melissa, melisse de calme, menta americana, oreganito, orégano de cerro/burro, poleo, prontoalivio, quita dolor, salva(e) limao/branca/real, salva vida, salvia, salvia trepadora, Santa María, sonora, teasam, toronjil. Its main secondary metabolites are volatile in nature and therefore make up many essential oils. For this reason, 32 chemical varieties (types/subtypes) have been reported for *L. alba*. Moreover, the EO have presented some interesting/particular bioproperties, e.g., antibacterial/antifungal, antiviral/antiherpetic, anticancer/antigenotoxic/cytotoxic, neurosedative, analgesic, antiinflammatory, cardiovascular, antiulcerogenic, anticonvulsive, antioxidant, acaricidal, repellent and insecticide (Agudelo-Gómez et al. 2010, Bernal et al. 2011, Celis et al. 2007, Ciccio and Ocampo 2006, Craveiro et al. 1981, de Souza et al. 2017, do Vale et al. 2002, Duke et al. 2009, Fun and Svendsen 1990, García-Barriga 1992a, Gómes et al. 2019, Hennebelle et al. 2006, 2008, HNC-ICN 2021, Lorenzo et al. 2001, Louchard and Araujo 2019, Matos et al. 1996, Mesa-Arango et al. 2009, 2010, Montero-Villegas et al. 2018, Olivero-Verbel et al. 2009, 2010, Pascual et al. 2001a, b, Peixoto et al. 2015a, b, Pino et al. 1996, Quattrocchi 2012, Rao et al. 2000, RBG Kew 2022, Ricciardi et al. 2009, Senatore and Rigano 2001, Shukla et al. 2009, 2011, The Plant List 2013, Viana et al. 1998, 2000, Zétola et al. 2002).

Based on Table 13, both the VF and the EO of *L. alba* from the northern region of Colombia were enriched with *cis*-piperitone oxide (17–56%)/germacrene D (11–32%)/limonene (8–18%). These compositions were different from those other Colombian chemotypes reported by Stashenko et al. (2004, 2014), López et al. (2011), Durán et al. (2007) and Mesa-Arango et al. (2009) which were mainly constituted by carvone (50%)/limonene (30%), citral (48%), myrcenone (63%) and/or bicyclosesquiphellandrene (27–36%)/carvone (25–27%). As mentioned earlier, there are 32 chemovars known for the *L. alba* EO; they are typified by one or combination of components like camphor, carvone, citral, citrol, dihydrocarvone, eucalyptol, limonene, linalool, lippione, myrcene, myrcenone, E/Z-β-ocimene, ocimenone, piperitone, γ-terpinene, estragole, bicyclosesquiphellandrene, β-caryophyllene, caryophyllene oxide, germacrene D and α-guaiene with percentage amounts between 10–91% and EO yields among 0.1–3.2%. The largest number of chemovars has been found in the South American countries with Argentina and Brazil as the main ones; each with 18 and 12 chemovars, respectively; the types of chemical structures that predominated in them were monoterpene derivatives with carbonyl groups (aldehydes and ketones) and terpene hydrocarbons (mono- and sesqui-). Thus, camphor (15–18%)/eucalyptol (16–35%) were reported for the Argentina/Guatemala/Uruguay EO; carvone (28–69%)/limonene (5–38%) were found in the Brazil/Uruguay/Costa Rica/Cuba/Argentina EO; carvone (18–70%) was present in the Brazil/Costa Rica/Cuba/Peru/Argentina EO; eucalyptol (32–47%)/limonene (18–35%) or limonene (31–47%)/piperitone (24–52%) were present in the Argentina/Brazil/Guatemala/Costa Rica EO; linalool (41–91%) was identified in the Argentina/Brazil/Cuba/Uruguay EO; citral (23–65%)/β-caryophyllene (6–24%) were characterized in the Brazil/Curasao/Argentina EO; citral (17–80%) was recognized in the Brazil/Argentina EO (Carvalho et al. 2017, Craveiro et al. 1981, de Souza et al. 2017, Dellacassa et al. 1990, Fischer et al. 2004, Frighetto et al. 1998, Gómes et al. 2019, Hennebelle et al. 2006, 2008, Leclercq et al. 1999, Lorenzo et al. 2001, Matos et al. 1996, Montero-Villegas et al. 2018, Peixoto et al. 2015a, b, Pino et al. 1996, Ricciardi et al. 2009, Senatore and Rigano 2001, Tavares et al. 2005, Zoghbi et al. 1998).

Other chemovars of Argentinian EO were dihydrocarvone (21–50%), linalool (40%)/citral (22%), myrcene (6–42%), myrcene (7–47%)/myrcenone (15–58%), ocimenone (20%); while other Brazilian EO contained β-caryophyllene (34%)/ caryophyllene oxide (18%), limonene (18–51%) and γ-terpinene (7–46%). One more chemovar of the EO were citral (22–72%)/citrol (6-16%), citral (36–47%)/ germacrene D (7–25%), citral (63–71%)/limonene (9–13%), citral (16–60%)/ myrcene (5–26%), citral (17–40%)/α-guaiene (10–15%), eucalyptol (20–25%)/citral (42–46%), myrcenone (63%), myrcenone (13%)/eucalyptol (18%) and myrcenone (47–71%)/ocimenone (7–34%) (Craveiro et al. 1981, do Vale et al. 2002, Fester et al. 1961, Fischer et al. 2004, Hennebelle et al. 2006, Matos et al. 2011, Montero-Villegas et al. 2018, Pereira-de-Morais et al. 2019, Pino et al. 1996, Riccardi et al. 2009, Tavares et al. 2005, Zétola et al. 2002, Zoghbi et al. 1998).

The Colombian chemovar enriched in "*cis*-piperitone oxide" showed a cytotoxicity (i) on human erythrocytes, moderate (>100 µg/mL < 1,000 µg/mL), (ii) on human lymphocytes, significant (>100 µg/mL < 250 µg/mL) and (iii) on Hep2 cell line which is high (> 10 µg/mL < 100 µg/mL); while that on AChE, the EO presented a moderate effect of inhibition. The other four Colombian chemovar demonstrated antiproliferative/cytotoxicity/antigenotoxic, antifungal/antibacterial, antiherpetic and antiparasitic activities.

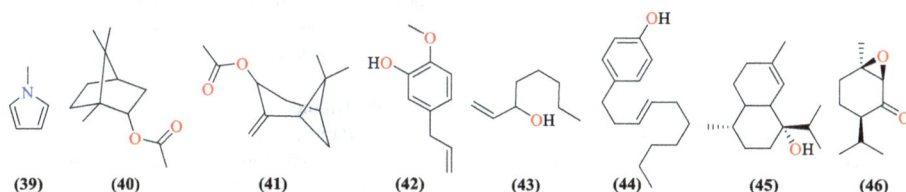

(39)   (40)   (41)   (42)   (43)   (44)   (45)   (46)

Conclusion and Perspective

The 23 plants included in this chapter presented volatile secondary metabolites, some of which could be playing an ecological role or defense/protection due to their chemical nature (e.g., mono-/sesqui-terpene hydrocarbons - α-/β-pinenes, myrcene, sabinene, limonene, (E)-β-ocimene, (E)-β-caryophyllene, germacrene D; oxygenated mono-/sesqui-terpenes- eucalyptol, linalool, mint furanone, guaiol, α-cadinol and *trans*-nerolidol). Some of these known species contained certain particular/prominent and/or biologically active molecules; e.g., germacrene B, β-bisabolol, β-eudesmol, *m*-mentha-2,8-diene, α-tolualdehyde, α-cadinol, δ-cadinene, cubebol, *trans*-pinocarvyl acetate, 1-octen-3-ol, gibbilimbol B, *cis*-cadin-4-en-7-ol, *cis*-piperitone oxide, etc. Meanwhile, other identified molecules were well-known (e.g., limonene, α-/β-pinenes/sabinene, myrcene/(E)-β-ocimene, eucalyptol, linalool, terpinen-4-ol, mint furanone, eugenol, methyleugenol) and with well-defined applications (e.g., perfumery and cosmetic). In other plants (e.g., *Ch. barranquillensis*, *Cy. cinereum*, *B. glabra*, *Cr. fragilis*, *Cr. fragrans* and *D. carthagenensis*), their volatile metabolites were identified/found/reported for the first time for Colombia/Latin America. In another case, the chemistry of volatiles for some plants (e.g., *B. simaruba*, *C. curassavica*, *L. camara*, *L. alba*, *Pl. alba* and *Pr. heptaphyllum*) differed from those found in the scientific literature consulted. The last but not least, the EO isolated from some Colombian plants was effective/active in the *in vitro* biological test/model (e.g., cyto-/toxic) cell line, against human lymphocytes and erythrocytes, *A. salina*), non-hemolytic, antibacterial, antifungal, antioxidant, fumigant/repellent). The expectation for the future of the Colombian species of interest is that with the interesting chemical compositions and the outstanding bioproperties determined, it would be necessary/imperative to protect such plant richness as well as to implement management plans and sustainable use of them that reduce the risk of deterioration of the biomes where the plants are located.

Acknowledgments

The authors thank Universidad Santo Tomás de Aquino (sede Bucaramanga) for using the Bibliometric Unit; Colciencias-SGR [Formación de Capital Humano (O.J.C.C, Convocatoria No. 1 Becas de Excelencia Bicentenario, 2019 – Departamento de Sucre; N.M.P.S. and C.P.G.G., Convocatoria No. 810, 2018 - Departamento de La Guajira)].

References

Abadia-Bejarano, L.P. and J. Chamorro-Crespo. 2011. Composición química y evaluación de la toxicidad y capacidades antioxidantes de las fracciones volátiles y aceites esenciales de las hojas y corteza de *Croton niveus* Jacq. Trabajo de pregrado, Profesional en Química y Farmacia. Universidad del Atlántico, Puerto Colombia, Colombia.

Afifi, M.S., O.M. Salama, A.A. Gohar and A.M. Marzouk. 2006. Iridoids with antimicrobial activity from *Plumeria alba* L. Bull. Pharm. Sci. (Assiut University) 29(Part 1): 215–223.

Agudelo-Gómez, L.S., G.A. Gómez Ríos, D.C. Durán, E.E. Stashenko and L. Betancur-Galvis. 2010. Chemical composition and evaluation *in vitro* of anti-herpetic activity of essential oils from *Lippia alba* (Mill) NE Brown and the main components. Salud UIS 42(3): 230–239.

Alhozaimy, G.A., E.S. Al-Sheddi and T.A. Ibrahim. 2017. Biological activity and isolation of compounds from stem bark of *Plumeria acutifolia*. Pharmacog. Mag. 13(Suppl 3): S505–S511.

Ali, S.N., O.H. Mustafvi, S. Begum, A. Ayub, R. Ghafoor and B.S. Siddiqui. 2019. Camaridin, a new triterpenoid from roots of *Lantana camara*. Chem. Nat. Comp. 55(2): 296–299.

Almeida, E.X., L.M. Conserva and R.P.L. Lemos. 2002. Coumarins, coumarinoligninoids and terpenes from *Protium heptaphyllum*. Biochem. Syst. Ecol. 30(7): 685–687.

Alonso-Castro, A.J., M.L. Villarreal, L.A. Salazar-Olivo, M. Gómez-Sánchez, F. Domínguez and A. García-Carranca. 2011. Mexican medicinal plants used for cancer treatment: Pharmacological, phytochemical and ethnobotanical studies. J. Ethnopharmacol. 133(3): 945–972.

Altamar, J.A., J.D. Rodríguez, A. Muñoz-Acevedo and R.G. Gutiérrez. 2016. Actividad hemolítica y cito-genotoxicidad del aceite esencial de *Croton malambo* H. Karst. p. 188. *In*: Memorias IV Congreso Latinoamericano de Plantas Medicinales, 17–19 agosto. Barranquilla, Colombia.

Álvarez, O.L. 1991. Recuperación de la medicina indígena. Memorias del I Encuentro de Médicos Tradicionales. Santiago - Putumayo. Cartilla de Educación Popular No. 8. Servicio Colombiano de Comunicación, Bogotá, Colombia.

Álvarez, A.L., S. Habtemariam and F. Parra. 2015. Inhibitory effects of lupene-derived pentacyclic triterpenoids from *Bursera simaruba* on HSV-1 and HSV-2 *in-vitro* replication. Nat. Prod. Res. 29(24): 2322–2327.

Álvarez, V.C., M.E. Niño and A. Muñoz. 2011a. Protective effect of the essential oils of flowers and leaves of *Chromolaena barranquillensis* from Sabanalarga (Atlántico). Vitae 18(S2): S46.

Álvarez, V.C., M.E. Niño and A. Muñoz. 2011b. Assessment of toxicity against *Artemia salina* and chemical composition of volatile oils of flowers/leaves of *Chromolaena barranquillensis* wild from Sabanalarga (Atlántico-Colombia). Abstract Book of 43rd World Chemistry Congress, 46th IUPAC General Assembly, 70th CQPR Annual Conference. No. 966.

Álvarez, V.C., M.J. Padilla, R.A. Jiménez and A. Muñoz-Acevedo. 2010. Chemical composition of the volatile fractions of flowers and leaves from *Chromolaena barranquillensis*. Libro de resúmenes XXIX Congreso Latinoamericano de Química (CLAQ), XVI Congreso Colombiano de Química y VI Congreso Colombiano de Cromatografía. PRN91.

Álvarez, V.C., S. Aristizábal-Córdoba, M.C. González, A.M. Molina, J. Rodríguez, E.E. Stashenko and A. Muñoz-Acevedo. 2015. Caracterización química y bioactividades del aceite esencial de hojas de *Dalea carthagenensis* (Jacq.) J.F. Macbr. (Atlántico, Colombia). Cienc. Amaz. (Iquitos) 5(2): 200.

Álvarez-Carillo, V.C. 2011. Determinación de la composición química de la fracción volátil y aceites esenciales de hojas y flores de *Chromolaena barranquillensis* aislados por destilación-extracción simultánea con solvente e hidrodestilación. Trabajo de pregrado, Profesional en Química. Universidad del Atlántico, Puerto Colombia, Colombia.

Amaral, M.P.M., F.A.V. Braga, F.F.B. Passos, F.R.C. Almeida, R.C.M. Oliveira, A.A. Carvalho, M.H. Chaves and F.A. Oliveira. 2009. Additional evidence for the anti-inflammatory properties of the essential oil of *Protium heptaphyllum* resin in mice and rat. Lat. Am. J. Pharm. 28(5): 775–782.

Amaya, C. and A.C. Acevedo. 2011. Effect of the extracts of *Piper cumanense* and *Piper eriopodon* in the behavior of genes involved in oxidation process of the skin. Planta Med. 77: PI6.

Andrade, E.H.A., L.M.M. Carreira, M.H.L. da Silva, J.D. da Silva, C.N. Bastos, P.J.C. Sousa, E.F. Guimarães and J.G.S. Maia. 2008. Variability in essential-oil composition of *Piper marginatum sensu lato*. Chem. Biodivers. 5(1): 197–208.

Anggoro, B., E. Istyastono and M. Hariono. 2020. Future molecular medicine from white frangipani (*Plumeria alba* L.): A review. J. Med. Plants Res. 14(10): 544–554.

Ankli, A., M. Heinrich, P. Bork, L. Wolfram, P. Bauerfeind, R. Brun, C. Schmid, C. Weiss, R. Bruggisser, J. Gertsch, M. Wasescha and O. Sticher. 2002. Yucatec Mayan medicinal plants: Evaluation based on indigenous uses. J. Ethnopharmacol. 79(1): 43–52.

Api, A.M., D. Belsito, D. Botelho, M. Bruze, G.A. Burton Jr., J. Buschmann, M.A. Cancellieri, M.L. Dagli, M. Date, W. Dekant, C. Deodhar, A.D. Fryer, L. Jones, K. Joshi, M. Kumar, A. Lapczynski, M. Lavelle, I. Lee, D.C. Liebler, H. Moustakas, M. Na, T.M. Penning, G. Ritacco, J. Romine, N. Sadekar, T.W. Schultz, D. Selechnik, F. Siddiqi, I.G. Sipes, G. Sullivan, Y. Thakkar and Y. Tokura. 2021. RIFM fragrance ingredient safety assessment, pinocarvyl acetate, CAS Registry Number 1078-95-1. Food Chem. Toxicol. 156(suppl. 1): 112451.

Aprotosoaie, A.C., M. Hăncianu, I.-I. Costache and A. Miron 2014. Linalool: A review on a key odorant molecule with valuable biological properties. Flavour Frag. J. 29(4): 193–219.

Aragão, G.F. 2004. Atividade Antiinflamatória, antiagregante plaquetária e efeitos centrais de alfa e beta amirina isolada de *Protium heptaphyllum* Albl March. Master Thesis. Universidade Federal do Ceará, Fortaleza, Brazil.

Araujo, D.A.O.V., C. Takayama, F.M. de Faria, E.A.R. Socca, R.J. Dunder, L.P. Manzo, A. Luiz-Ferreira and A.R.M. Souza-Brito. 2011. Gastroprotective effects of essential oil from *Protium heptaphyllum* on experimental gastric ulcer models in rats. Braz. J. Pharmacogn. 21(4): 721–729.

Arbeláez-Cortés, E. 2013. Knowledge of Colombian biodiversity: Published and indexed. Biodivers. Conserv. 22: 2875–2906.

Aristizabal-Córdoba, S. and J.L. Pacheco-Castro. 2014. Caracterización química y evaluación de la capacidad antioxidante y citotóxicidad de los metabolitos secundarios volátiles de hojas y flores frescas de *Croton fragrans* Kunth. Trabajo de pregrado, Profesional en Química y Farmacia. Universidad del Atlántico, Puerto Colombia, Colombia.

Aristizabal-Córdoba, S., J. Pacheco, A.L. Méndez, J.D. Rodríguez, R.G. Gutiérrez De Aguas and A. Muñoz-Acevedo. 2014. Capacidad antioxidante, citotoxicidad y composición química del aceite esencial de las hojas de *Croton fragrans* Kunth. Rev. Prod. Nat. 4(1): 57.

Autran, E.S., I.A. Neves, C.S.B. da Silva, G.K.N. Santos, C.A.G. da Câmara and D.M.A.F. Navarro. 2009. Chemical composition, oviposition deterrent and larvicidal activities against *Aedes aegypti* of essential oils from *Piper marginatum* Jacq. (Piperaceae). Bioresource Technol. 100(7): 2284–2288.

Ávila Ayala, M.J. 2014. Estudio fitoquímico de las hojas de la planta *Croton malambo* Karst. Trabajo de pregrado, Profesional en Licenciatura en Química. Universidad Central de Venezuela, Caracas.

Ayalew, A.A. 2020. Insecticidal activity of *Lantana camara* extract oil on controlling maize grain weevils. Toxicol. Res. Appl. 4: 1–10.

Ayub, A., S. Begum, S.N. Ali, S.T. Ali and B.S. Siddiqui. 2019. Triterpenoids from the aerial parts of *Lantana camara*. J. Asian Nat. Prod. Res. 21(2): 141–149.

Ayub, A., S. Begum, S.T. Ali Sara and B.S. Siddiqui. 2019. Isolation and spectral studies of a new lactone triterpenoid from *Lantana camara*. Chem. Nat. Comp. 55(3): 478–481.

Baghel, A.S., C.K. Mishra, A. Rani, D. Sasmal and R.K. Nema. 2010. Antibacterial activity of *Plumeria rubra* Linn. plant extract. J. Chem. Pharm. Res. 2(6): 435–440.

Balick, M.J. and H. O'Brian. 2004. Ethnobotanical and floristic research in Belize: Accomplishments, challenges and lessons learned. Ethnobot. Res. Applic. 2: 77–78.

Bandeira, P.N., A.M. Fonseca, S.M.O. Costa, M.U.D.S. Lins, O.D.L. Pessoa, F.J.Q. Monte, N.A.P. Nogueira and T.L.G. Lemos. 2006. Antimicrobial and antioxidant activities of the essential oil of resin of *Protium heptaphyllum*. Nat. Prod. Commun. 1(2): 117–120.

Bandeira, P.N., M.I.L. Machado, F.S. Cavalcanti and T.L.G. Lemos. 2001. Essential oil composition of leaves, fruits and resin of *Protium heptaphyllum* (Aubl.) March. J. Essent. Oil Res. 13(1): 33–34.

Bandeira, P.N., O.D.L. Pessoa, M.T.S. Trevisan and T.L.G. Lemos. 2002. Secondary metabolites of *Protium heptaphyllum* March. Quim. Nova 25(6B): 1078–1080.

Barriga, H.G. 1992. Flora medicinal de Colombia: Botánica médica: Vol II. Instituto de Ciencias Naturales, Universidad Nacional. Tercer Mundo, Bogotá.

Barros, J.A., M.C. Garcés, J. Cabrera, J. Fuentes, J.D. Rodríguez, A. Muñoz-Acevedo and R.G. Gutiérrez. 2015. Actividad hemolítica y citotoxicidad en linfocitos humanos de aceites esenciales y extractos totales de *Piper marginatum, Croton niveus* e *Hyptis suaveolens*. Cienc. Amaz. (Iquitos) 5(2): 218.

Bawa, E., K.L. Kagoro and J.A. Wapwera. 2019. Study on the phytochemical and antimicrobial screening of ethyl acetate extract of *Plumeria rubra* leaves and stems bark. Agr. Res. Tech. Open Acc. J. 21(4): ID 556170.

Bayeux, M.C., A.T. Fernandes, M.A. Foglio and J.E. Carvalho. 2002. Evaluation of the antiedematogenic activity of artemetin isolated from *Cordia curassavica* DC. Braz. J. Med. Biol. Res. 35(10): 1229–1232.

Begum, S., A. Ayub, S.Q. Zehra and B.S. Siddiqui. 2013. Chemical constituents of the aerial parts of *Lantana camara*. Chem. Nat. Comp. 49(3): 566–567.

Begum, S., A. Ayub, S.Q. Zehra, B.S. Siddiqui, M.I. Choudhary and Samreen. 2014. Leishmanicidal triterpenes from *Lantana camara*. Chem. Biodivers. 11(5): 709–718.

Begum, S., A. Wahab and B. Siddiqui. 2002. Ursethoxy acid, a new triterpene from *Lantana camara*. Nat. Prod. Lett. 16(4): 235–238.

Begum, S., A. Wahab and B.S. Siddiqui. 2008. Antimycobacterial activity of flavonoids from *Lantana camara*. Nat. Prod. Res. 22(6): 467–470.

Begum, S., A. Wahab, B.S. Siddiqui and F. Qamar. 2000. Nematicidal constituents of the aerial parts of Lantana camara. J. Nat. Prod. 63(6): 765–767.

Begum, S., S.M. Raza, B.S. Siddiqui and S. Siddiqui. 1995. Triterpenoids from the aerial parts of *Lantana camara*. J. Nat. Prod. 58(10): 1570–1574.

Benites, J., C. Moiteiro, G. Miguel, L. Rojo, J. López, F. Venâncio, L. Ramalho, S. Feio, S. Dandlen, H. Casanova and I. Torres. 2009. Composition and biological activity of the essential oil of Peruvian *Lantana camara*. J. Chil. Chem. Soc. 54(4): 379–384.

Benítez, J.D., J.D. Rodríguez, A.L. Méndez, A.M. Molina, E.E. Stashenko and A. Muñoz-Acevedo. 2014a. Metabolitos secundarios volátiles de hojas y ramas de *Bursera glabra* (Jacq.) Triana and Planch. del cerro "pan de azúcar" (Barranquilla). Rev. Prod. Nat. 4(1): 76.

Benítez, J.D., J.D. Rodríguez, A.L. Méndez, A.M. Molina, E.E. Stashenko and A. Muñoz-Acevedo. 2014b. Análisis composicional de los metabolitos secundarios volátiles de hojas y ramas de *Bursera simaruba* (L.) Sarg. del cerro pan de azúcar (Barranquilla). Rev. Prod. Nat. 4(1): 158.

Bernadi, W.A., J.C. Zanotelli, E.M. de Lima, T.D. de Souza, D.C. Edringer and V.R.C. de Souza. 2015. Effects of topical application of essential oil from resin of almescar (*Protium heptaphyllum* (Aubl.) Marchand) in experimentally induced skin wounds in rats Rev. Bras. Cien. Vet. 22(1): 10–15.

Bernal, H.Y., H. García-Martínez and F. Quevedo-Sánchez. 2011. Pautas para el conocimiento, conservación y uso sostenible de las plantas medicinales nativas en Colombia: Estrategia nacional para la conservación de plantas. Ministerio de Ambiente, Vivienda y Desarrollo Territorial e Instituto de Investigación de Recursos Biológicos Alexander von Humboldt. Bogotá, Colombia.

Bernal, R., S.R. Gradstein and M. Celis Morales. 2019. Catálogo de plantas y líquenes de Colombia. Instituto de Ciencias Naturales, Universidad Nacional de Colombia, Bogotá. http://catalogoplantasdecolombia.unal.edu.co.

Bihani, T. 2021. *Plumeria rubra* L.—A review on its ethnopharmacological, morphological, phytochemical, pharmacological and toxicological studies. J. Ethnopharmacol. 264: 113291.

Borrero-Meza, E. and L.Y. Hurtado Palacio. 2013. Caracterización química de los metabolitos secundarios volátiles presentes en las hojas de *Piper marginatum* Jacq. recolectadas en Tubará - Atlántico. Trabajo de pregrado, Profesional en Química y Farmacia. Universidad del Atlántico, Puerto Colombia, Colombia.

Bristot, S.F., M.P.D. Colle, A.E. Rossato and V. Citadini-Zanette. 2020. Medicinal use of *Varronia curassavica* Jacq. "erva-baleeira" (Boraginaceae): Case study in Southern Brazil. Braz. J. Animal Environm. Res. 4(1): 170–182.

Bru Hernández, J.C. and Y. Fernández-Aparicio. 2014. Caracterización química del efecto protector a la oxidación y citotoxicidad de los metabolitos secundarios volátiles de las hojas frescas de *Eugenia procera* (Sw.) Poir. Trabajo de pregrado, Profesional en Química y Farmacia. Universidad del Atlántico, Puerto Colombia, Colombia.

Brú, J. and J.D. Guzman. 2016. Folk medicine, phytochemistry and pharmacological application of *Piper marginatum*. Braz. J. Pharmacogn. 26: 767–779.

Bru, J., Y. Fernández-Aparicio, A.L. Méndez, S. Aristizabal-Córdoba, J.D. Rodríguez, R.G. Gutiérrez De Aguas and A. Muñoz-Acevedo. 2014. Composición química y evaluaciones de la citotoxicidad y capacidad antiradicalaria del aceite esencial de las hojas de *Eugenia procera* (Sw.) Poir. Rev. Prod. Nat. 4(1): 81.

Bussmann, R.W., N.Y. Paniagua-Zambrana, C. Romero and R.E. Hart. 2018. Astonishing diversity—The medicinal plant markets of Bogotá, Colombia. J. Ethnobiol. Ethnomed. 14: 43.

Cabral, R.S.C., C.C.S. Alves, H.R.F. Batista, W.C. Sousa, I.S. Abrahão, A.E.M. Crotti, M.B. Santiago, C.H.G. Martins and M.L.D. Miranda. 2020. Chemical composition of essential oils from different parts of *Protium heptaphyllum* (Aubl.) Marchand and their *in vitro* antibacterial activity. Nat. Prod. Res. 34(16): 2378–2383.

Cabral, R.S.C., C.C. Fernandes, A.L.B. Dias, H.R.F. Batista, L.G. Magalhães, M.C. Pagotti and M.L.D. Miranda. 2021. Essential oils from *Protium heptaphyllum* fresh young and adult leaves (Burseraceae): Chemical composition, *in-vitro* leishmanicidal and cytotoxic effects. J. Essent. Oil Res. 33(3): 276–282.

Cáceres-Ferreira, W., M. Rengifo-Carrillo, L. Rojas and C. Rosquete-Porcar. 2019. Chemical composition of essential oils from *B. simaruba* (L.) Sarg. fruits and the resins from three *Bursera* species: *B. simaruba* (L.) Sarg, *B. glabra* Jack and *B. inversa* Daly. Avances Quím. 14(1): 25–29.

Cárdenas, D.L., J.A. Lora, R.L. Márquez and P.J. Blanco. 2005. Actividad leishmanicida de *Annona purpurea*. Actual. Biol. 27(Supl. 1): 35–37.

Carmona, R., C.E. Quijano-Celís and J.A. Pino. 2009. Leaf oil composition of *Bursera graveolens* (Kunth) Triana et Planch. J. Essent. Oil Res. 21(5): 387–389.

Carretero, M.E., J.L. López-Pérez, M.J. Abad, P. Bermejo, S. Tillet, A. Israel and B. Noguera-P. 2008. Preliminary study of the anti-inflammatory activity of hexane extract and fractions from *Bursera simaruba* (Linneo) Sarg. (Burseraceae) leaves. J. Ethnopharmacol. 116: 11–15.

Caroprese Araque, J.F., M.I. Parra Garcés, D. Arrieta Prieto and E. Stashenko. 2011. Anatomía microscópica y metabolitos secundarios volátiles en tres estadios del desarrollo de las inflorescencias de *Lantana camara* (Verbenaceae). Rev. Biol. Trop. 59(1): 473–486.

Carvalho, P.M.M., C.A.F. Macêdo, T.F. Ribeiro, A.A. Silva, R.E.R. Da Silva, L.P. de Morais, M.R. Kerntopf, I.R.A. Menezes and R. Barbosa. 2017. Effect of the *Lippia alba* (Mill.) N.E. Brown essential oil and its main constituents, citral and limonene, on the tracheal smooth muscle of rats. Biotechnol. Rep. 17: 31–34.

Case, R.J., A.O. Tucker, M.J. Maciarello and K.A. Wheeler. 2003. Chemistry and ethnobotany of commercial incense copals, copal blanco, copal oro, and copal negro, of North America. Econ. Bot. 57(2): 189–202.

Castañeda Muñoz, M.L. 2007. Estudio de la composición química y la actividad biológica de los aceites esenciales de diez plantas aromáticas colombianas. Trabajo de pregrado, Profesional en Química. Universidad Industrial de Santander, Bucaramanga.

Castañeda, M.L., A. Muñoz, J.R. Martínez and E.E. Stashenko. 2007. Estudio de la composición química y la actividad biológica de los aceites esenciales de diez plantas aromáticas colombianas. Sci. Techn. 33(1): 165–166.

Castro, S. 2012. Análisis florístico y fitogeográfico de ambientes asociados al complejo de ciénagas de Zapatosa (Cesar) en el caribe colombiano. Tesis de Maestría en Ciencias Biológicas, Universidad Nacional de Colombia. Bogotá, Colombia.

Castro-Laportte, M. 2013. Estudio taxonómico del género *Bursera* Jacq. ex L. (Burseraceae) en Venezuela. ERNSTIA 23(2): 125–169.

Celis, C.N., P.E. Rivero, J.H. Isaza, E.E. Stashenko and J.R. Martínez. 2007. Estudio comparativo de la composición y actividad biológica de los aceites esenciales extraídos de *Lippia alba*, *Lippia origanoides* y *Phyla dulcis*, especies de la familia Verbenaceae. Sci. Techn. 33(1): 103–105.

Celotto, A.C., D.Z. Nazario, M.A. Spessoto, C.H.G. Martins and W.R. Cunha. 2003. Evaluation of the *in vitro* antimicrobial activity of crude extracts of three *Miconia* especies. Braz. J. Microbiol. 34(4): 339–340.

Cepleanu, F., K. Ohtani, M. Hamburger, K. Hostettmann, M.P. Gupta and P. Solis. 1993. Novel acetogenins from the leaves of *Annona purpurea*. Helv. Chim. Acta 76(3): 1379–1388.

Chamorro, J., L. Abadia, C. Trillos, M.E. Niño and A. Muñoz. 2011. Evaluación de la toxicidad frente *Artemia salina* y composición química de los aceites esenciales de hojas y corteza de la raíz de *Croton niveus* Jacq. Vitae 18(Supl. 2): S212.

Chang, F.R., C.Y. Chen, P.H. Wu, R.Y. Kuo, Y.C. Chang and Y.C. Wu. 2000. New alkaloids from *Annona purpurea*. J. Nat. Prod. 63(6): 746–748.

Chang, F.R., J.L. Wei, C.M. Teng and Y.C. Wu. 1998a. Two new 7-dehydroaporphine alkaloids and antiplatelet action aporphines from the leaves of *Annona purpurea*. Phytochemistry 49(7): 2015–2018.

Chang, F.R., J.L. Wei, C.M. Teng and Y.C. Wu. 1998b. Antiplatelet aggregation constituents from *Annona purpurea*. J. Nat. Prod. 61(12): 1457–1461.

Chávez, D. and R. Mata. 1999. Purpuracenin: A new cytotoxic adjacent bis-tetrahydrofuran annonaceous acetogenin from the seeds of *Annona purpurea*. Phytochemistry 50(5): 823–828.

Chen, T.K., D.F. Wiemer and J.J. Howard. 1984. A volatile leafcutter ant repellent from *Astronium graveolens*. Naturwissenschaften 71: 97–98.

Ciccio, J.F. and R.A. Ocampo. 2006. Variación anual de la composición química del aceite esencial de *Lippia alba* (Verbenaceae) cultivada en Costa Rica. Lankesteriana Int. J. Orch. 6(3): 149–154.

Citó, A.M.G.L., F.B. Costa, J.A.D. Lopes, V.M.M. Oliveira and M.H. Chaves. 2006. Identificação de constituintes voláteis de frutos e folhas de *Protium heptaphyllum* Aubl (March). Rev. Bras. Plant. Med. 8(4): 4–7.

Cohen, S.M., S. Fukushima, N.J. Gooderham, S.S. Hecht, L.J. Marnett, I.M.C.M. Rietjens, R.L. Smith, M. Bastaki, M.M. McGowen, C. Harman and S.V. Taylor. 2015. GRAS flavoring substances 27. Food Technol. 69(8): 40–59.

Comisión Técnica Subregional para la Política de Acceso a Medicamentos. 2014. Plantas medicinales de la Subregión Andina. Organismo Andino de Salud/Convenio Hipólito Unanue, Lima, Perú.

Correa, J.M. and H.Y. Bernal. 1990. Especies vegetales promisorias de los países del Convenio Andrés Bello. Tomo III. Programa de Recursos Vegetales del Convenio Andrés Bello. Editora Secretaria Ejecutiva del Convenio Andrés Bello, Bogotá.

Correa-Barrera, L. 2006. Estudio fitoquímico de hojas y cortezas de *Bursera graveolens*. Tesis de Maestría en Ciencias Biológicas, Pontificia Universidad Javeriana. Bogotá, Colombia.

Correa Navarro, Y.M., L.R. Palomino García and O. Marino Mosquera. 2001. Actividad antioxidante y antifúngica de Piperaceas de la flora colombiana. Rev. Cubana Plant. Med. 20(2): 167–181.

Costa, J.G.M., F.F.G. Rodrigues, E.O. Sousa, D.M.S. Junior, A.R. Campos, H.D.M. Coutinho and S.G. de Lima. 2010. Composition and larvicidal activity of the essential oils of *Lantana camara* and *Lantana montevidensis*. Chem. Nat. Comp. 46(2): 313–315.

Craveiro, A.A., J.W. Alencar, F.J.A. Matos, C.H.S. Andrade and M.I.L. Machado. 1981. Essential oils from Brazilian Verbenaceae. Genus *Lippia*. J. Nat. Prod. 44(5): 598–601.

Cruz, J.P., V.O. Vasconcelos, P.C. Medeiros, J.C. Figueiredo, Y.R.F. Nunes, E.R. Duarte and A.P. Venuto. 2021. Acaricide potential of *Xylopia emarginata* Mart. leaf extract against *Rhipicephalus microplus* larvae. Agrarian Sci. J. 13: 1–5.

Cuervo Cuervo, A., D. García Vásquez, L. Orozco Gómez and D. Ramírez Ospino. 2017. Efectividad del aceite esencial de *Piper eriopodon* en inhibición del crecimiento de *Trichophyton rubrum* y *Trichophyton mentagrophytes*. Anuario de Investigación. Fundación Universitaria Juan. M. Corpas. 2017: 30–31.

da Silva, E.R., D.R. de Oliveira, M.F.F. Melo, H.R. Bizzo and S.G. Leitão. 2016. Report on the Malungo expedition to the Erepecuru river, Oriximiná, Brazil. Part I: Is there a difference between black and white breu? Braz. J. Pharmacogn. 26: 647–656.

da Silva, J.K.R., N.N.S. Silva, J.F.S. Santana, E.H.A. Andrade, J.G.S. Maia and W.N. Setzer. 2016. Phenylpropanoid-rich essential oils of *Piper* species from the Amazon and their antifungal and anti-cholinesterase activities. Nat. Prod. Comm. 11(12): 1907–1911.

da Silva, M.H.L., E.H.A. Andrade, M.G.B. Zoghbi, A.I.R. Luz, J.D. da Silva and J.G.S. Maia. 1999. The essential oils of *Lantana camara* L. occurring in North Brazil. Flavour Frag. J. 14(4): 208–210.

Dabhadkar. D. and V. Zade. 2012. Abortifacient activity of *Plumeria rubra* (Linn) pod extract in female albino rats. Indian J. Exp. Biol. 50: 702–707.

David Higuita, H., O. Díaz Vasco, L.M. Urrea and F. Cardona Naranjo. 2014. Guía ilustrada flora cañón del río Porce, Antioquia. EPM E.S.P. Universidad de Antioquia, Herbario Universidad de Antioquia - Medellín, Colombia.

de Lima, E.M., D.S.P. Cazelli, F.E. Pinto, R.A. Mazuco, I.C. Kalil, D. Lenz, R. Scherer, T.U. de Andrade and D.C. Endringer. 2016. Essential oil from the resin of *Protium heptaphyllum*: Chemical composition, cytotoxicity, antimicrobial activity, and antimutagenicity. Pharmacogn. Mag. 12(Suppl. 1): S42–S46.

de Mesquita, M.L., P. Grellier, L. Mambu, J.E. de Paula and L.S. Espindola. 2007. *In vitro* antiplasmodial activity of Brazilian cerrado plants used as traditional remedies. J. Ethnopharmacol. 110(1): 165–170.

De Solla Price, D.J. 1976. A general theory of bibliometric and other cumulative advantage pro-cesses. J. Assoc. Inf. Sci. Technol. 27: 292–306.

De Sousa, E.O., A.S. Lima, S.G. Lopes, L.M. Costa-Junior and J.G.M. Da Costa. 2020. Chemical composition and acaricidal activity of *Lantana camara* L. and *Lantana montevidensis* Briq. essential oils on the tick *Rhipicephalus microplus*. J. Essent. Oil Res. 32(4): 316–322.

de Souza, M.T., M.T. de Souza, D. Bernardi, D. Krinski, D.J. de Melo, D.C. Oliveira, M. Rakes, P.H.G. Zarbin, B.H.L.N.S. Maia and M.A.C. Zawadneak. 2020. Chemical composition of essential oils of selected species of *Piper* and their insecticidal activity against *Drosophila suzukii* and *Trichopria anastrephae*. Env. Sci. Poll. Res. 27: 13056–13065.

de Souza, R.C., M.M. da Costa, B. Baldisserotto, B.M. Heinzmann, D. Schmidt, B.O. Caron and C.E. Copatti. 2017. Antimicrobial and synergistic activity of essential oils of *Aloysia triphylla* and *Lippia alba* against *Aeromonas* spp. Microb. Pathog. 113: 29–33.

DeCarlo, A., N.S. Dosoky, P. Satyal, A. Sorensen and W.N. Setzer. 2019. The essential oils of the Burseraceae. pp. 61–145. *In*: S. Malik (ed.). Essential Oil Research. Trends in Biosynthesis, Analytics, Industrial Applications and Biotechnological Production. Springer Nature, Switzerland.

Delgado-Altamirano, R., M.E. García-Aguilera, J. Delgado-Domínguez, I. Becker, E. Rodríguez de San Miguel, A. Rojas-Molina and N. Esturau-Escofet. 2021. ^{1}H NMR profiling and chemometric analysis as an approach to predict the leishmanicidal activity of dichloromethane extracts from *Lantana camara* (L.). J. Pharm. Biomed. Anal. 199: 114060.

Dellacassa, E., E. Soler, P. Menéndez and P. Moyna. 1990. Essential oils from *Lippia alba* (Mill.) N.E. Brown and *Aloysia chamaedrifolia* Cham. (verbenaceae) from Uruguay. Flavour Frag. J. 5(2): 107–108.

Devprakash, D., T. Rohan, G. Suhas, G.P. Senthil Kumarand and M.T. Tamizh. 2012. An review of phytochemical constituents and pharmacological activity of *Plumeria* species. Int. J. Curr. Pharm. Res. 4(1): 1–6.

Dey, A., T. Das and S. Mukherjee. 2011. *In vitro* antibacterial activity of n-hexane fraction of methanolic extract of *Plumeria rubra* L. (Apocynaceae) stem bark. J. Plant Sci. 6: 135–142.

Dey, A. and A. Mukherjee. 2015. *Plumeria rubra* L. (Apocynaceae): Ethnobotany, phytochemistry and pharmacology: A mini review. J. Plant Sci. 10: 54–62.

do Vale, T.G., E. Couto Furtado, J.G. Santos Jr. and G.S.B. Viana. 2002. Central effects of citral, myrcene and limonene, constituents of essential oil chemotypes from *Lippia alba* (Mill.) N.E. Brown. Phytomedicine 9(8): 709–714.

Dos Santos, R.C., A.A. De Melo Filho, E.A. Chagas, I.M. Fernández, J.A. Takahashi and V.P. Ferraz. 2019. Influence of diurnal variation in the chemical composition and bioactivities of the essential oil from fresh and dried leaves of *Lantana camara*. J. Essent. Oil Res. 31(3): 228–234.

Dougnon, G. and M. Ito. 2020. Sedative efects of the essential oil from the leaves of *Lantana camara* occurring in the Republic of Benin via inhalation in mice. J. Nat. Med. 74: 159–169.

Duke, J.A., M.J. Bogenschutz-Godwin and A.R. Ottesen. 2009. Duke's Handbook of Medicinal Plants of Latin America. CRC Press, Boca Ratón.

Durán, D.C.D., L.A. Monsalve, J.R. Martínez and E.E. Stashenko. 2007. Estudio comparativo de la composición química de aceites esenciales de *Lippia alba* provenientes de diferentes regiones de Colombia, y efecto del tiempo de destilación sobre la composición del aceite. Sci. Techn. 33(1): 435–438.

Dutra, K., V. Wanderley-Teixeira, C. Guedes, G. Cruz, D. Navarro, A. Monteiro, A. Agra, C.L. Neto and Á. Teixeira. 2020. Toxicity of essential oils of leaves of plants from the genus *Piper* with influence on the nutritional parameters of *Spodoptera frugiperda* (J.E. Smith) (Lepidoptera: Noctuidae). J. Essent. Oil Bear. Pl. 23(2): 213–229.

Fahlbusch, K.-G., F.-J. Hammerschmidt, J. Panten, W. Pickenhagen and D. Schatkowski. 2005. Flavors and fragrances. pp. 1–127. *In*: F. Ullmann and M. Bohnet (eds.). Ullmann's Encyclopedia of Industrial Chemistry. Wiley-VCH Verlag GmbH and Co, Weinheim.

Fatope, M.O., L. Salihu, S.K. Asante and Y. Takeda. 2002. Larvicidal activity of extracts and triterpenoids from *Lantana camara*. Pharm. Biol. 40(8): 564–567.

Fester, G., E. Martinuzzi, J. Retamar and A. Ricciardi. 1961. Aceites esenciales de la República Argentina. Acad Nac. Cienc. 36–38.

Fischer, D.C.H., N.C.A. Gualda, D. Bachiega, C.S. Carvalho, F.N. Lupo, S.V. Bonotto, M.O. Alves, A. Yogi, S.M. Di Santi, P.E. Avila, K. Kirchgatter and P.R.H. Moreno. 2004. *In vitro* screening for antiplasmodial activity of isoquinoline alkaloids from Brazilian plant species. Acta Trop. 92(3): 261–266.

Fischer, U., R. Lopez, E. Pöll, S. Vetter, J. Novak and C.M. Franz. 2004. Two chemotypes within *Lippia alba* populations in Guatemala. Flavour Frag. J. 19(4): 333–335.

Fon-Fay, F.M., J.A. Pino, I. Hernández, I. Rodeiro and M.D. Fernández. 2019. Chemical composition and antioxidant activity of *Bursera graveolens* (Kunth) Triana et Planch essential oil from Manabí, Ecuador. J. Essent. Oil Res. 31(3): 211–216.

Frighetto, N., J.G. de Oliveira, A.C. Siani and K.C. das Chagas. 1998. *Lippia alba* Mill N.E. Br. (Verbenaceae) as a source of linalool. J. Essent. Oil Res. 10(5): 578–580.

Frischkorn, C.G.B., H.E. Frischkorn and E. Carrazzoni. 1948. Cercaricidal acttivity of some essential oils of plants from Brazil. Naturwissenschaften 65(2): 480–483.

Fun, C.E. and A.B. Svendsen. 1990. The essential oil of *Lippia alba* (Mill.) N.E. Br. J. Essent. Oil Res. 2(5): 265–267.

Fundación BioColombia. 2013. Estudio básico para la declaratoria de un área natural protegida en la zona de bañaderos - municipio de Riohacha - departamento de La Guajira, y formulación de su plan de manejo. Corpoguajira, Riohacha, Colombia.

García-Barriga, H. 1992a. Flora medicinal de Colombia. Botánica médica. Tomo II. Tercer Mundo Editores, Bogotá, Colombia.

García-Barriga, H. 1992b. Flora medicinal de Colombia. Botánica médica. Tomo I. Tercer Mundo Editores, Bogotá, Colombia.

García, C.C., E.G. Acosta, A.C. Carro, M.C. Fernández Belmonte, R. Bomben, C.B. Duschatzky, M. Perotti, C. Schuff and E.B. Damonte. 2010. Virucidal activity and chemical composition of essential oils from aromatic plants of central west Argentina. Nat. Prod. Comm. 5(8): 1307–1310.

García, M., L. Monzote, A.M. Montalvo and R. Scull. 2010. Screening of medicinal plants against *Leishmania amazonensis*. Pharm. Biol. 48: 1053–1058.

Gil-Otaiza, R., J. Carmona and A. Rodríguez. 2006. Plantas medicinales de la Mesa de Los Indios, Municipio Campo Elías (estado Mérida. Venezuela). Plántula 4: 55–67.

Gomes, A.F., M.P. Almeida, M.P. Leite, S. Schwaiger, H. Stuppner, M. Halabalaki, J.G. Amaral and J.M. David. 2019. Seasonal variation in the chemical composition of two chemotypes of *Lippia alba*. Food Chem. 273: 186–193.

Gómez Chimá, L.G. and J.A. Vargas Campos. 2018. Caracterización de especies arbóreas con propiedades medicinales del bosque natural primario del Centro de Investigaciones Santa Lucía del Instituto Universitario de La Paz Barrancabermeja, Santander. Trabajo de pregrado, Programa de Ingeniería Ambiental y de Saneamiento. Universidad de La Paz, Barrancabermeja.

Gómez-Estrada, H., F. Díaz-Castillo, L. Franco-Ospina, J. Mercado-Camargo, J. Guzmán-Ledezma, J.D. Medina and R. Gaitán-Ibarra. 2011. Folk medicine in the northern coast of Colombia: An overview. J. Ethnobiol. Ethnomed. 7: 27–37.

Gonçalves, R., V.F.S. Ayres, L.G. Magalhães, A.E.M. Crotti, G.M. Corrêa, A.C. Guimarães and R. Takeara. 2019. Chemical composition and schistosomicidal activity of essential oils of two *Piper* species from the Amazon region. J. Essent. Oil Bear. Pl. 22(3): 811–820.

González, M.C. 2019. Nueva quimiovariedad de *Lippia alba* (Mill.) y reactividad de su componente mayoritario. Trabajo de postgrado, Maestría en Ciencias Naturales. Universidad del Norte, Puerto Colombia, Colombia.

González, M.C., J.D. Rodríguez, S. Aristizabal, V.C. Álvarez, Y.Sh. De Moya, H.G. San Juan, R.G. Gutiérrez and A. Muñoz-Acevedo. 2016a. Nuevo quimiotipo de *Lippia alba* - composición química del aceite esencial de hojas y determinación de sus actividades *in vitro* citotóxicas, anticolinesterásica/ repelencia y reactividad frente ABTS⁺. p. 76. *In*: Memorias IV Congreso Latinoamericano de Plantas Medicinales, 17–19 agosto. Barranquilla, Colombia.

González, M.C., J.D. Rodríguez, Y.Sh. De Moya, A. Barrios, V.C. Álvarez, H.G. San Juan, R.G. Gutiérrez and A. Muñoz-Acevedo. 2016b. Composición química del aceite esencial de hojas de *Piper eriopodon* y estimación de sus actividades *in vitro* citotóxicas, antibacterianas, anticolinesterásica/repelencia y antiradicalaria. p. 66. *In*: Memorias IV Congreso Latinoamericano de Plantas Medicinales, 17–19 agosto. Barranquilla, Colombia.

González-Rocha, E. and R. Cerros-Tlatilpa. 2015. La familia Apocynaceae (Apocynoideae y Rauvolfioideae) en el Estado de Morelos, México. Acta Bot. Mex. 110: 21–70.

Goswami, P., A. Chauhan, R.S. Verma and R.C. Padalia. 2016. Chemical constituents of floral volatiles of *Plumeria rubra* L. from India. Med. Aromat. Plants S3: 005.

Gotyal, B.S., C. Srivastava and S. Walia. 2016. Fumigant toxicity of essential oil from *Lantana camara* against almond moth, *Cadra cautella* (Walker). J. Essent. Oil Bear. Pl. 19(6): 1521–1526.

Grandtner, M.M. 2005. Elsevier's Dictionary of Trees, Volume 1. North America: With Names in Latin, English, French, Spanish and other Languages. Elsevier B.V., Amsterdam, The Netherlands.

Grandtner, M.M. and J. Chevrette. 2014. Dictionary of Trees, Volume 2: South America: Nomenclature, Taxonomy and Ecology (Elsevier's Dictionary of Trees). Academic Press, Elsevier Inc., Oxford, UK.

Gupta, H., K.R. Sharma and J.N. Sharma. 2017. Fungal inhibition in wood treated with *Lantana camara* L. extract. pp. 269–276. *In*: K.K. Pandey, V. Ramakantha, S.S. Chauhan and A.N.A. Kumar (eds.). Wood is Good, Current Trends and Future Prospects in Wood Utilization. Springer Nature, Singapore.

Gupta, M., Rakhi, N., Yadaf, Saroj, Pinky, Siksha, Manisha, Priyanka, Amit, Rahul, Sumit and Ankit. 2016. Phytochemical screening of leaves of *Plumeria alba* and *Plumeria acuminata*. J. Chem. Pharm. Res. 8(5): 354–358.

Gupta, M.P. 1995. 270 plantas medicinales iberoamericanas. CYTED Programa Iberoamericano de Ciencia y Tecnología para el Desarrollo. Subprograma de Química Fina Farmacéutica, Convenio Andrés Bello, Bogotá.

Gutiérrez, C.C., M.E. Niño, R.A. Jiménez and A. Muñoz-Acevedo. 2010. Estudio comparativo de la fracción volátil de flores de *Plumeria rubra* y *Plumeria alba*. Libro de resúmenes XXIX Congreso Latinoamericano de Química (CLAQ), XVI Congreso Colombiano de Química y VI Congreso Colombiano de Cromatografía. PRN100.

Guzman, J.D., A. Gupta, D. Evangelopoulos, C. Basavannacharya, L.C. Pabon, E.A. Plazas, D.R. Muñoz, W.A. Delgado, L.E. Cuca, W. Ribon, S. Gibbons and S. Bhakta. 2010. Anti-tubercular screening of natural products from Colombian plants: 3-methoxynordomesticine, an inhibitor of MurE ligase of *Mycobacterium tuberculosis*. J. Antimicrob. Chemother. 65(10): 2101–2107.

Hennebelle, T., S. Sahpaz, C. Dermont, H. Joseph and F. Bailleul. 2006. The essential oil of *Lippia alba*: Analysis of samples from French overseas departments and review of previous works. Chem. Biodivers. 3(10): 1116–1125.

Hennebelle, T., S. Sahpaz, H. Joseph and F. Bailleul. 2008. Ethnopharmacology of *Lippia alba*. J. Ethnopharmacol. 116(2): 211–222.

Herbert, J.M., J.P. Maffrand, K. Taoubi, J.M. Augereau, I. Fouraste and J. Gleye. 1991. Verbascoside isolated from *Lantana camara*, an Inhibitor of Protein Kinase C. J. Nat. Prod. 54(6): 1595–1600.

Hernández, D., J. Orozco, R. Serrano, A. Durán, S. Meraz, M. Jiménez-Estrada, A. García-Bores, J.G. Ávila and T. Hernández. 2014. Temporal variation of chemical composition and antimicrobial activity of the essential oil of *Cordia curassavica* (Jacq.) Roemer and Schultes: Boraginaceae. Bol. Latinoam. Caribe Plant. Med. Aromat. 13(1): 100–108.

Hernández, T., M. Canales, J.G. Ávila, A. Durán, J. Caballero, A. Romo de Vivar and R. Lira. 2003. Ethnobotany and antibacterial activity of some plants used in traditional medicine of Zapotitlán de las Salinas, Puebla (México). J. Ethnopharmacol. 88(1): 181–188.

Hernández, T., M. Canales, B. Terán, O. Ávila, A. Durán, A.M. García, H. Hernández, O. Ángeles-López, M. Fernández-Araiza and G. Ávila. 2007. Antimicrobial activity of the essential oil and extracts of *Cordia curassavica* (Boraginaceae). J. Ethnopharmacol. 111(1): 137–141.

Hernández, V., F. Mora and P. Meléndez. 2012. A study of medicinal plant species and their ethnomedicinal values in Caparo Barinas, Venezuela. Emir. J. Food Agric. 24(2): 128–132.

Hernández, V., F. Mora, M. Araque, S. De Montijo, L. Rojas, P. Meléndez and N. De Tommasi. 2013. Chemical composition and antibacterial activity of *Astronium graveolens* Jacq. essential oil. Rev. Latinoamer. Quím. 41(2): 89–94.

Hernández, V., N. Malafronte, F. Mora, M.S. Pesca, R.P. Aquino and T. Mencherini. 2014. Antioxidant and antiangiogenic activity of *Astronium graveolens* Jacq. leaves. Nat. Prod. Res. 28(12): 917–922.

Hernández, Y., R. Lizcano, J. Rodríguez, M. Quintana, A. Muñoz-Acevedo and R. Gutiérrez. 2015. Cito-genotoxicidad *in vitro* del aceite esencial de hojas de *Croton niveus* Jacq. en linfocitos y espermatozoides humanos. Cienc. Amaz. (Iquitos) 5(2): 174.

Hiremath, K.Y., N. Jagadeesh, S. Belur, S.S. Kulkarni and S.R. Inamdar. 2020. A lectin with anti-microbial and anti proliferative activities from *Lantana camara*, a medicinal plant. Protein Exp. Purif. 170: 105574.

HNC-ICN - Herbario Nacional de Colombia - Instituto de Ciencias Naturales. 2021. Universidad Nacional de Colombia. URL: http://www.biovirtual.unal.edu.co/es/colecciones/search/plants/.

Hung, N.H., D.N. Dai, P. Satyal, L.T. Huong, B.T. Chinh, D.Q. Hung, T.A. Tai and W.N. Setzer. 2021. *Lantana camara essential* oils from Vietnam: Chemical composition, molluscicidal, and mosquito larvicidal activity. Chem. Biodivers. 18: e2100145.

IFRA - The International Fragrance Association. 2015. IFRA Standards. 48th amendment. URL: https://ifrafragrance.org/docs/default-source/ifra-code-of-practice-and-standards/ifra-standards---48th-amendment/ifra-standards-in-full---booklet.pdf.

IFRA - The International Fragrance Association. 2021. IFRA transparency list. URL: https://ifrafragrance.org/priorities/ingredients/ifra-transparency-list.

Ioset, J.R., A. Marston, M.P. Gupta and K. Hostettmann. 2000. Antifungal and larvicidal cordiaquinones from the roots of *Cordia curassavica*. Phytochemistry 53(5): 613–617.

Jaramillo, B.E., E. Duarte, K. Muñoz and E. Stashenko. 2010. Composición química volátil del aceite esencial de *Croton malambo* H. Karst. colombiano y determinación de su actividad antioxidante. Rev. Cubana Plant. Med. 15(3): 133–142.

Jaramillo, B.E. and K. Muñoz. 2007. Composición química volátil y toxicidad aguda (CL$_{50}$) frente a *Artemia salina* del aceite esencial del *Croton malambo* colectado en la costa norte Colombiana. Sci. Techn. 33(1): 299–302.

Jaramillo-Colorado, B., J. Julio-Torres, E. Duarte-Restrepo, A. González-Coloma and L.F. Julio-Torres. 2015. Estudio comparativo de la composición volátil y las actividades biológicas del aceite esencial de *Piper marginatum* Jacq Colombiano. Bol. Latinoam. Caribe Plant. Med. Aromat. 14(5): 343–354.

Jaramillo-Colorado, B.E., K. Muñoz, E. Duarte, E. Stashenko and J. Olivero. 2014. Volatile secondary metabolites from Colombian *Croton malambo* (Karst) by different extraction methods and repellent activity of its essential oil. J. Essent. Oil Bear. Pl. 17(5): 992–1001.

Jaramillo-Colorado, B.E., S. Suarez-López and V. Marrugo-Santander. 2019. Volatile chemical composition of essential oil from *Bursera graveolens* (Kunth) Triana and Planch and their fumigant and repellent activities. Acta Scient. Biol. Sci. 41(1): e46822.

Jiménez-Escobar, N.D. 2012. Uso y conocimiento de árboles en la comunidad campesina de la bahía de Cispatá, departamento de Córdoba-Colombia. Tesis de Maestría en Ciencias - Biología. Universidad Nacional de Colombia, Bogotá, Colombia.

Jiménez-Medina, D. 2002. Estudio fitoquímico de la resina, hojas y tallos de *Protium heptaphyllum* (Aubl.) March. Tesis de Maestría, Facultad de Farmacia. Universidad de los Andes, Mérida, Venezuela.

Junor, G.A.O., R.B.R. Porter, P.C. Facey and T.H. Yee. 2007. Investigation of essential oil extracts from four native Jamaican species of *Bursera* for antibacterial activity. West Indian Med. J. 56: 22–25.

Junor, G.A.O., R.B.R. Porter and T.H. Yee. 2008. The chemical composition of the essential oils from the leaves, bark and fruits of *Bursera simaruba* (L.) Sarg. from Jamaica. J. Essent. Oil Res. 20: 426–429.

Kamatou, G.P.P. and Viljoen, A.M. 2008. Linalool—A review of a biologically active compound of commercial importance. Nat. Prod. Comm. 3(7): 1183–1192.

Khan, A.S. 2017. Medicinally Important Trees. Springer Nature, Switzerland.

Khan, M., A. Mahmood and H.Z. Alkhathlan. 2016. Characterization of leaves and flowers volatile constituents of *Lantana camara* growing in central region of Saudi Arabia. Arabian J. Chem. 9(6): 764–774.

Khan, M., S.K. Srivastava, N. Jain, K.V. Syamasundar and A.K. Yadav. 2003. Chemical composition of fruit and stem essential oils of *Lantana camara* from northern India. Flavour Frag. J. 18(5): 376–379.

Khan, M., S.K. Srivastava, K.V. Syamasundar, M. Singh and A.A. Naqvi. 2002. Chemical composition of leaf and flower essential oil of *Lantana camara* from India. Flavour Frag. J. 17(1): 75–77.

Kinjo, J., D. Nakano, T. Fujioka and H. Okabe. 2016. Screening of promising chemotherapeutic candidates from plants extracts. J. Nat. Med. 70: 335–360.

Knudsen, J.T. and L. Tollsten. 1993. Trends in floral scent chemistry in pollination syndromes: Floral scent composition in moth-pollinated taxa. Bot. J. Linnean Soc. 113(3): 263–284.

Kurade, N.P., V. Jaitak, V.K. Kaul and O.P. Sharma. 2010. Chemical composition and antibacterial activity of essential oils of *Lantana camara*, *Ageratum houstonianum* and *Eupatorium adenophorum*. Pharm. Biol. 48(5): 539–544.

Lago, J.H.G., A.A. Reis, D. Martins, F.G. Cruz and N.F. Roque. 2005. Composition of the leaf oil of *Xylopia emarginata* Mart. (Annonaceae). J. Essent. Oil Res. 17(6): 622–623.

Lawal, O.A., I.A. Ogunwande and A.R. Opoku. 2014. Constituents of essential oils from the leaf and flower of *Plumeria alba* L. grown in Nigeria. Nat. Prod. Commun. 9(11): 1613–1614.

Lawal, O.A., I.A. Ogunwande and A.R. Opoku. 2015. Chemical composition of essential oils of *Plumeria rubra* L. grown in Nigeria. Eur. J. Med. Plants 6(1): 55–61.

Leal, S.M., N. Pino, E.E. Stashenko, J.R. Martínez and P. Escobar. 2013. Antiprotozoal activity of essential oils derived from *Piper* spp. grown in Colombia. J. Essent. Oil Res. 25(6): 512–519.

Leclercq, P.A., H. Silva Delgado, J. García, J.E. Hidalgo, T. Cerruttti, M. Mestanza, F. Ríos, E. Nina, L. Nonato, R. Alvarado and R. Menéndez. 1999. Aromatic plant oils of the Peruvian Amazon. Part 1. *Lippia alba* (Mill.) N.E. Br. and *Cornutia odorata* (Poeppig) Poeppig ex Schauer, Verbenaceae. J. Essent. Oil Res. 11(6): 753–756.

Leyva, M., M.C. Marquetti, D. Montada, J. Payroll, R. Scull, G. Morejón and O. Pino. 2020. Essential oils of *Eucalyptus globulus* (Labill) and *Bursera graveolens* (Kunth) Triana and Planch for the control of mosquitoes of medical importance. Biologist 18(2): 239–250.

Leyva, M.A., J.R. Martínez and E.E. Stashenko. 2007. Composición química del aceite esencial de hojas y tallos de *Bursera graveolens* (burseraceae) de Colombia. Sci. Techn. 33(1): 201–202.

Lim, T.K. 2014. Edible Medicinal and Non-medicinal Plants. Volume 7, Flowers. Springer Science+Business Media B.V., New York.

Lima, T.A.A.C., L.P. Cunha, J.E.L.S. Ribeiro, M.O.M. Marques and M.P. Lima. 2021. Evaluation of volatile constituents, exudation of resin and occurrence of galls of *Protium aracouchini* (Aubl.) Marchand. Acta Brasil. 5(3): 88–91.

Lima-Junior, R.C.P., F.A. Oliveira, L.A. Gurgel, I.J.M. Cavalcante, K.A. Santos, D.A. Campos, C.A.L. Vale, R.M. Silva, M.H. Chaves, V.S.N. Rao and F.A. Santos. 2006. Attenuation of visceral nociception by α- and β-amyrin, a triterpenoid mixture isolated from the resin of *Protium heptaphyllum*, in mice. Planta Med. 72(1): 34–39.

Litaudon, M., C. Jolly, C. Le Callonec, D.D. Cuong, P. Retailleau, O. Nosjean, V.H. Nguyen, B. Pfeiffer, J.A. Boutin and F. Guéritte. 2009. Cytotoxic pentacyclic triterpenoids from *Combretum sundaicum* and *Lantana camara* as inhibitors of Bcl-xL/BakBH3 domain peptide interaction. J. Nat. Prod. 72(7): 1314–1320.

Liu, Y., H. Wang, S. Wei and Z. Yan. 2012. Chemical composition and antimicrobial activity of the essential oils extracted by microwave-assisted hydrodistillation from the flowers of two *Plumeria* species. Anal. Lett. 45(16): 2389–2397.

López, M.A., E.E. Stashenko and J.L. Fuentes. 2011. Chemical composition and antigenotoxic properties of *Lippia alba* essential oils. Genet. Mol. Biol. 34(3): 479–488.

Lorenzo, D., P. Davies, R. Vila, S. Cañigueral and E. Dellacassa. 2001. Composition of a new essential oil type of *Lippia alba* (Mill.) N.E. Brown from Uruguay. Flavour Frag. J. 16(5): 356–359.

Louchard, B.O. and T.G. de Araújo. 2019. Pharmacological effects of different chemotypes of *Lippia alba* (Mill.) N.E. Brown. Bol. Latinoam. Caribe Plant. Med. Aromat. 18(2): 95–105.

Loureiro, R.M. 2014. Atividade anti-*Helicobacter pylori* de extratos hidroetanólicos de caule e folhas de *Astronium fraxinifolium* e *Astronium graveolens*. Trabajo de pregrado, Profesional en Farmácia-Bioquímica, Universidade Estadual Paulista.

Luján, M., Y. León and R. Riina. 2015. Sinopsis de *Croton* (Euphorbiaceae) en los Andes de Mérida, Venezuela. Caldasia 37(1): 73–90.

Luna-Cazáres, L.M. and A.R. González-Esquinca. 2008. Actividad antibacteriana de extractos de *Annona diversifolia* Safford y *Annona purpurea* Mociño and Sessé ex Dunal. Polibotánica 25: 120–125.

Maia, J.G.S., E.H.A. Andrade, A.C.M. da Silva, J. Oliveira, L.M.M. Carreira and J.S. Araújo. 2005. Leaf volatile oils from four Brazilian *Xylopia* species. Flavour Frag. J. 20(5): 474–477.

Maia, R.M., P.R. Barbosa, F.G. Cruz, N.F. Roque and M. Fascio. 2000. Triterpenes from the resin of *Protium heptaphyllum* March (Burseraceae): Characterization in binary mixtures. Quim. Nova 23(5): 623–626.

Majolo, C., P.C. Monteiro, A.V.P. do Nascimento, F.C.M. Chaves, P.E. Gama, H.R. Bizzo and E.C. Chagas. 2019. Essential oils from five Brazilian *Piper* species as antimicrobials against strains of *Aeromonas hydrophila*. J. Essent. Oil Bear. Pl. 22(3): 746–761.

Manzano-Santana, P., M. Miranda, Y. Gutiérrez, G. García, T. Orellana and A. Orellana. 2009. Antinflammatory effect and chemical composition of *Bursera graveolens* Triana and Planch. branch oil (palo santo) from Ecuador. Rev. Cubana Plant. Med. 14(3): 45–53.

Marcano, E.E., A. Padilla-Baretic and L. Rojas-Fermín. 2013. Aceite esencial extraído por hidrodestilación del tejido xilemático de ramas de *Bursera simaruba* (L.) Sarg. Rev. Forestal Latinoamer 28: 27–36.

Marin, W.A. and E.M. Flores. 2002. *Astronium graveolens* Jacq. pp. 311–314. *In*: J.A. Vozzo (ed.). Tropical Tree Seed Manual. Part II—Species Descriptions. 2002. USDA Forest Service, Washington.

Marongiu, B., A. Piras, S. Porcedda, E. Tuveri, A. Deriu and S. Zanetti. 2007. Extraction of *Lantana camara* essential oil by supercritical carbon dioxide: influence of the grinding and biological activity. Nat. Prod. Res. 21(1): 33–36.

Martínez Gordillo, M.J. 1988. Contribución al conocimiento del género *Croton* (Euphorbiaceae) en el estado de Guerrero. Trabajo de pregrado, Profesional en Biología. Universidad Nacional Autónoma de México, Ciudad de México.

Matos, F.J.A., M.I.L. Machado, A.A. Craveiro and J.W. Alencar. 1996. Essential oil composition of two chemotypes of *Lippia alba* grown in Northeast Brazil. J. Essent. Oil Res. 8(6): 695–698.

Meccia, G., L.B. Rojas, J. Velasco, T. Díaz, A. Usubillaga, J. Carmona Arzola and S. Ramos. 2009. Chemical composition and antibacterial activity of the essential oil of *Cordia verbenacea* from the Venezuelan Andes. Nat. Prod. Comm. 4(8): 1119–1122.

Medeiros, L.B.P., M.S. Rocha, S.G. de Lima, G.R. de Sousa Júnior, A.M.G.L. Citó, D. da Silva, J.A.D. Lopes, D.J. Moura, J. Saffi, M. Mobin and J.G.M. da Costa. 2012. Chemical constituents and evaluation of cytotoxic and antifungal activity of *Lantana camara* essential oils. Braz. J. Pharmacogn. 22(6): 1259–1267.

Medeiros, P.C., Y.R.F. Nunes, J.P. Cruz, D.M. de Souza, M.A. Ávila, F. Morais-Costa, S.R. Arrudas, V.O. Vasconcelos, T.M. Vieira and A.P.V. Moura. 2021. Seasonal variation in the content of condensed tannins in leaves of *Xylopia emarginata* Mart. (Annonaceae) in response to phenology and climate. J. Agric. Sci. 13(10): 142–151.

Melanie, M., F.Y. Kosasih, H. Kasmara, D.M. Malini, C. Panatarani, I.M. Joni, T. Husodo and W. Hermawan. 2020. Antifeedant activity of *Lantana camara* nano suspension prepared by reverse emulsion of ethyl acetate active fraction at various surfactant organic-phase ratio. Biocat. Agric. Biotech. 29: 101805.

Méndez, A.L., S. Aristizabal, E. Torres, J.D. Rodríguez, A.M. Molina and A. Muñoz-Acevedo. 2014a. Metabolitos secundarios volátiles de frutos, hojas y ramas de *Croton fragilis* Kunth. del cerro "pan de azúcar" (Barranquilla). Rev. Prod. Nat. 4(1): 177.

Méndez, A.L., J.D. Rodríguez, S. Aristizabal, S. Vargas, L. Yamaguchi, M.J. Kato and A. Muñoz-Acevedo. 2014b. HPLC/DAD/MS/ESI/TOF analysis and evaluation of the antioxidant capacities and cytotoxicities of the ethanolic extracts of the leaves of *Chromolaena barranquillensis* (Hieron.) and *Cyanthillium cinereum* (L.). Rev. Prod. Nat. 4(1): 64.

Méndez Beltrán, A.L. and E.M. Ortega Morales. 2012. Caracterización química, efecto protector y letalidad de los metabolitos secundarios volátiles de las flores y hojas de *Cyanthillium cinereum* recolectadas en los municipios de Juan de Acosta y Piojó, Atlántico. Trabajo de pregrado, Profesional en Química y Farmacia. Universidad del Atlántico, Puerto Colombia, Colombia.

Mendoza-Meza, D.L., H. Benavides-Henríquez and M.E. Taborda-Martínez. 2014. Actividad acaricida del aceite esencial de la corteza de *Croton malambo* H. Karst, metil-eugenol y metil-isoeugenol contra *Dermatophagoides farinae* Hughes, 1961. Bol. Latinoam. Caribe Plant. Med. Aromat. 13(6): 537–544.

Mesa-Arango, A.C., L. Betancur-Galvis, J. Montiel, J.G. Bueno, A. Baena, D.C. Durán, J.R. Martínez and E.E. Stashenko. 2010. Antifungal activity and chemical composition of the essential oils of *Lippia alba* (Miller) N.E. Brown grown in different regions of Colombia. J. Essent. Oil Res. 22(6): 568–574.

Mesa-Arango, A.C., J. Montiel-Ramos, T. Zapata, C. Durán, L. Betancur-Galvis and E.E. Stashenko. 2009. Citral and carvone chemotypes from the essential oils of Colombian *Lippia alba* (Mill.) N.E. Brown: Composition, cytotoxicity and antifungal activity. Mem. Inst. Oswaldo Cruz 104(6): 878–884.

Misra, L. and A.K. Saikia. 2011. Chemotypic variation in Indian *Lantana camara* essential oil. J. Essent. Oil Res. 23(3): 1–5.

Mobin, M., S.G. de Lima, L.T.G. Almeida, J.C. Silva Filho, M.S. Rocha, A.P. Oliveira, M.B. Mendes, F.A.A. Carvalho, M.S.C. Melhem and J.G.M. Costa. 2017. Gas chromatography-triple quadrupole mass spectrometry analysis and vasorelaxant effect of essential oil from *Protium heptaphyllum* (Aubl.) March. BioMed. Res. Int. 2017: 1928171.

Mobin, M., S.G. de Lima, L.T.G. Almeida, J.P. Takahashi, J.B. Teles, M.W. Szeszs, M.A. Martins, A.A. Carvalho and M.S.C. Melhem. 2016. MDGC-MS analysis of essential oils from *Protium heptaphyllum* (Aubl.) and their antifungal activity against *Candida* specie. Rev. Bras. Pl. Med. 18(2): 531–538.

Molina, A.M., A.L. Méndez, A. Muñoz-Acevedo, R. Gutiérrez and J.D. Guzmán. 2014. Antimicrobial activity of the ethanolic extracts of five plant species used in traditional medicine in the departamento del Atlántico. Rev. Prod. Nat. 4(1): 77.

Moncayo, S., M. Rondón, L. Araujo, L. Rojas, X. Cornejo and W. Guamán. 2021. Composición química y actividad biológica de los aceites esenciales de *Piper marginatum* Jacq. y *Piper tuberculatum* Jacq. de Ecuador. Rev. Fac. Farm. 63(1): 14.

Montero-Villegas, S., R. Crespo, B. Rodenak-Kladniew, M.A. Castro and M. Galle. 2018. Cytotoxic effects of essential oils from four *Lippia alba* chemotypes in human liver and lung cancer cell lines. J. Essent. Oil Res. 30(3): 167–181.

Montes-de-Oca, C., C.T. Hernández-Delgado, J. Orozco-Martínez, A.M. García-Bores, J.G. Ávila-Acevedo, M.T. Ortiz-Melo, I. Peñalosa-Castro, G. López-Moreno and R. Serrano-Parrales. 2017. Antibacterial and antifungal activity of *Dalea carthagenensis* (Jacq.) J.F. Macbr. Rev. Fitotec. Mex. 40: 161–168.

Monzote, L., G.M. Hill, A. Cuellar, R. Scull and W.N. Setzer. 2012. Chemical composition and anti-proliferative properties of *Bursera graveolens* essential oil. Nat. Prod. Comm. 7(11): 1531–1534.

Morales, A., P. Pérez, R. Mendoza, R. Compagnone, A.I. Suárez, F. Arvelo, J.L. Ramírez and I. Galindo-Castro. 2005. Cytotoxic and proapoptotic activity of ent-16β-17α-dihydroxykaurane on human mammary carcinoma cell line MCF-7. Cancer Lett. 218(1): 109–116.

Moreira, I.C., J.H.G. Lago and N.F. Roque. 2003. Alkaloid, flavonoids and terpenoids from leaves and fruits of *Xylopia emarginata* (Annonaceae). Biochem. Syst. Ecol. 31(5): 535–537.

Moreira, I.C., N.F. Roque and G. Lago. 2006. Diterpene adducts from branches of *Xylopia emarginata*. Biochem. Syst. Ecol. 34(11): 833–837.

Moreira, I.C., N.F. Roque, K. Contini and J.H.G. Lago. 2007. Sesquiterpenos e hidrocarbonetos dos frutos de *Xylopia emarginata* (Annonaceae). Braz. J. Pharmacogn. 17: 55–58.

Muñoz, D.R. 2008. Estudio fitoquímico y evaluación de la actividad fungicida e insecticida de la especie *Piper eriopodon* (Piperaceae). Trabajo de pregrado, Profesional en Química. Universidad Nacional de Colombia, Bogotá.

Muñoz, D.R., A.G. Sandoval-Hernandez, W.A. Delgado, G.H. Arboleda and L.E. Cuca. 2018. *In vitro* anticancer screening of Colombian plants from *Piper* genus (Piperaceae). J. Pharmacogn. Phytother. 10(9): 174–181.

Muñoz-Acevedo, A. and O. Castillo-Contreras. 2022. Chemical characterization by GC-FID/MS of the volatile secondary metabolites (SDE/HS-SPME/MWHD) isolated/identified in different parts of *Protium heptaphyllum* (Aubl.) Marchand from Sucre (Colombia) (*manuscript in preparation*).

Muñoz-Acevedo, A., A. Serrano-Uribe, X.J. Parra, L.A. Olivares and M.E. Niño. 2013. Multivariate analysis and chemical variability of the volatile metabolites present in the aerial parts and the resin of *Bursera graveolens* (Kunth) Triana and Planch from Soledad (Atlántico, Colombia). Bol. Latinoam. Caribe Plant. Med. Aromat. 12(3): 322–337.

Muñoz-Acevedo, A., A.L. Méndez, E.M. Ortega and M.E. Niño. 2012. Chemical study of the volatile secondary metabolites of flowers and leaves of *Cyanthillium cinereum* (L.) H. Rob from Juan de Acosta (Atlántico, Colombia). Bol. Latinoam. Caribe Plant. Med. Aromat. 11(4): 331–340.

Muñoz-Acevedo, A., C.E. Puerto, J.D. Rodríguez, S. Aristizábal-Córdoba, R.G. Gutiérrez and V.V. Kouznetsov. 2014b. Estudio químico-biólogico de los aceites esenciales de *Croton malambo* H. Karst y su componente mayoritario, metileugenol. Bol. Latinoam. Caribe Plant. Med. Aromat. 13(4): 336–343.

Muñoz-Acevedo, A., C.E. Puerto, J.D. Rodríguez, R.G. Gutiérrez and V.V. Kouznetsov. 2014a. Estudio de la citotoxicidad *in vitro* e *in vivo* del aceite esencial de *Croton malambo* y sus transformaciones químicas. Rev. Prod. Nat. 4(1): 67.

Muñoz-Acevedo, A., C.L. Trillos, J. Chamorro-Crespo, L.P. Abadia and M.E. Niño. 2011c. Oxidation inhibition effects and composition of the volatile oils of leaves/bark of *Croton niveus* Jacq. from Tubará (Colombia). Conference paper No. 612. 43rd IUPAC World Chemistry Congress/46th IUPAC General Assembly/70th CQPR Annual Conference and Exhibition. July 31–August 5, San Juan, Puerto Rico.

Muñoz-Acevedo, A., E. Borrero-Meza, E. Vitola-Méndez, M.E. Niño and C.L. Trillos. 2011b. Metabolitos secundarios volátiles aislados por SDE e hidrodestilación de hojas de *Piper peltatum* encontrada en Tubará (Atlántico). Vitae 18(S2): S210.

Muñoz-Acevedo, A., M.C. González and E.E. Stashenko. 2019a. Volatile fractions and essential oils of the leaves and branches of *Dalea carthagenensis* (Jacq.) J.F. Macbr. from northern region of Colombia. J. Essent. Oil Bear. Pl. 22(3): 774–788.

Muñoz-Acevedo, A., M.C. González and E.E. Stashenko. 2019b. Composición de metabolitos secundarios volátiles de dos especies de Bursera encontradas en el bosque seco tropical de la costa norte colombiana. VI Congreso Latinoamericano de Plantas Medicinales. Med. Plant. Comm. 2(2): 52.

Muñoz-Acevedo, A., M.C. González, J.D. Rodríguez and Y.Sh. De Moya. 2019. New chemovariety of *Lippia alba* from Colombia: Compositional analysis of the volatile secondary metabolites and some *in vitro* biological activities of the essential oil from plant leaves. Nat. Prod. Comm. July 2019(1–7).

Muñoz-Acevedo, A., M.C. González, J.D. Rodríguez, Y.Sh. De Moya and R.G. Gutiérrez. 2016. Composición química y actividades biológicas de aceites esenciales de algunas especies vegetales del bosque seco tropical de la costa norte colombiana. Dominguezia 32: 12.

Muñoz-Acevedo, A., M.C. González, J.D. Rodríguez, Y.Sh. De Moya and R.G. Gutiérrez. 2017. Aceites esenciales de especies vegetales del bosque seco tropical del Caribe colombiano: Composición química y actividades biológicas. Libro resúmenes VCOLAPLAMED, La Paz, Bolivia.

Muñoz-Acevedo, A., S. Aristizabal-Córdoba, J.D. Rodríguez, E.A. Torres, R.G. Gutiérrez and V.V. Kouznetsov. 2016. *In-vitro* cytotoxicity/anti-radical capability and structural characterization by

GC-MS/^{1}H-^{13}C-NMR of the leaves essential oils of young/old trees of *Annona purpurea* Moc. and Sessé ex Dunal from Repelón (Atlántico, Colombia). Bol. Latinoam. Caribe Plant. Med. Aromat. 15(2): 99–111.

Muñoz-Acevedo, A., V.C. Álvarez and M.E. Niño. 2011a. Chemical characterization of the volatile fractions and essential oils of leaves and flowers of *Chromolaena barranquillensis* from Sabanalarga (Atlántico, Colombia). Bol. Latinoam. Caribe Plant. Med. Aromat. 10(6): 440–448.

Muñoz-Acevedo, A., Y.Sh. De Moya, M.C. González and J.D. Rodríguez. 2022. Composition of the volatile metabolites of *Piper eriopodon* (Miq.) C. DC. from Northern Region of Colombia and assessment of *in vitro* bioactivities of the leaf essential oil (*manuscript submitted/in evaluation*).

Murillo, J. 2001. Las Annonaceae de Colombia. Biota Colomb. 2(1): 49–58.

Murillo Aldana, J.C. 1999. Composición y distribución del género *Croton* (Euphorbiaceae) en Colombia, con cuatro especies nuevas. Caldasia 21(2): 141–166.

Nakanishi, T., Y. Inatomi, H. Murata, K. Shigeta, N. Iida, A. Inada, J. Murata, M.A. Perez Farrera, M. Iinuma, T. Tanaka, S. Tajima and N. Oku. 2005. A new and known cytotoxic aryltetralin-type lignans from stems of *Bursera graveolens*. Chem. Pharm. Bull. 53(2): 229–231.

Ngassoum, M.B., S. Yonkeu, L. Jirovetz, G. Buchbauer, G. Schmaus and F.-J. Hammerschmidt. 1999. Chemical composition of essential oils of *Lantana camara* leaves and flowers from Cameroon and Madagascar. Flavour Frag. J. 14(4): 245–250.

Noel-Martinez, K.C., G.J.F. Cruz and R.L. Solis-Castro. 2021. *Bursera graveolens* essential oil: Physiochemical characterization and antimicrobial activity in pathogenic microorganisms found in *Kajikia audax*. Scientia Agrop. 12(3): 303–309.

Oliveira, F.A., C.L.S. Costa, M.H. Chaves, F.R.C. Almeida, I.J.M. Cavalcante, A.F. Lima, R.C.P. Lima-Junior, R.M. Silva, A.R. Campos, F.A. Santos and V.S.N. Rao. 2005b. Attenuation of capsaicin-induced acute and visceral nociceptive pain by α- and β-amyrin, a triterpene isolated from *Protium heptaphyllum* resin in mice. Life Sci. 77(23): 2942–2952.

Oliveira, F.A., G.M. Vieira-Junior, M.H. Chaves, F.R.C. Almeida, M.G. Florêncio, R.C.P. Lima-Junior, R.M. Silva, F.A. Santos and V.S.N. Rao. 2004a. Gastroprotective and anti-inflammatory effects of resin from *Protium heptaphyllum* in mice and rats. Pharmacol. Res. 49(2): 105–111.

Oliveira, F.A., G.M. Vieira-Junior, M.H. Chaves, F.R.C. Almeida, K.A. Santos, F.S. Martins, R.M. Silva, F.A. Santos and V.S.N. Rao. 2004b. Gastroprotective effect of the mixture of alpha- and beta-amyrin from Protium *heptaphyllum*: Role of capsaicin-sensitive primary afferent neurons. Planta Med. 70(8): 780–782.

Oliveira, F.A., M.H. Chaves, F.R.C. Almeida, R.C.P. Lima-Junior, R.M. Silva, J.L. Maia, G.A.A.C. Brito, F.A. Santos and V.S.M. Rao. 2005a. Protective effect of α- and β-amyrin, a triterpene mixture from *Protium heptaphyllum* (Aubl.) March. trunk wood resin, against acetaminophen-induced liver injury in mice. J. Ethnopharmacol. 98(1-2): 103–108.

Olivero-Verbel, J., J. Güete-Fernández and E.E. Stashenko. 2009. Acute toxicity against *Artemia franciscana* of essential oils isolated from plants of the genus *Lippia* and *Piper* collected in Colombia. Bol. Latinoam. Caribe Plant. Med. Aromát. 8(5): 419–427.

Olivero-Verbel, J., T. González-Cervera, J. Güette-Fernandez, B. Jaramillo-Colorado and E. Stashenko. 2010. Chemical composition and antioxidant activity of essential oils isolated from Colombian plants. Braz. J. Pharmacogn. 20(4): 568–574.

Omata, A., K. Yomogida, S. Nakamura, S. Hashimoto, T. Arai and K. Furukawa. 1991. Volatile components of *Plumeria* flowers. Part 1. *Plumeria rubra* forma *acutifolia* (Poir.) Woodson cv. 'Common Yellow'. Flavour Frag. J. 6(4): 277–279.

Omata, A., S. Nakamura, S. Hashimoto and K. Furukawa. 1992. Volatile components of *Plumeria* flowers. Part 2. *Plumeria rubra* L. cv. 'Irma Bryan'. Flavour Frag. J. 7(1): 33–35.

O'Neill, M.J., J.A. Lewis, H.M. Noble, S. Holland, C. Mansat, J.E. Farthing, G. Foster, D. Noble, S.J. Lane, P.J. Sidebottom, S.M. Lynn, M.V. Hayes and C.J. Dix. 1998. Isolation of translactone-containing triterpenes with thrombin inhibitory activities from the leaves of *Lantana camara*. J. Nat. Prod. 61(11): 1328–1331.

Ono, M., A. Hashimoto, M. Miyajima, A. Sakata, C. Furusawa, M. Shimode, S. Tsutsumi, S. Yasuda, M. Okawa, J. Kinjo, H. Yoshimitsu and T. Nohara. 2020. Two new triterpenoids from the leaves and stems of *Lantana camara*. Nat. Prod. Res. 35(21): 3757–3765.

Ontiveros-Rodríguez, J.C., E. Burgueño-Tapia, J. Porras-Ramírez, P. Joseph-Nathan and L.G. Zepeda. 2018. Configurational study of an aporphine alkaloid from *Annona purpurea*. Nat. Prod. Comm. 13(7): 831–836.

Oyedeji, O.A., O. Ekundayo and W.A. König. 2003. Volatile leaf oil constituents of *Lantana camara* L. from Nigeria. Flavour Frag. J. 18(5): 384–386.

Oza, M.J. and Y.A. Kulkarni. 2017. Traditional uses, phytochemistry and pharmacology of the medicinal species of the genus *Cordia* (Boraginaceae). J. Pharm. Pharmacol. 69(7): 755–789.

Padalia, R.C., R.S. Verma and A. Chauhan. 2015. Variation in the essential oils constituents of two chemovariant of *Lantana camara* L. J. Essent. Oil Bear. Pl. 18(4): 775–784.

Páez Aranzalez, D.M. and A. Sánchez-Corredor. 2018. Extracción de aceite esencial de las especies aromáticas venturosa *Lantana camara* L., Escobillo *Xylopia emarginata* Mart., clavo de monte *Miconia* sp. Ruiz and Pav, para la determinación de actividad biológica en el Centro de Investigación Santa Lucía Barrancabermeja, Santander. Trabajo de pregrado, Profesional en Ingeniería Ambiental y de Saneamiento. Universidad de La Paz, Barrancabermeja, Colombia.

Pasaribu, T., R.D.D.M. Tobing, T. Kostaman and B. Dewantoro. 2020. Active substance compounds and antibacterial activity of extract flower and leaves of *Plumeria rubra* and *Plumeria alba* against *Escherichia coli*. AIP Conference Proceedings 2296: 020105.

Pascual, M.E., K. Slowing, E. Carretero, D. Sánchez Mata and A. Villar. 2001a. *Lippia*: Traditional uses, chemistry and pharmacology: A review. J. Ethnopharmacol. 76(3): 201–214.

Pascual, M.E., K. Slowing, M.E. Carretero and Á. Villar. 2001b. Antiulcerogenic activity of *Lippia alba* (Mill.) N.E. Brown (Verbenaceae). Farmaco 56(5-7): 501–504.

Passos, J.L., L.C.A. Barbosa, A.J. Demuner, E.S. Alvarenga, C.M. da Silva and R.W. Barreto. 2012. Chemical characterization of volatile compounds of *Lantana camara* L. and *L. radula* Sw. and their antifungal activity. Molecules 17: 11447–11455.

Paz, R.F., E.F. Guimarães and C.S. Ramos. 2017. The occurrence of phenylpropanoids in the saps of six *Piper* species (Piperaceae) from Brazil. Gayana Bot. 74(1): 236–239.

Peixoto, M.G., L. Bacci, A.F. Blank, A.P.A. Araújo, P.B. Alves, J.H.S. Silva, A.A. Santos, A.P. Oliveira, A.S. da Costa and A.F. Arrigoni-Blank. 2015a. Toxicity and repellency of essential oils of *Lippia alba* chemotypes and their major monoterpenes against stored grain insects. Ind. Crop. Prod. 71: 31–36.

Peixoto, M.G., L.M. Costa-Júnior, A.F. Blank, A.S. Lima, T.S.A. Menezes, D.A. Santos, P.B. Alves, S.C.H. Cavalcanti, L. Bacci and M.F. Arrigoni-Blank. 2015b. Acaricidal activity of essential oils from *Lippia alba* genotypes and its major components carvone, limonene, and citral against *Rhipicephalus microplus*. Vet. Parasitol. 210(1-2): 118–122.

Peraza-Sánchez, S.R., N.E. Salazar-Aguilar and L.M. Peña-Rodríguez. 1995. A new triterpene from the resin of *Bursera simaruba*. J. Nat. Prod. 58(2): 271–274.

Pereira, K.L.G., P.C.L. Nogueira, M.F. Arrigoni-Blank, D.A.C. Nizio, D.C. Silva, J.A.O. Pinto, T.S. Sampaio and A.F. Blank. 2020. Chemical diversity of essential oils of *Lantana camara* L. native populations. J. Essent. Oil Res. 32(1): 32–47.

Pereira-de-Morais, L., A.A. Silva, R.E.R. da Silva, R.H.S. da Costa, A.B. Monteiro, C.R.S. Barbosa, T.S. Amorim, I.R.A. de Menezes and M.R. Kerntopf. 2019. Tocolytic activity of the *Lippia alba* essential oil and its major constituents, citral and limonene, on the isolated uterus of rats. Chem. Biol. Interact. 297: 155–159.

Pérez Arbeláez, E. 1996. Plantas útiles de Colombia. DAMA/Fondo FEN Colombia, Bogotá.

Pino, J.A., A. Ferrer, D. Álvarez and A. Rosado. 1994. Volatiles of an alcoholic extract of flowers from *Plumeria rubra* L. var. *acutifolia*. Flavour Frag. J. 9(6): 343–345.

Pino, J.A., A. Ortega and A. Rosado. 1996. Chemical composition of the essential oil of *Lippia alba* (Mill.) N.E. Brown from Cuba. J. Essent. Oil Res. 8(4): 445–446.

Pino, J.A., R. Marbot, A. Rosado, C. Romeu and M.P. Mart. 2004. Chemical composition of the essential oil of *Lantana camara* L. from Cuba. J. Essent. Oil Res. 16(3): 216–218.

Pinto, F.E., O.A. Heringer, M.A. Silva, T.U. Andrade, J.S. Ribeiro, D. Lenz, F.C.R. Lessa and D.C. Endringer. 2015. Stability and disinfecting proprieties of the toothbrush rinse of the essential oil of *Protium heptaphyllum*. African J. Pharm. Pharmacol. 9(6): 173–181.

Piñeros, L.P. and F. González. 2020. Taxonomic revision of *Dalea* (Leguminosae: Papilionoideae) in Colombia. Caldasia 42(2): 220–240.

PNN - Parques Nacionales Naturales. 2005. Plan de manejo Parque Nacional Natural Macuira. Unidad administrativa especial del sistema de Parques Nacionales Naturales, Ministerio de Ambiente y Desarrollo Sostenible, Colombia.

Pontes, W.J.T., J.C.G. de Oliveira, C.A.G. da Câmara, A.C.H.R. Lopes, M.G.C. Gondim Jr, J.V. de Oliveira, R. Barros and M.O.E. Schwartz. 2007. Chemical composition and acaricidal activity of the leaf and fruit essential oils of *Protium heptaphyllum* (Aubl.) Marchand (Burseraceae). Acta Amazon. 37(1): 103–110.

Pour, B.M. and S. Sasidharan. 2011. *In vivo* toxicity study of *Lantana camara*. Asian Pac. J. Trop. Biomed. 1(3): 230–232.

Qamar, F., S. Begum, S.M. Raza, A. Wahab and B.S. Siddiqui. 2005. Nematicidal natural products from the aerial parts of *Lantana camara* Linn. Nat. Prod. Res. 19(6): 609–613.

Quattrocchi, U. 2012. CRC World Dictionary of Medicinal and Poisonous Plants: Common Names, Scientific Names, Eponyms, Synonyms, and Etymology (5 volume set). CRC Press, Boca Ratón.

Raguso, R.A. and E. Pichersky. 1999. A day in the life of a linalool molecule: Chemical communication in a plant-pollinator system. Part 1: Linalool biosynthesis in flowering plants. Plant Spec. Biol. 14: 95–120.

Rahalison, L., M. Hamburger, K. Hostettmann, E. Monod, E. Frenk, M.P. Gupta, A.I. Santana, M.D. Correa and A.G. Gonzalez. 1993. Screening for antifungal activity of Panamanian plants. Int. J. Pharmacog. 31: 68–76.

Rajakrishnan, R., P. Kuppusamy, R. Sathya, P. Nandhakumari, A.D.V. Bensy and G.D. Biji. 2022. Antifungal phytochemicals from the methanol and aqueous extract of *Acacia concinna* and *Lantana camara* and synergistic biological control of the hibiscus mealybug (*Maconellicoccus hirsutus*). Physiol. Mol. Plant Pathol. 2022: 101813.

Ramos, L.S., M.L. da Silva, A.I.R. Luz, M.G.B. Zoghbi and J.G.S. Maia. 1986. Essential oil of *Piper marginatum*. J. Nat. Prod. 49(4): 712–713.

Rana, V.S., D. Prasad and M.A. Blazquez. 2005. Chemical composition of the leaf oil of *Lantana camara*. J. Essent. Oil Res. 17(2): 198–200.

Randrianalijaona, J.-A., P.A.R. Ramanoelina, J.R.E. Rasoarahona and E.M. Gaydou. 2006. Chemical compositions of aerial part essential oils of *Lantana camara* L. chemotypes from Madagascar. J. Essent. Oil Res. 18(4): 405–407.

Rao, G.P., M. Singh, P. Singh, S.P. Singh, C. Catalán, I.P.S. Kapoor, O.P. Singh and G. Singh. 2000. Studies on chemical constituents and antifungal activity of leaf essential oil of *Lippia alba* (Mill.). Indian J. Chem. Technol. 7: 332–335.

Rao, V.S., J.L. Maia, F.A. Oliveira, T.L.G. Lemos, M.H. Chaves and F.A. Santos. 2007. Composition and antinociceptive activity of the essential oil from *Protium heptaphyllum* resin. Nat. Prod. Commun. 2(12): 1199–1202.

RBG Kew - Royal Botanic Gardens Kew. 2022. Useful Plants of Colombia. https://colplanta.org/.

Reigada, J.B., C.M. Tcacenco, L.H. Andrade, M.J. Kato, A.L.M. Porto and J.H.G. Lago. 2007. Chemical constituents from *Piper marginatum* Jacq. (Piperaceae) - antifungal activities and kinetic resolution of (RS)-marginatumol by *Candida antarctica* lipase (Novozym 435). Tetrahedron (Asymm) 18(9): 1054–1058.

Rey-Valeirón, C., L. Guzmán, L.R. Saa, J. López-Vargas and E. Valarezo. 2017. Acaricidal activity of essential oils of *Bursera graveolens* (Kunth) Triana and Planch and *Schinus molle* L. on unengorged larvae of cattle tick *Rhipicephalus* (*Boophilus*) *microplus* (Acari:Ixodidae). J. Essent. Oil Res. 29(4): 525–526.

Ricciardi, G., J.F. Cicció, R. Ocampo, D. Lorenzo, A. Ricciardi, A. Bandoni and E. Dellacassa. 2009. Chemical variability of essential oils of *Lippia alba* (Miller) N.E. Brown growing in Costa Rica and Argentina. Nat. Prod. Commun. 4(6): 853–858.

Robles, J., R. Torrenegra, A.I. Gray, C. Piñeros, L. Ortiz and M. Sierra. 2005. Triterpenos aislados de corteza de *Bursera graveolens* (Burseraceae) y su actividad biológica. Braz. J. Pharmacogn. 15(4): 283–286.

Rodríguez-Burbano, D., C.E. Quijano-Celis and J.A. Pino. 2010. Composition of the essential oil from leaves of *Astronium graveolens* Jacq grown in Colombia. J. Essent. Oil Res. 22(6): 488–489.

Rodríguez, G.M., K. Banda, S.P. Reyes and A.C. Estupiñán González. 2012. Lista comentada de las plantas vasculares de bosques secos prioritarios para la conservación en los departamentos de Atlántico y Bolívar (Caribe colombiano). Biota Colomb. 13(2): 7–39.

Rodríguez, J.D., A. Muñoz-Acevedo, A.L. Méndez, R.A. Jiménez and R.G. Gutiérrez. 2014. Analysis of germination rate, mitotic index, and karyotype of *Chromolaena barranquillensis* (Hieron.) R.M. King and H. Rob. - Asteraceae. South Afr. J. Bot. 94: 149–154.

Rodríguez, J.D., A.L. Méndez, A.M. Molina, M.J. Bermejo and A. Muñoz-Acevedo. 2014. Metabolitos secundarios volátiles de flores, hojas y ramas de *Cordia curassavica* Jacq. del manantial "la sierra" (Sabanalarga, Atlántico). Rev. Prod. Nat. 4(1): 182.

Rojas, E.T. and L. Rodriguez-Hahn. 1978. Nivenolide, a diterpene lactone from *Croton niveus*. Phytochemistry 17(3): 574–575.

Rosales-Ovares, K.M. and J.F. Cicció-Alberti. 2002. The volatile oil of the fruits of *Bursera simaruba* (L.) Sarg. (Burseraceae) from Costa Rica. Ing. Cienc. Quim. 20: 60–61.

Roth, I. and H. Lindorf. 2002. South American Medicinal Plants. Botany, Remedial Properties and General Use. Springer-Verlag, Heidelberg.

Rüdiger, A.L., A.C. Siani and V.F. Veiga Junior. 2007. The chemistry and pharmacology of the South America genus *Protium* Burm. f. (Burseraceae). Phcog. Rev. 1(1): 93–104.

Ruíz-Baquero, J., O. Camacho-Romero, S. Bolívar-González and A. Castro-Zafra. 2021. Actividad anticoagulante *in vitro* del extracto etanólico de las hojas de dos especies de la familia Euphorbiaceae. Rev. U.D.C.A Act. Divulg. Cient. 24(2): e1681.

Saavedra Barrera, R.A.S. 2015. Análisis y estudio comparativo de los extractos obtenidos con fluido supercrítico y de los aceites esenciales de diferentes plantas del género *Piper*. Trabajo de pregrado, Profesional en Química. Universidad Industrial de Santander, Bucaramanga, Colombia.

Sahoo, A., B. Dash, S. Jena, A. Ray, P. Chandra Panda and S. Nayak. 2021. Phytochemical composition of flower essential oil of *Plumeria alba* grown in India. J. Essent. Oil Bear. Pl. 24(4): 671–676.

Saleh, M., A. Kamel, X. Li and J. Swaray. 1999. Antibacterial triterpenoids isolated from *Lantana camara*. Pharm. Biol. 37(1): 63–66.

Sánchez, O., L.P. Kvist and Z. Aguirre. 2006. Bosques secos en Ecuador y sus plantas útiles. *In*: Botánica Económica de los Andes Centrales. Universidad Mayor de San Andrés, La Paz, Bolivia.

Santos, R.P., E.P. Nunes, R.F. Nascimento, G.M.P. Santiago, G.H.A. Menezes, E.R. Silveira and O.D.L. Pessoa. 2006. Chemical composition and larvicidal activity of the essential oils of *Cordia leucomalloides* and *Cordia curassavica* from the Northeast of Brazil. J. Braz. Chem. Soc. 17(5): 1027–1030.

Sarma, N., T. Begum, S.K. Pandey, R. Gogoi, S. Munda and M. Lal. 2020. Chemical profiling of leaf essential oil of *Lantana camara* Linn. from North-East India. J. Essent. Oil Bear. Pl. 23(5): 1035–1041.

Sarmiento Bernal, D.C., L.P. Espitia Palencia and R. López Camacho. 2017. Caracterización de los productos forestales no maderables del bosque seco tropical asociado a las comunidades del Caribe colombiano. Braz. J. Biosci. 15(4): 187–198.

Satyal, P., R.A. Crouch, L. Monzote, P. Cos, N.A.A. Ali, M.A. Alhaj and W.N. Setzer. 2016. The chemical diversity of *Lantana camara*: Analyses of essential oil samples from Cuba, Nepal, and Yemen. Chem. Biodivers. 13(3): 336–342.

Schmidt Werka, J., A.K. Boehme and W.N. Setzer. 2007. Biological activities of essential oils from Monteverde, Costa Rica. Nat. Prod. Commun. 2(12): 1215–1219.

Scott, P. and R. Martin. 1984. Avian consumers of *Bursera*, *Ficus* and *Ehretia* fruit in Yucatan. Biotropica 16: 319–323.

Sefidkon, F. 2002. Essential oil of *Lantana camara* L. occurring in Iran. Flavour Frag. J. 17(1): 78–80.

Senatore, F. and D. Rigano. 2001. Essential oil of two *Lippia* spp. (Verbenaceae) growing wild in Guatemala. Flavour Frag. J. 16(3): 169–171.

Setzer, W.N. 2006. Chemical compositions of the bark essential oils of *Croton monteverdensis* and *Croton niveus* from Monteverde, Costa Rica. Nat. Prod. Commun. 1(7): 567–572.

Setzer, W.N. 2014. Leaf and bark essential oil compositions of *Bursera simaruba* from Monteverde, Costa Rica. American J. Essent. Oil Nat. Prod. 1: 34–36.

Sharma, M., A. Alexander, S. Sara, S. Saraf, U.K. Vishwakarma, K.T. Nakhate and Ajazuddin. 2021. Mosquito repellent and larvicidal perspectives of weeds *Lantana camara* L. and *Ocimum gratissimum* L. found in central India. Biocat. Agric. Biotechnol. 34: 102040.

Sharma, M., P.D. Sharma and M.P. Bansal. 2008. Lantadenes and their esters as potential antitumor agents. J. Nat. Prod. 71(7): 1222–1227.

Shinde, P.R., P.S. Patil and V.A. Bairagi. 2014. Phytopharmacological review of *Plumeria* species. Sch. Acad. J. Pharm. 3(2): 217–227.

Shukla, R., A. Kumar, P. Singh and N.K. Dubey. 2009. Efficacy of *Lippia alba* (Mill.) N.E. Brown essential oil and its monoterpene aldehyde constituents against fungi isolated from some edible legume seeds and aflatoxin B1 production. Int. J. Food Microbiol. 135(2): 165–170.

Shukla, R., P. Singh, B. Prakash, A. Kumar, P.K. Mishra and N.K. Dubey. 2011. Efficacy of essential oils of *Lippia alba* (Mill.) N.E. Brown and *Callistemon lanceolatus* (Sm.) Sweet and their major constituents on mortality, oviposition and feeding behaviour of *Pulse beetle, Callosobruchus chinensis* L. J. Sci. Food Agric. 91(12): 2277–2283.

Siani, A.C., M.F.S. Ramos, A.C. Guimarães, G.S. Susunaga and M.G.B. Zoghbi. 1999. Volatile constituents from oleoresin of *Protium heptaphyllum* (Aubl.) March. J. Essent. Oil Res. 11(1): 72–74.

Siani, A.C., M.F.S. Ramos, O.M. Lima Junior, R.R. Santos, E.F. Ferreira, R.O.A. Soares, E.C. Rosas, G.S. Susunaga, G.S. Guimarães, M.G.B. Zoghbi and M.G.M.O. Henriques. 1999. Evaluation of anti-inflammatory-related activities of essential oils from the leaves and resin of species of *Protium*. J. Ethnopharmacol. 66(1): 57–69.

Siani, A.C., M.F.S. Ramos, S.S. Monteiro, R. Ribeiro-dos-Santos and R.O.A. Soares. 2011. Essential oils of the oleoresins from *Protium heptaphyllum* growing in the Brazilian southeastern and their cytotoxicity to neoplastic cell lines. J. Essent. Oil Bear. Pl. 14(3): 373–378.

Sibi, G., A. Venkategowda and L. Gowda. 2014. Isolation and characterization of antimicrobial alkaloids from *Plumeria alba* flowers against food borne pathogens. Am. J. Life Sci. 2(6-1): 1–6.

Silva, J.R.D.A., M.G.B. Zoghbi, A.C. Pinto, R.L.O. Godoy and A.C.F. Amaral. 2009. Analysis of the hexane extracts from seven oleoresins of *Protium* species. J. Essent. Oil Res. 21(4): 305–308.

Sirisha, K., Y. Rajendra, P. Gomathi, K. Soujanya and N. Yasmeen. 2013. Antioxidant and anti-inflammatory activities of flowers of *Plumeria rubra* L. f. *rubra* and *Plumeria rubra* f. *lutea*: A comparative study. Res. J. Pharm. Biol. Chem. Sci. 4(4): 743–756.

Sotelo-Méndez, A.H., C.G. Figueroa Cornejo, M.F. Césare Coral and M.C. Alegría Arnedo. 2017. Chemical composition, antimicrobial and antioxidant activities of the essential oil of *Bursera graveolens* (burseraceae) from Perú. Indian J. Pharm. Ed. Res. 51(3): S429–S436.

Soukup, J. 1970. Vocabulario de los nombres vulgares de la flora peruana. Imprenta Salesiana, Lima, Perú.

Sousa, E.O., C.M.B.A. Miranda, C.B. Nobre, A.A. Boligon, M.L. Athayde and J.G.M. Costa. 2015. Phytochemical analysis and antioxidant activities of *Lantana camara* and *Lantana montevidensis* extracts. Ind. Crop. Prod. 70: 7–15.

Sousa, E.O., N.F. Silva, F.F.G. Rodrigues, A.R. Campos, S.G. Lima and J.G.M. Costa. 2010. Chemical composition and resistance-modifying effect of the essential oil of *Lantana camara* Linn. Pharmacogn. Mag. 6(22): 79–82.

Sousa, E.O., T.S. Almeida, I.R.A. Menezes, F.F.G. Rodrigues, A.R. Campos, S.G. Lima and J.G.M. da Costa. 2012. Chemical composition of essential oil of *Lantana camara* L. (Verbenaceae) and synergistic effect of the aminoglycosides gentamicin and amikacin. Rec. Nat. Prod. 6(2): 144–150.

Souto, R.N.P., A.Y. Harada, E.H.A. Andrade and J.G.S. Maia. 2012. Insecticidal activity of Piper essential oils from the Amazon against the fire ant *Solenopsis saevissima* (Smith) (Hymenoptera: Formicidae). Neotrop. Entomol. 41: 510–517.

Srivastava, G., A. Gupta, M.P. Singh and A. Mishra. 2017. Pharmacognostic standardization and chromatographic fingerprint analysis on triterpenoids constituents of the medicinally important plant *Plumeria rubra* f. *rubra* by HPTLC technique. Pharmacogn. J. 9(2): 135–141.

Stashenko, E.E., B.E. Jaramillo and J.R. Martínez. 2004. Comparison of different extraction methods for the analysis of volatile secondary metabolites of *Lippia alba* (Mill.) N.E. Brown, grown in Colombia, and evaluation of its *in vitro* antioxidant activity. J. Chromatogr. A 1025(1): 93–103.

Stashenko, E.E., J.R. Martínez, D.C. Durán, Y. Córdoba and D. Caballero. 2014. Estudio comparativo de la composición química y la actividad antioxidante de los aceites esenciales de algunas plantas del género *Lippia* (Verbenaceae) cultivadas en Colombia. Rev. Acad. Colombia. Cienc. Exact. 38: 89–105.

Suárez, A.I., L.J. Vásquez, M.A. Manzano and R.S. Compagnone. 2005. Essential oil composition of *Croton cuneatus* and *Croton malambo* growing in Venezuela. Flavour Frag. J. 20(6): 611–614.

Suárez, A.I., L.J. Vásquez, A. Taddei, F. Arvelo and R.S. Compagnone. 2008. Antibacterial and cytotoxic activity of leaf essential oil of *Croton malambo*. J. Essent. Oil Bear. Pl. 11(2): 208–213.

Suárez, A.I., R.S. Compagnone, M.M. Salazar-Bookaman, S. Tillet, F. Delle Monache, C. Di Giulio and G. Bruges. 2003. Antinociceptive and anti-inflammatory effects of *Croton malambo* bark aqueous extract. J. Ethnopharmacol. 88(1): 11–14.

Sundufu, A.J. and H. Shoushan. 2004. Chemical composition of the essential oils of *Lantana camara* L. occurring in south China. Flavour Frag. J. 19(3): 229–232.

Sura, J., S. Dwivedi and R. Dubey. 2018. Pharmacological, phytochemical, and traditional uses of *Plumeria alba* Linn. an Indian medicinal plant. J. Pharm. BioSci. 6(1): 1–4.

Susunaga, G.S., A.C. Siani, M.G. Pizzolatti, R.A. Yunes and F.D. Monache. 2001. Triterpenes from resin of *Protium heptaphyllum*. Fitoterapia 72(6): 709–711.

Swart, J.J. 1942. Monograph of the genus *Protium* and some allied genera (Burseraceae). Gouda (Ed). Drukkerij Koch in Knuttel/Smithsonian Libraries.

Syakira, M.H. and L. Brenda. 2010. Antibacterial capacity of *Plumeria alba* petals. World Aca. Sci. Eng. Tech. 4(8): 1202–1205.

Sylvestre, M., A.P.A. Longtin and J. Legault. 2007. Volatile leaf constituents and anticancer activity of *Bursera simaruba* (L.) Sarg. essential oil. Nat. Prod. Commun. 2(12): 1273–1276.

Tangarife-Castaño, V., J.B. Correa-Royero, V.C. Roa-Linares, N. Pino-Benitez, L.A. Betancur-Galvis, D.C. Durán, E.E. Stashenko and A.C. Mesa-Arango. 2014. Anti-*Dermatophyte*, anti-*Fusarium* and cytotoxic activity of essential oils and plant extracts of *Piper* genus. J. Essent. Oil Res. 26(3): 221–227.

Tafurt-García, G. and A. Muñoz-Acevedo. 2012. Metabolitos volátiles presentes en *Protium heptaphyllum* (Aubl.) March. colectado en Tame (Arauca - Colombia). Bol. Latinoam. Caribe Plant. Med. Aromat. 11(3): 223–232.

Tavares, E.S., L.S. Julião, D. Lopes, H.R. Bizzo, C.L.S. Lage and S.G. Leitão. 2005. Analysis of the essential oil from leaves of three *Lippia alba* (Mill.) N.E. Br. (Verbenaceae) chemotypes cultivated on the same conditions. Braz. J. Pharmacogn. 15(1): 1–5.

Tene, V., O. Malagón, P. Vita Finzi, G. Vidari, Ch. Armijos and T. Zaragoza. 2007. An ethnobotanical survey of medicinal plants used in Loja and Zamora-Chinchipe, Ecuador. J. Ethnopharmacol. 111(1): 63–81.

The Plant List. 2013. Version 1.1. Published on the Internet; http://www.theplantlist.org/ (accessed 10 november 2021).

Tohar, N., K. Awang, M.A. Mohd and I. Jantan. 2006a. Chemical composition of the essential oils of four *Plumeria* species grown on peninsular Malaysia. J. Essent. Oil Res. 18(6): 613–617.

Tohar, N., M.A. Mohd, I. Jantan and K. Awang. 2006b. A comparative study of the essential oils of the genus *Plumeria* Linn. from Malaysia. Flavour Frag. J. 21(6): 856–863.

Ustáriz Fajardo, F.J., M.E. Lucena de Ustáriz, F.G. Urbina Carmona, D.M. Villamizar Sánchez, L.B. Rojas Fermín, Y.E. Cordero de Rojas, J.E. Ustáriz Lucena, L.C. González Ramírez and L.M. Araujo Baptista. 2020. Composition and antibacterial activity of the *Piper eriopodon* (Miq.) C.DC. Essential oil from the Venezuelan Andes. PharmacologyOnLine 2: 13–22.

Uter, W., K. Yazar, E.-M. Kratz, G. Mildau and C. Lidén. 2013. Coupled exposure to ingredients of cosmetic products: I. Fragrances. Contact Derm. 69(6): 335–341.

Valenzuela-Vergara, E.E., G. Tafurt-García and E. Stashenko. 2014. Análisis de la composición química de aceites esenciales de especies de *Piper* (Fam. Piperaceae). Rev. Prod. Nat. 4(1): 159.

Vásquez-Londoño, C.A. 2012. Clasificación por categorías térmicas de las plantas medicinales, en el sistema tradicional de salud de la comunidad afrodescendiente de Palenque San Basilio, Bolivar, Colombia. Tesis de Maestría en Medicina Alternativa, Medicina. Universidad Nacional de Colombia, Bogotá, Colombia.

Velandia, S.A., E. Quintero, E.E. Stashenko and R.E. Ocazionez. 2018. Actividad antiproliferativa de aceites esenciales de plantas cultivadas en Colombia. Acta Biol. Colomb. 23(2): 189–198.

Viana, G.S.B., T.G. do Vale, V.S.N. Rao and F.J.A. Matos. 1998. Analgesic and antiinflammatory effects of two chemotypes of *Lippia alba*: A comparative study. Pharm. Biol. 36(5): 347–351.

Viana, G.S.B., T.G. do Vale, C.C.M. Silva and F.J.A. Matos. 2000. Anticolvulsant activity of essential oils and active principles from chemotypes of *Lippia alba* (Mill.) N.E. Brown. Biol. Pharm. Bull. 23(5): 1314–1317.

Vieira-Junior, G.M., C.M.L. Souza and M.H. Chaves. 2005. The *Protium heptaphyllum* resin: Isolation, structural characterization and evaluation of thermal properties. Quim. Nova 28(2): 183–187.

Vogler, B., J.A. Noletto, W.A. Haber and W.N. Setzer. 2006. Chemical constituents of the essential oils of three *Piper* species from Monteverde, Costa Rica. J. Essent. Oil Bear. Pl. 9(3): 230–238.

Volpato, G., D. Godínez, A. Beyra and A. Barreto. 2009. Uses of medicinal plants by Haitian immigrants and their descendants in the Province of Camagüey, Cuba. J. Ethnobiol. Ethnomed. 5: 16.

Wangrawa, D.W., A. Badolo, W.M. Guelbéogo, R.C.H. Nébié, N'F. Sagnon, D. Borovsky and A. Sanon. 2018. Insecticidal activity of local plants essential oils against laboratory and field strains of *Anopheles gambiae* s. l. (Diptera: Culicidae) from Burkina Faso. J. Econ. Entomol. 111(6): 2844–2853.

Wangrawa, D.W., A. Badolo, Z. Ilboudo, W.M. Guelbéogo, M. Kiendrébeogo, R.C.H. Nébié, N.'F. Sagnon and A. Sanon. 2018. Insecticidal activity of local plants essential oils against laboratory and field strains of *Anopheles gambiae* s. l. (Diptera: Culicidae) from Burkina Faso. J. Econ. Entomol. 111(6): 2844–2853.

Wedler, E. 2013. Atlas de las plantas medicinales silvestres y cultivadas en la zona tropical. Especial Impresores SAS, Colombia.

Wu, P., Z. Song, X. Wang, Y. Li, Y. Li, J. Cui, M. Tuerhong, D.-Q. Jin, M. Abudukeremu, D. Lee, J. Xu and Y. Guo. 2020. Bioactive triterpenoids from *Lantana camara* showing anti-inflammatory activities *in vitro* and *in vivo*. Bioorg. Chem. 101: 104004.

Xia, Y.-Y., C.-Z. Lin, X.-J. Lu, F.-L. Liu, A.-Z. Wu, L. Zhang and C.-C. Zhu. 2018. New iridoids from the flowers of *Plumeria rubra* "*acutifolia*". Phytochem. Lett. 25: 81–85.

Yasunaka, K., F. Abe, A. Nagayama, H. Okabe, L. Lozada-Pérez, E. López-Villafranco, E. Estrada-Muñiz, A. Aguilar and R. Reyes-Chilpa. 2005. Antibacterial activity of crude extracts from Mexican medicinal plants and purified coumarins and xanthones. J. Ethnopharmacol. 97(2): 293–299.

Young, D.G., S. Chao, H. Casabianca, M.-C. Bertrand and D. Minga. 2007. Essential oil of *Bursera graveolens* (Kunth) Triana et Planch from Ecuador. J. Essent. Oil Res. 19(6): 525–526.

Yukawa, C. and H. Iwabuchi. 2003. Terpenoids of the volatile oil of *Bursera graveolens*. J. Oleo Sci. 52(9): 483–489.

Yukawa, C., H. Iwabuchi, T. Kamikawa, S. Komemushi and A. Sawabe. 2004a. Terpenoids of the volatile oil of *Bursera graveolens*. Flavour Frag. J. 19(6): 565–570.

Yukawa, C., H. Iwabuchi, S. Komemushi and A. Sawabe. 2004b. Eudesmane-type sesquiterpenoids in the volatile oil from *Bursera graveolens*. J. Oleo Sci. 53(7): 343–348.

Yukawa, C., H. Iwabuchi, S. Komemushi and A. Sawabe. 2005. Mono and sesquiterpenoids of the volatile oil of *Bursera graveolens*. Flavour Frag. J. 20(6): 635–658.

Zachrisson, B., A. Santana and M. Gupta. 2019. Effects of essential oils from two species of Piperaceae on parasitized and unparasitized eggs of *Oebalus insularis* (Heteroptera: Pentatomidae) by *Telenomus podisi* (Hymenoptera: Platygastridae). Nat. Prod. Commun. 14(1): 83–84.

Zaheer, Z., A.G. Konale, K.A. Patel, S. Khan and R.Z. Ahmed. 2010. Comparative phytochemical screening of flowers of *Plumeria alba* and *Plumeria rubra*. Asian J. Pharm. Clinical Res. 3(4): 167.

Zamora-Villalobos, N., Q. Jiménez-Madrigal and L.J. Poveda. 2000. Árboles de Costa Rica Vol II. Centro Científico Tropical, Conservación Internacional and Instituto Nacional de Biodiversidad. Ed. INBio, San José.

Zandi-Sohani, Nooshin, M. Hojjati and Á.A. Carbonell-Barrachina. 2012. Bioactivity of *Lantana camara* L. essential oil against *Callosobruchus maculatus* (Fabricius). Chilean J. Agric. Res. 72(4): 502–506.

Zétola, M., T.C.M. de Lima, D. Sonaglio, G. González-Ortega, R.P. Limberger, P.R. Petrovick and V.L. Bassani. 2002. CNS activities of liquid and spray-dried extracts from *Lippia alba*-Verbenaceae (Brazilian false melissa). J. Ethnopharmacol. 82(2-3): 207–215.

Zhao, M., Z. Liang, Z. Xie, D. Yang and X. Xu. 2015. Separation and purification of 15-demethylplumieride, cerberic acid B, and kaempferol-3-rutinoside from *Plumeria rubra* '*acutifolia*' by high-speed counter-current chromatography. Sep. Sci. Technol. 50(15): 2360–2366.

Zoghbi, M.G.B., J.G.S. Maia and A.I.R. Luz. 1995. Volatile constituents from leaves and stems of *Protium heptaphyllum* (Aubl.) March. J. Essent. Oil Res. 7(5): 541–543.

Zoghbi, M.G.B., E.H.A. Andrade, A.S. Santos, M.H.L. Silva and J.G.S. Maia. 1998. Essential oils of *Lippia alba* (Mill.) N.E. Br. growing wild in the Brazilian Amazon. Flavour Frag. J. 13(1): 47–48.

Zoubiri, S. and A. Baaliouamer. 2012. Chemical composition and insecticidal properties of *Lantana camara* L. leaf essential oils from Algeria. J. Essent. Oil Res. 24(4): 377–383.

Zúñiga, B., P. Guevara-Fefer, J. Herrera, J.L. Contreras, L. Velasco, F.J. Pérez and B. Esquivel. 2005. Chemical composition and anti-inflammatory activity of the volatile fractions from the bark of eight Mexican *Bursera* species. Planta Med. 71(9): 825–828.

4

Essential Oils' Chemical Characterization and Investigation of Some Biological Activities

A Demanding Review

Prerna Sharma,[1,*] *Kumar Guarve,*[1] *Sumeet Gupta,*[2]
Garima Malik[2] *and Kashish Wilson*[2]

Introduction

One of the most important products of agriculture is essential oils of plant origin. They are widely employed as flavoring ingredients in food, beverages, perfumes, medicines, and cosmetics (Guenther 1948). Around 3,000 essential oils have been utilizing at least 2,000 plant species, 300 of which are commercially relevant. The production of 40,000–60,000 tons per year with an estimated market worth of 700 million US dollars indicates that the production and consumption of essential oils are expanding all over the world (Association Française de Normalisation (AFNOR) 2000). Many factors influence the chemistry of EOs, including genetic diversity, plant ecotype or variety, plant nutrition, fertilizer application, geographic location of the plants, surrounding climate, seasonal fluctuations, stress during growth or maturity, and post-harvest drying and storage. Furthermore, the type of plant material utilized and the method of extraction define the quantity and composition (constituents) of an EO, hence it has distinctive biological qualities (Carette Delacour 2000, Sell 2006, Vainstein et al. 2001). For example, EO derived from various plant components like flowers, leaves, stems, roots, fruits, and fruit peels demonstrate a variety of biological and therapeutic effects. Similarly, various polarities of solvents

[1] Guru Gobind Singh College of Pharmacy, Yamunanagar (135002), India.
[2] MM College of Pharmacy, Maharishi Markandeshwar (Deemed to be University), Mullana-Ambala (133207), India.
* Corresponding author: presharma31@yahoo.com

extract distinct groups of chemicals. It is difficult to distinguish and study the effects of these components since they influence one another (Pophof et al. 2005, Anwar et al. 2009, Cocking and Middleton 1935).

Essential oils are complex combinations of low molecular weight molecules (typically fewer than 500 daltons) extracted using steam distillation, hydrodistillation, or solvent extraction. They are typically kept in the plant's oil ducts, resin ducts, glands, or trichomes (glandular hairs) (Nickerson and Likens 1996). On a commercial basis, steam distillation is the primary method for essential oil extraction. EOs can be composed of 20–100 different plant secondary metabolites from various chemical classes. The main ingredients of essential oils are terpenoids and phenylpropanoids (Shelef 1983). There are also a few aromatic and aliphatic components. The most abundant chemical entities in Eos are monoterpenes, sesquiterpenes, and oxygenated derivatives (Nychas 1995, Lambert et al. 2001, Sikkema et al. 1994). EO is determined by one or two of its primary components. However, there are occasions when overall activity cannot be assigned to any of the major constituents, and the presence of a combination of molecules modifies the activity sufficiently to have a substantial influence (Gustafson et al. 1998, Cox et al. 2000). For example, it has been found that the inhibitory activity of rosemary oil against insect larvae (lepidopteron larvae) is due to the synergistic effects of many chemical ingredients, even though no single compound exhibits the activity (Carson and Riley 1995).

Natural products and their derivatives are rich in new medicinal compounds. Plant essential oils have a wide range of applications, mostly in the health, agriculture, cosmetic, and food industries. The use of essential oils in traditional medical systems has been used since prehistoric times in human history (Ultee et al. 2002, Denyer and Hugo 1991). Researchers from all around the world are working to describe Eos' antibacterial, antiviral, antimutagenic, anticancer, antioxidant, anti-inflammatory, immunomodulatory, and antiprotozoal characteristics. The concentrations necessary to suppress the growth of target organisms are used to compare the efficacy of various Eos (Farag et al. 1989). For bioactivity comparison, minimum growth inhibitory concentrations (MICs), minimum lethal concentrations (MBCs or MFCs), MIC50, and LD50 values are commonly utilized (Cosentino et al. 2002). These values are obtained by the use of defined procedures. For antimicrobial susceptibility testing, for example, Clinical Laboratory Standards Institute (CLSI) methods and cell viability assessment by MTT or XTT assays are used (Dorman and Deans 2000).

The emergence of drug-resistant pathogen strains, an increase in the immunocompromised population, and the limitations of current antibiotics/drugs have prompted people to seek out complementary and alternative medicines, including the use of essential oils (Davidson 1997). Secondary metabolites, which are spontaneously generated by plants in response to insect pests and herbivore attacks, form a complex blend of EOs. These small molecule metabolites, both alone and in combination (such as EOs), have important therapeutic effects and may thus be employed for chemotherapy of infectious and non-infectious disorders (Knobloch et al. 1986).

Taxonomy of Essential Oil Producing Plants

Plants that produce EOs come from a variety of taxa that are divided into around 60 families. Alliaceae, Apiaceae, Asteraceae, Lamiaceae, Myrtaceae, Poaceae, and Rutaceae are widely known for their potential to create medicinal and industrial EOs. Terpenoids are abundant in all of the plant families that produce EO (Pauli 2001). Plant groups that contain phenylpropanoids more frequently include Apiaceae (Umbelliferae), Lamiaceae, Myrtaceae, Piperaceae, and Rutaceae. Plants from these families are utilized in the commercial synthesis of EO. Coriander, anise, dill, and fennel oils, for example, are derived from the herbs such as *Coriandrum sativum*, *Pimpinella anisum*, *Anethum graveolens*, and *Foeniculum vulgare*, respectively (Fabian et al 2006). All of these plants are members of the Apiaceae family and are widely known for their antibacterial, antifungal, anticancer, and antiviral properties. The Lamiaceae family also contains numerous genera renowned for their chemotherapeutic, antiviral, antibacterial, antimutagenic, antioxidant, and anti-inflammatory capabilities (Marino et al. 1999). These are also beneficial in the treatment of intestinal problems and bronchitis. Some popular Lamiaceae EO generating plants are *Mentha piperita*, *Rosmarinus officinalis*, *Ocimum basilicum*, *Salvia officinalis*, *Origanum vulgare*, *Melissa officinalis*, *Satureja hortensis*, *Thymus vulgaris*, and *Lavandula angustifolia*. Cinnamon oil, which is high in eugenol and derived from Cinnamomum verum, is a good example from the Lauraceae family. It has antibacterial and anticancer properties (Senatore et al. 2000). The Myrtaceae family contains many commercially important plants. *Melaleuca altrnifolia*, *Eucalyptus globulus*, *Syzygium aromaticum* (*Eugenia caryophyllus*), and *Myrtus communis*, for example, create EOs having antibacterial, antifungal, antitumor, anticancer, and antiviral effects (Canillac and Mourey 2001). Poaceae is a grass family that produces lemongrass oil (from Cymbopogon citrates), citronella oil (from *C. nardus*), and palmarosa (*C. martini*) oils. These Eos' medicinally active components, such as citral, geraniol, and geranyl acetate, have antibacterial and anticancer effects. Citrus oils, which contain limonene and linalool, are extracted from the fruit peel of plants of the Rutaceae family (Cimanga et al. 2002). These ingredients have antibacterial properties. Pelargonium graveolens and Santalum spp. of the Geraniaceae and Santalaceae families, respectively, provide two significant oils, geranium and sandalwood oil. Several other families, including Cupressaceae, Hypericaceae (Clusiaceae), Fabaceae (also known as Leguminosae), Liliaceae, Pinaceae, and Zygophyllaceae, may yield EOs with substantial biological activities that should be investigated (Delaquis et al. 2002).

Chemistry of Essential Oils

Essential oils are created by a variety of structures, the quantity, and properties of which are quite diverse. Essential oils are found in the cytoplasm of specific plant cell secretions, which can be found in one or more plant organs, such as secretory hairs or trichomes, epidermal cells, interior secretory cells, and secretory pockets. These oils are complicated combinations containing up to 300 distinct components. They are made up of organic volatile chemicals with molecular weights below 300. Their vapor pressure is sufficiently high at atmospheric pressure and room temperature for them to be found partially in the vapor state (Ratledge and Wilkinson 1988). Alcohols, ethers or oxides, aldehydes, ketones, esters, amines, amides, phenols, heterocycles, and, most notably, terpenes are among the volatile chemicals. Fruity ((E)-nerolidol), floral (Linalool), citrus (Limonene), herbal (-selinene), and other aromatic notes can be found in alcohols, aldehydes, and ketones. Furthermore, the vast majority of essential oil components are members of the terpene family. Many thousands of terpenes have been identified in essential oils, including functionalized alcohols (geraniol and -bisabolol), ketones (menthone and p-vetivone), aldehydes (citronellal and sinensal), esters (-tepinyl acetate and cedryl acetate), and phenols (thymol). Non-terpenic chemicals produced by the phenylpropanoids route, such as eugenol, cinnamaldehyde, and safrole are also found in essential oils shown in Figure 1 (Davidson and Parish 1989).

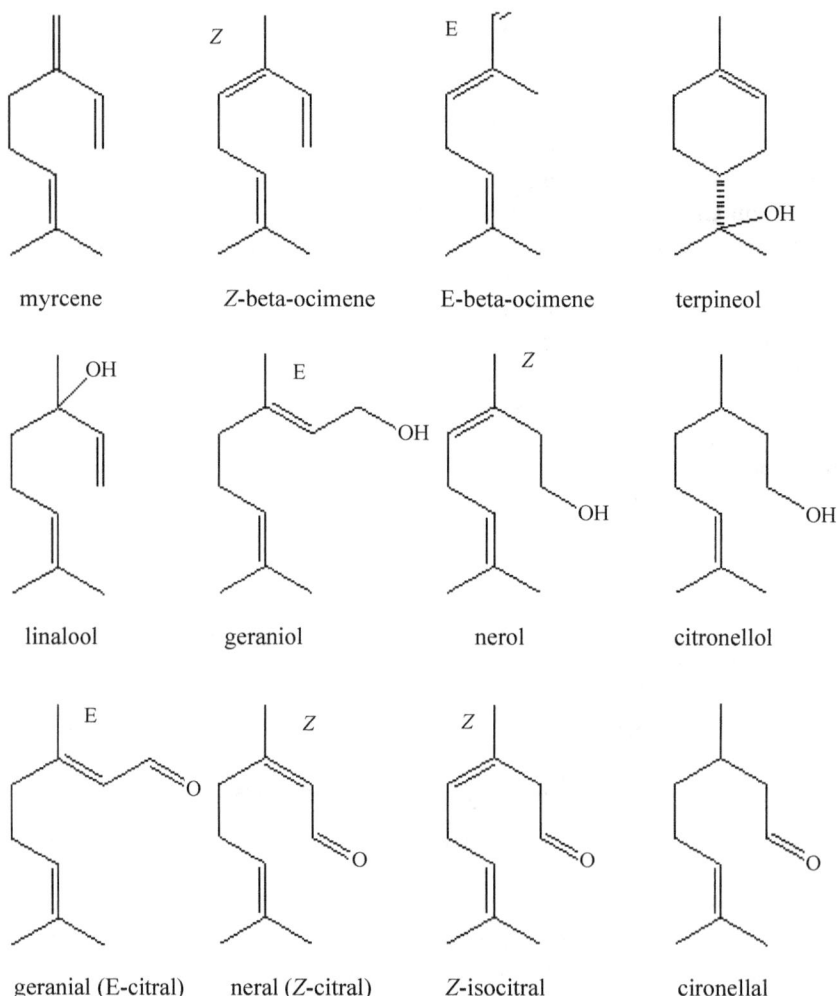

Figure 1: Structures of some terpenes.

Terpenoids and phenylpropanoids are biogenetically distinct with different fundamental metabolic precursors and biosynthesis pathways. The mevalonate and mevalonate-independent (deoxyxylulose phosphate) routes are involved in terpenoids, whereas the shikimate pathway is involved in phenylpropanoids (Gill et al. 2002). The biosynthesis routes of terpenoids and phenylpropanoids as well as the enzymes and enzyme processes involved have been reviewed by several publications, as well as the information on genes encoding for these enzymes. Essential oils have a lot of variation in terms of composition, both qualitatively and quantitatively. This variability is caused by a number of causes that can be divided into two categories:

• Even at harvest time during the day, intrinsic characteristics connected to the plant as well as interaction with the environment (such as soil type and climate) and the age of the plant in question are the elements that need to be considered.

• Extrinsic factors related to the extraction method and the environment.

There are various elements that influence essential oil yield and composition. It can be difficult to separate these aspects from one another in some circumstances because they are interconnected and influence one another. Seasonal fluctuations, plant organs, degree of maturity, geographic origin, and genetics are among these characteristics. Volatiles from aromatic plants is trapped using a variety of approaches. The circulatory distillation apparatus described by Cocking and Middleton and included in the European Pharmacopoeia and several other pharmacopeias is the most commonly used device (Reichling et al. 2009). This equipment comprises a heated round-bottom flask containing chopped plant material and water, which is coupled to a vertical condenser and a graduated tube for volumetric oil determination. The essential oil is extracted from the aqueous phase at the end of the distillation process for further research. The length of the distillation process is determined by the plant material being studied. It is commonly set at 3–4 hours. Using a closed circuit distillation system, the device allows for continuous concentration of volatiles during hydrodistillation in one step shown in Figure 2 (Burt 2004).

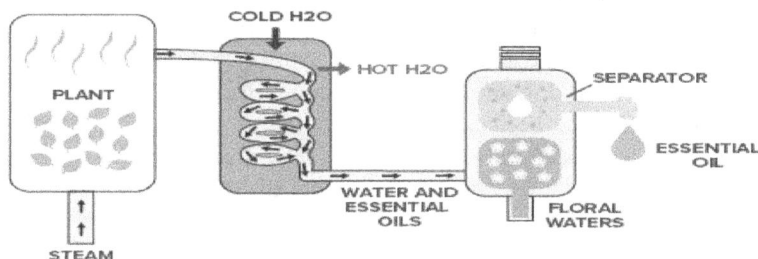

Figure 2: Hydrodistillation process.

Biological Activities of Essential Oils

Antibacterial Activity

The antibacterial effects of essential oils and their constituents have been investigated in depth as well as the mechanism of action. Essential oils have a high hydrophobicity, which allows them to partition into the lipids of bacteria's cell membrane, altering the structure and making it more permeable. This can lead to ion and other cellular molecule leaks. Although a limited amount of bacterial cell leakage can be tolerated without compromising viability, a greater loss of cell contents or essential molecule and ion production can result in cell death (Edris 2007). The activity of EOs and/ or their constituents can have single or numerous targets. Trans-cinnamaldehyde, for example, can stop *E. coli* and *Salmonella typhimirium* from growing without destroying the OM or depleting intracellular ATP. Trans-cinnamaldehyde, like thymol and carvacrol, is thought to gain access to the periplasm and deeper parts of the cell. Carvone does not affect the cellular ATP pool and is ineffective against the OM. Antibacterial activity was shown to be strongest in EOs containing mostly aldehydes or phenols, such as cinnamaldehyde, citral, carvacrol, eugenol, or thymol followed by EOs containing terpene alcohols (Braga et al. 2006).

Other EOs with ketones or esters, such as myrcene, thujone, or geranyl acetate, had substantially lower activity, while volatile oils with terpene hydrocarbons were almost always inert. Essential oils with high levels of phenolic compounds, such as carvacrol, eugenol, and thymol, have antibacterial properties in general. The disruption of the cytoplasmic membrane, the driving force of protons, electron movement, active transport, and cell coagulation are all caused by these chemicals. The mode of action of essential oils in terms of antibacterial activity is influenced by their chemical structure (Aruoma 1998). The significance of the hydroxyl group in phenolic compounds, like carvacrol and thymol, has been confirmed. The intensity of the antibacterial activity appears to be unaffected by the relative position of the phenolic hydroxyl group on the ring. Thymol appears to have a similar effect on *Bacillus cereus*, Staphylococcus aureus, and Pseudomonas aeruginosa as carvacrol. Carvacrol and thymol, on the other hand, have differing effects on Gram-positive and Gram-negative bacteria. Thymol, eugenol, and carvacrol show antibacterial properties against *Escherichia coli, Bacillus cereus, Listeria monocytogenes,* Salmonella enterica, Clostridium jejuni, Lactobacillus sake, Staphylococcus aureus, and Helicobacter pyroli bacteria. Certain alcohols, aldehydes, and ketones, monoterpenes (geraniol, linalol, menthol, terpineol, thujanol, myrcenol, citronelîaî, neral, thujone, camphor, carvone, and others), phenylpropanes (cinnamaldehyde), and monoterpenes (γ-terpinene) (Kamatou and Viljoen 2010). Carvacrol is the most active of these chemicals. It is used as a preservative and culinary flavoring in drinks, desserts, and other dishes because it is non-toxic. It is worth noting that essential oils are more effective against Gram-positive bacteria than they are against Gram-negative bacteria. The outer membrane enclosing the cell wall, which inhibits the diffusion of hydrophobic chemicals through its lipopolysaccharide film, makes them less vulnerable to the action of essential oils (Maruyama et al. 2005).

Furthermore, essential oils' antibacterial activity was linked to their chemical composition, volatile molecule quantities, and interactions. When the sum of the individual impacts equals the sum of the combined effects, an additive effect is observed. When one or both substances are examined together, their effect is less important than when they are evaluated separately, this is known as antagonism. When the combined effects of two or more substances are higher than the sum of their separate effects, a synergistic effect is noticed (Koh et al. 2002). According to several studies, using the entire essential oil produces a stronger benefit than using the primary components separately. This implies that minor components are required for activity and may work in tandem. Using checkerboard and time-kill assays, the combinations of 1,8-cineole and aromadendrene were found to have additive and synergistic effects against methicillin-resistant *Staphylococcus aureus* (MRSA), vancomycin-resistant enterococci (VRE), and *Enterococcus faecalis*, respectively. The combination of inhibitory effects of plant volatile oils and benzoic acid derivatives against *L. monocytogenes* and *S. enteritidis* is deemed synergistic because the combined components enabled a $\log^{10}$ stronger inhibition than the total of the inhibitory effects of the components employed individually (Caldefie-Chézet et al. 2004).

Combinations (1:5, 1:7, and 1:9) of essential oils of *S. aromaticum* (clove) and *Rosmarinus officinalis* have increased antifungal activity against *C. albicans*. Furthermore, Lambert et al. found that utilizing half-fold dilutions within the Bioscreen mat, carvacrol and thymol had additive effects against *S. aureus* and *P. aeruginosa* (Caldefie-Chézet et al. 2006). To explain the synergistic effects of cinnamaldehyde/thymol or cinnamaldehyde/carvacrol against *S. typhimurium*, two hypotheses have been proposed. On one hand, it is to prove that thymol or carvacrol can increase the permeability of the cytoplasmic membrane, allowing cinnamaldehyde to be more easily transported into the cell. On the other hand, these facts support the existence of a synergistic impact when these two components are combined. The mechanisms of interaction that result in antagonistic consequences have received little attention. Furthermore, essential oils are beneficial in inhibiting growth and reducing the populations of more dangerous foodborne pathogens, like *Salmonella* spp., *E. coli* O157:H7, and *Listeria monocytogenes* (Hart et al. 2000).

Antioxidant Activity

Essential oils have been shown to have antioxidant effects in numerous research. An essential oil's antioxidant capacity is determined by its composition. It is generally known that phenolics and secondary metabolites containing conjugated double bonds have strong antioxidant capabilities. Alcohols (Achillea filipendulina), aldehydes (Galagania fragrantissima), ketones (Anethum graveolens, Artemisia rutifolia, Hyssopus seravschanicus, Mentha longifolia, and Ziziphora clinopodioides), and esters (Anethum graveolens, Artemisia rutifolia, Hyssopus sera, and Salvia sclarea). Monoterpene hydrocarbons characterize Artemisia absinthium and Artemisia scoparia EOs, but phenolic terpenoids, such as thymol or carvacrol, characterize Origanum tyttanthum and Mentha longifolia EOs, which could explain why both plants had the highest antioxidant activity (Sharififar et al. 2012).

Thymol and carvacrol, which are abundant in Origanum tyttanthum, are also responsible for the antioxidant action of Mentha longifolia and Thymus serpyllus essential oils. The most important antioxidant qualities are found in the essential oils of cinnamon, nutmeg, clove, basil, parsley, oregano, and thyme (Manjamalai et al. 2012). The most active chemicals are thymol and carvacrol. Their phenolic structure is linked to their action. These phenolic compounds have redox characteristics and so play a key role in the neutralization of free radicals as well as the decomposition of peroxide. Certain alcohols, ethers, ketones, aldehydes, and monoterpenes—such as linalool, 1,8-cineole, geranial/neral, citronellal, isomenthone, menthone, and other monoterpenes like Terpinene, Terpinene, and Terpinolene—are essential oils with high free radical scavenging potential that may be useful in the prevention of diseases like brain dysfunction, cancer, heart disease, and immune system decline. In reality, free radical damage to cells may be the cause of several disorders. In aging polyunsaturated fatty acids mammals, EOs have been demonstrated to (Yoon et al. 2000) operate as hepatoprotective agents, and it has been proven that they have a positive effect on PUFAs, particularly long-chain C20 and C22 acids. Furthermore, the ability of essential oils to scavenge free radicals may have a role in the prevention of diseases, such as brain dysfunction, cancer, heart disease, and immune system decline. The antioxidant activity of Zataria multiflora Boiss was studied by Sharififar et al. (Lamiaceae) essential oil in rats. Antioxidant activity was assessed using the 1,1-diphenyl-2-picrylhydrazyl (DPPH) radical inhibition assay and the index of thiobarbituric acid reactive compounds to assess lipid peroxidation inhibition (TBARs) (Pyun and Shin 2006). For ten days, animals were given three doses of 100, 200, and 400 L/kg by intragastric intubation (i.g) route. The blood was drawn on the eleventh day via direct puncture, and the liver was quickly removed. The animals' histopathology investigations were compared to (Milner 2001) those of the butylated hydroxyl toluene (BHT) group. According to the authors, all doses of Zataria multiflora oils ZMO examined were able to scavenge DPPH radicals (p 0.05).

TBARs were also reduced by ZMO in a dose-dependent manner. In the ZMO-treated groups, no changes in liver function test LFT enzymes or alterations in liver histology were observed. The findings suggested that ZMO might be used in human health and food production. According to Manjamalai and Grace, the essential oil of Wedelia chinensis (Osbeck) raises catalase and glutathione peroxidase levels in the lung and liver tissues; however, catalase levels in the serum dropped on the 22nd day (2.32 0.016 Lung tissue, 6.47 0.060 liver tissue and 0.94 0.007 serum). Furthermore, the level of Glutathione Peroxidase GPx in the liver (the range) was found to be lower in the EO-treated group than in the cancer-induced and control groups, although the level of GPx in the lung tissue was found to be low (Milner 2006).

Anti-Inflammatory Activity

Inflammation is a natural protective response to tissue injury or infection that helps the body fight intruders (microorganisms and non-self cells) and remove dead or damaged host cells. Increased permeability of endothelial lining cells and influxes of blood leukocytes into the interstitium, oxidative burst, and the production of cytokines, such as interleukins and tumor necrosis factor- (TNF-) are all part of the inflammatory response. It also increases the activity of various enzymes (oxygenases, nitric oxide synthases, peroxidases, and so on), as well as the metabolism of arachidonic acid (Wu et al. 2002). Essential oils have recently been used to treat inflammatory disorders, like rheumatism, allergies, and arthritis in clinical settings. Melaleuca alternifolia EO has been shown to have significant anti-inflammatory properties. Its main component, terpineol, is linked to this activity (Ahmad et al. 1997). The active chemicals work by lowering the generation of inflammatory mediators or limiting the release of histamine. Another example is geranium essential oil. Linalool and linalyl acetate were found to have anti-inflammatory properties in paw-induced mouse carrageenan oedema. According to Yoon et al. (2000) the oils of Torreya nucifera Siebold et Zucc., which are predominantly made up of limonene, -3-carene, and -pinene, inhibit COX-2, resulting in a considerable reduction in prostaglandin (PGE2) synthesis. Furthermore, 1,8-cineole, which is found in many essential oils, has been found to block the biosynthesis of leukotrienes (LTB4) and PGE2, which are both produced by arachidonic acid metabolic pathways (Schnitzler et al. 2007).

Essential oils' anti-inflammatory properties can be linked to their antioxidant properties, as well as their interactions with signaling pathways including cytokines and regulatory transcription factors, as well as their effects on the expression of pro-inflammatory genes. As a result, essential oils constitute a novel therapy alternative for inflammatory illnesses (Parveen et al. 2004).

Cancer Chemoprotective Activity

Essential oils' diverse medicinal potential has piqued researchers' interest in recent years due to their possible anti-cancer activities. The research, as well as their volatile ingredients, are aimed at discovering new anticancer natural compounds. Essential oils could be used to prevent cancer as well as to treat it. Certain foods, such as garlic and turmeric, are widely known for their anticancer properties (Hong et al. 2004). Garlic essential oil contains sulfur compounds that have been shown to help prevent cancer. Examples include diallylsulfide, diallyldisulfide, and diallyltrisulfide. According to Wu et al. (2002) these chemicals stimulate enzymes involved in the hepatic phase 1 (disintegration of chemical bonds that bind carcinogenic poisons to each other) and phase 2 detoxification processes in rats (bonds to toxins released detoxifying enzymes, such as glutathione S-transferase) (Rota et al. 2004). The liver, the body's largest internal organ, is primarily responsible for metabolism. The portal vein transports blood directly from the small intestine to the liver. Hepatic cells make up about 60% of the tissue in the liver. Chemical reactions occur more often in these cells than in any other type of cell in the body. Chemical reactions such as oxidation (the most prevalent), reduction, and hydrolysis are all part of phase 1 metabolism. Phase 1 metabolism can have three different outcomes (Si et al. 2006). The medication is rendered fully inactive. To put it another way, the metabolites have no pharmacological activity. One or more metabolites are pharmacologically active, although not to the same extent as the original substance (Sonboli et al. 2006).

One of the metabolites of the original drug is pharmacologically active. A prodrug is a name given to the original substance. Phase 2 metabolism entails chemical reactions that convert the medication or phase 1 metabolites into soluble molecules that can be eliminated in urine. The molecule (drug or metabolite) is connected to an ionisable cluster in these processes. The process is known as conjugation, and the result is known as a conjugate. Phase 2 metabolites are unlikely to be pharmacologically active (Bruni et al. 2003, Sacchetti et al. 2005). Some medications go through phase 1 or phase 2 metabolism, although the majority go through phase 1 and then phase 2. Myristicin, an allylbenzene found in several essential oils, particularly nutmeg, is another example (Myristica fragrans). In mice, this chemical activates glutathione S-transferase and inhibits carcinogenesis induced by benzo(a)pyrene in the lungs. Myristicin has recently been discovered to induce apoptosis in human neuroblastoma (SK-N-SH). Other volatile chemicals have been shown to have a cytotoxic effect on cancer cell types (Basile et al. 2006). Geraniol reduces the resistance of colon cancer cells (TC118) to the anticancer drug 5-fluorouracil. As a result, geraniol boosts the inhibitory effect of 5-fluorouracil on tumor growth. In various cell lines, the essential oil of balsam fir and Humulene demonstrated substantial anticancer action with negligible harm to healthy cells (Duschatzky et al. 2005). Furthermore, D-limonene, the major component of Citrus essential oil, has been shown to have anticancer properties, particularly in cases of stomach cancer and liver cancer. The sesquiterpene alcohol, such as bisabolol, is prevalent in chamomile essential oil (Matricaria) and has antigliomale action. *Melissa officinalis*, *Melaleuca alternifolia*, *Artemisia annua*, and *Comptonia peregrina* are just a few of the essential oils that exhibit cytotoxic properties (El Hadri et al. 2010).

Cytotoxicity

Essential oils lack specific cellular ligands due to their complicated chemical composition. They can cross the cell membrane and break down the layers of polysaccharides, phospholipids, and fatty acids as well as permeabilize because they are lipophilic mixtures. Such membrane damage appears to be part of this cytotoxicity (Zeytinoglu et al. 2003). Membrane permeabilization in bacteria is linked to ion loss, a decrease in membrane potential, the collapse of the proton pump, and the depletion of the ATP pool. Essential oils have the potential to coagulate the cytoplasm and cause lipid and protein damage (Asekun and Adeniyi 2004). Damage to the cell wall and membrane can result in macromolecule leaking and lysis. Furthermore, essential oils alter membrane fluidity, making it unusually porous, allowing radicals, cytochrome C, Ca2+ ions, and proteins to leak out, similar to oxidative stress. Cell death occurs as a result of the permeabilization of the outer and inner membranes, which results in apoptosis and necrosis. In a number of compartments, ultrastructural changes in the cell can be seen. Electron microscopy has also shown that essential oils disrupt the viral envelope of the herpes simplex virus HSV. An investigation revealed that treatment with terpinene affects microtubule *Saccharomyces cerevisiae* genes, which are present in ergosterol production, sterol absorption, lipid metabolism, cell wall construction and function, cellular detoxification, and transport (Yu et al. 2007). Recent research on the yeast *Saccharomyces cerevisiae* has revealed that the cytotoxicity of particular essential oils can fluctuate greatly depending on their chemical composition. The presence of phenols, alcohols, and monoterpene aldehydes in essential oils is generally associated with cytotoxicity. Essential oils' cytotoxic characteristics are important because they allow them to be used not just against some human infections and animal parasites, but to also protect agricultural and marine products from microbial attack. Certain components of essential oils are efficient against bacteria, viruses, fungi, protozoa, parasites, mites, and other microbes (Ravizza et al. 2008).

In addition, γ-humulene has been shown to be cytotoxic to breast cancer cells *in vitro* humulene was found to be the cause of cytotoxicity (CI50 55 mM). It reduced cellular glutathione (GSH) levels and increased reactive oxygen species (ROS) generation in a dose- and time-dependent manner (Kilani et al. 2008). Furthermore, Zeytinoglu et al. (2003) investigated the effects of carvacrol—one of the main compounds in oregano EO—on DNA synthesis in N-ras transformed mouse myoblast CO25 cells and discovered that this monoterpenic phenol was able to inhibit DNA synthesis in both the growth medium and the ras-activating medium, which contained dexamethasone. Because it inhibits the proliferation of myoblast cells even after activation of the mutant N-ras oncogene, they believe it could be useful in cancer therapy. In the cancer cell line, the EO of the Anonaceae Xylopia aethiopica (Ethiopian pepper), a plant grown in Nigeria, demonstrated a cytotoxic action at a dosage of 5 mg/mL. (Hep-2). Yu et al. also investigated the cytotoxicity of the essential oil of the Aristolochiaceae Aristolochia mollissima rhizome on four human cancer cell lines (ACHN, Bel-7402, Hep G2 and HeLa) (Moon et al. 2006). The rhizome oil possessed a significantly greater cytotoxic effect on these cell lines than the oil extracted from the aerial plant. Linalool inhibits only mild cell growth, but it potentiates doxorubicin-induced cytotoxicity and proapoptotic effects in both MCF7 WT and MCF7 AdrR cell lines at subtoxic concentrations. This monoterpene enhances the therapeutic index in the treatment of breast cancer, particularly tumors with multidrug resistance (MDR). The EO of Cyperus rotundus (Cyperaceae), which is characterized by the predominance of cyperene, γ-Cyperone, isolongifolen-5-one, rotundene, and cyperorotundene were very effective against L1210 leukaemia cells. According to an *in vitro* cytotoxicity assay, it correlates with significantly increased apoptotic DNA fragmentation (Priestley et al. 2006).

Allelopathic Activity

Allelopathy was described as "the science that examines any process involving secondary metabolites produced by plants, algae, bacteria, and fungi that influences the growth and development of agricultural and biological systems" by the International Allelopathy Society (IAS) in 1996. The synthesis of secondary metabolites causes allelopathic interactions. Plants and bacteria produce secondary metabolites for a variety of defenses. Allelochemicals are the secondary metabolites in question (Rim and Jee 2006). Volatile oils and their constituents are being studied for weed and pest control, and they are thought to be a major source of lead compounds in agriculture. Bioactive terpenoids are an important part of many species' defensive mechanisms, and they represent a relatively unexplored supply of active chemicals with potential applications in both agriculture and medicine. In reality, the terpenoid pathway produces a vast number of very phytotoxic allelochemicals, and the phytotoxicity of essential oils has been studied. Angelini et al. (2003) studied the allelopathic activity of *Melaleuca alternifolia* (Maiden and Betche) Cheel (tea tree) essential oil against *Trichoderma harzianum*, a fungal contaminant that causes considerable losses in *Pleurotus* species cultivation. During *in vitro* process, this essential oil exerts an allelopathic effect on *Trichoderma harzianum* (Macías et al. 2006). In dual-culture studies with varying doses, the antifungal activity of *M. alternifolia* essential oil and antagonist activities between *Pleurotus* species against three *T. Harzianum* strains were investigated. The EOs of leaves and rhizomes induced a decrease in dry matter, according to Santos et al. (2011). In lettuce seedlings, they also noticed a drop in branch length. They found a reduction in these parameters when evaluating the effect of these Eos on lettuce seedling germination and vigor and determined that the oil from the rhizomes produced a higher reduction in all of the variables than the oil from the leaves (Singh et al. 2009).

Dudai et al., reported that monoterpenes act on seeds at very low levels. In particular, among the Lamiaceae family, many species release phytotoxic monoterpenes that hinder the development of herbaceous species, including pinene, limonene, p-Cymene, and 1,8-cineole. Furthermore, monoterpenes in essential oils are known to have phytotoxic effects in plants, causing an accumulation of lipid globules in the cytoplasm as well as a reduction in some organelles, such as mitochondria, possibly due to inhibition of DNA synthesis or disruption of membranes surrounding mitochondria and nuclei. Because the ongoing use of synthetic herbicides may jeopardize agricultural sustainability and cause major ecological and environmental concerns; essential oils having allelopathic qualities could be used as part of alternative tactics to develop biodegradable and non-toxic chemicals (Tellez et al. 2002).

Repellent and Insecticidal Activity

Essential oils are a rich source of structurally varied molecules with a wide range of insecticidal and repellant properties. Numerous studies have shown that these chemicals, as well as their parent blends, have biological activity that might cause arthropod pests to become ill. Regulatory restrictions, intellectual property value, biological activity, product performance, and product quality are all elements that influence the marketing of plant essential oil extracts as repellents (Angelini et al. 2003). Essential oils have a great killing power on insects, like the rice weevil Sitophilus oryzae, beetles Callosobruchus chinensis (Coleoptera: Bruchidae), and *S. paniceum* as well as *M. domestica*. Essential oils have a wide range of effects on different species. Essential oils of Mentha, Lavandula (Lamiaceae), or Pinus (Pinaceae) were found to be harmful to Myzus persicae (Homoptera: Aphididae) and Trialeurodes vaporariorum (Homoptera: Aleyrodidae) as well as the Colorado beetle Leptinotarsa decemlineata (Coleoptera: Chrysomelidae) and the (Hymenoptera: Stephanidae). Essential oils are commonly breathed, eaten, or absorbed through the skin by insects. Essential oils and their major components, volatile monoterpenes, have been shown to have fumigant toxicity (Santos et al. 2011). Insects were also quite susceptible to topical treatments. Citrus (Rutacae) essential oils reacted with Sitophilus zea-mais (Coleoptera: Curculionidae), Tribolium castaneum, and Prostephanus truncatus (Coleoptera: Bostrychidae). Contact with Eucalyptus saligna (Myrtaceae) oil killed Pediculus capitis (Anoplura: Pediculidae), Anopheles funestus (Diptera: Culicidae), Cimex lectularius (Hemiptera: Cimicidae), and Periplaneta orientalis (Dictyoptera: Blattidae) within 2 to 30 minutes.

Essential oils belonging to plants in the citronella genus (Poaceae) are commonly used as ingredients of plant-based mosquito repellents, mainly Cymbopogon nardus, which are sold in Europe and North America in commercial preparations as shown in Table 1 (Dudai et al. 1999).

Table 1: List of essential oil and its uses.

Essential Oil		Uses/Benefits
Balsam Peru		Hydration, clear airways, and mental awareness.
Bergamot		Relaxation and digestive system.
Clary Sage		Relaxation, calming, stress relief, and warming.
Cedarwood		Calming, stress relief, restful sleep, circulatory system, and clear airways.
Cinnamon Leaf		Revitalizing, refreshing, and warming.
Geranium		Relaxation, calming, and mood improvement.
Ginger		Digestive system.
Grapefruit		Refreshing, detoxifying, and astringent.
Lavender		Restful sleep, mood improvement, and soothes sun-damaged skin.

Table 1 contd. ...

...Table 1 contd.

Essential Oil		Uses/Benefits
Lemon		Uplifting, clarifying, and astringent.
Lemongrass		Clarifying and mental cleansing.
Lime		Uplifting, refreshing, and revitalizing.
Patchouli		Relaxation.
Peppermint		Energizing, stimulating, and revitalizing.
Rosemary		Cleansing, clarifying and invigoration; reduce the appearance of water retention.
Sweet Orange		Uplifting, calming and digestive system
Tangerine		Cheering and energizing, clears the mind and help in balancing.
Tea Tree		Burns and deodorizing.
White Camphor		Depression and improves mood.
Ylang Ylang		Sensual, relaxing and soothing, hydrating, tension and stress relief

Conclusion

Essential oils have a wide range of biological activity and possible industrial applications, which are being investigated further. Multiple drug resistance associated with infectious and noninfectious disorders necessitates novel methods of chemotherapy and chemoprevention. There is a need to raise knowledge among medical and healthcare professionals, as well as patients who use EOs, about the risks and advantages related to their medicinal usage. The antimalarial medication artemisinin (derived from Artemisia annua) and the anticancer drug taxol (derived from Taxus brevifolia) are two well-known instances of this approach's success. Many of the essential oils extracted from herbs and spices are regularly utilized in cooking. The Food and Drug Administration of the United States has given GRAS (Generally Recognized as Safe) certification to selected compounds from several of these EOs. Phytochemical scientists have used a variety of analytical techniques to uncover the chemical diversity of essential oils and their constituent molecules. These molecules have the potential to serve as scaffolds for the development of novel therapeutic molecules, and there is a lot of room for more research. To find novel bioactivities of EOs, efforts should be directed toward automation and high-throughput screening. Furthermore, the vast amounts of data obtained by *in vitro* assays must be verified through rigorous animal research and clinical examinations.

References

Ahmad, H., M.T. Tijerina and A.S. Tobola. 1997. Preferential overexpression of a class MU glutathione S-transferase subunit in mouse liver by myristicin. Biochem. Biophys. Res. Commun. 236: 825–828.

Angelini, L.G., G. Carpanese, P.L. Cioni, I. Morelli, M. Macchia and G. Flamini. 2003. Essential oils from Mediterranean Lamiaceae as weed germination inhibitors. J. Agric. Food Chem. 51: 6158–6164.

Anwar, F., A.I. Hussain, S.T.H. Sherazi and M.I. Bhanger. 2009. Changes in composition and antioxidant and antimicrobial activities of essential oil of fennel (Foeniculum vulgare Mill.) fruit at different stages of maturity. J. Herbs Spices Med. Plants 15: 1–16.

Aruoma, O.I. 1998. Free radicals, oxidative stress, and antioxidants in human health and disease. J. Am. Oil Chem. Soc. 75: 199–212.

Asekun, O.T. and B.A. Adeniyi. 2004. Antimicrobial and cytotoxic activities of the fruit essential oil of Xylopia aethiopica from Nigeria. Fitoterapia 75: 368–370.

Association Française de Normalisation (AFNOR). 2000. Huiles Essentielles, Tome 2, Monographies Relatives Aux Huiles Essentielles, 6th ed.; AFNOR, Association Française de Normalisation: Paris, France.

Basile, A., F. Senatore, R. Gargano, S. Sorbo, M. Del Pezzo, A. Lavitola, A. Ritieni, M. Bruno, D. Spatuzzi, D. Rigano et al. 2006. Antibacterial and antioxidant activities in Sideritis italica (Miller) Greuter et Burdet essential oils. J. Ethnopharmacol. 107: 240–248.

Braga, P.C., M. dal Sasso, M. Culici, L. Gasastri, M.X. Marceca and E.E. Guffanti. 2006. Antioxidant potential of thymol determined by chemiluminescence inhibition in human neutrophils and cell-free systems. Pharmacology 76: 61–68.

Bruni, R., A. Medici, E. Andreotti, C. Fantin, M. Muzzoli and M. Dehesa. 2003. Chemical composition and biological activities of Isphingo essential oil, a traditional Ecuadorian spice from Ocotea quixos (Lam.) Kosterm. (Lauraceae) flower calices. Food Chem. 85: 415–421.

Burt, S. 2004. Essential oils: Their antibacterial properties and potential applications in foods. Int. J. Food Microbiol. 94: 223–253.

Caldefie-Chézet, F., C. Fusillier, T. Jarde, H. Laroye, M. Damez and M.P. Vasson. 2006. Potential antiinflammatory effects of Malaleuca alternifolia essential oil on human peripheral blood leukocytes. Phytother. Res. 20: 364–370.

Caldefie-Chézet, F., M. Guerry, J.C. Chalchat, C. Fusillier, M.P. Vasson and J. Guillot. 2004. Antiinflammatory effects of Malaleuca alternifolia essential oil on human polymorphonuciear neutrophils and monocytes. Free Radic. Res. 38: 805–811.

Canillac, N. and A. Mourey. 2001. Antibacterial activity of the essential oil of Picea excelsa on Listeria, Staphylococcus aureus and coliform bacteria. Food Microbiol. 18: 261–268.

Carette Delacour, A.S. 2000. La Lavande et son Huile Essentielle. PhD. Thesis, Universite Lille 2, Lille, France.

Carson, C.F. and T.V. Riley. 1995. Antimicrobial activity of the major components of the essential oil of Melaleuca alternifolia. J. Appl. Bacteriol. 78: 264–269.

Cimanga, K., K. Kambu, L. Tona, S. Apers, T. de Bruyne, N. Hermans, J. Totté, L. Pieters and A.J. Vlietinck. 2002. Correlation between chemical composition and antibacterial activity of essential oils of some aromatic medicinal plants growing in the Democratic Republic of Congo. J. Ethnopharmacol. 79: 213–220.

Cocking, T.T. and G. Middleton. 1935. Improved method for the estimation of the essential oil content of drugs. Q. J. Pharm. Pharmacol. 8: 435–442.

Cosentino, S., C.I.G. Tuberoso, B. Pisano, M. Satta, V. Mascia, E. Arzedi and F. Palmas. 2002. *In vitro* antimicrobial activity and chemical composition of Sardinian Thymus essential oils. Lett. Appl. Microbiol. 29: 130–135.

Cox, S.D., C.M. Mann, J.L. Markham, H.C. Bell, J.E. Gustafson, J.R. Warmington and S.G. Wyllie. 2000. The mode of antimicrobial action of essential oil of Melaleuca alternifola (tea tree oil). J. Appl. Microbiol. 88: 170–175.

Davidson, P.M. 1997. Chemical preservatives and natural antimicrobial compounds. pp. 520–556. *In*: M.P. Doyle, L.R. Beuchat and T.J. Montville (eds.). Food Microbiology: Fundamentals and Frontiers; ASM Press: Washington, DC, USA.

Davidson, P.M. and M.E. Parish. 1989. Methods for testing the efficacy of food antimicrobials. Food Technol. 43: 148–155.

Delaquis, P.J., K. Stanich, B. Girard and G. Mazza. 2002. Antimicrobial activity of individual and mixed fractions of dill, cilantro, coriander and eucalyptus essential oils. Int. J. Food Microbiol. 74: 101–109.

Denyer, S.P. and W.B. Hugo. 1991. Biocide-induced damage to the bacterial cytoplasmic membrane. pp. 171–188. *In*: S.P. Denyer and W.B. Hugo (eds.). Mechanisms of Action of Chemical Biocides, the Society for Applied Bacteriology, Technical Series No 27; Oxford Blackwell Scientific Publication: Oxford, UK.

Dorman, H.J.D. and S.G. Deans. 2000. Antimicrobial agents from plants: Antibacterial activity of plant volatile oils. J. Appl. Microbiol. 88: 308–316.

Dudai, N., A. Poljakoff-Mayber, A.M. Mayer, E. Putievsky and H.R. Lerne. 1999. Essential oils as allelochemicals and their potential use as bioherbicides. J. Chem. Ecol. 25: 1079–1089.

Duschatzky, C.B., M.L. Possetto, L.B. Talarico, C.C. Garcia, F. Michis, N.V. Almeida, M.P. de Lampasona, C. Schuff and E.B. Damonte. 2005. Evaluation of chemical and antiviral properties of essential oils from South American plants. Antivir. Chem. Chemother. 16: 247–251.

Edris, A.E. 2007. Pharmaceutical and therapeutic potentials of essential oils and their individual volatile constituents. Phytother. Res. 21: 308–323.

El Hadri, A., M.A. Gómez Del Río, J. Sanz, A. González Coloma, M. Idaomar, B. Ribas Ozonas, J. Benedí González and M.I. Sánchez Reus. 2010. Cytotoxic activity of α-humulene and transcaryophyllene from *Salvia officinalis* in animal and human tumor cells. An. Real Acad. Nac. Farm. 76: 343–356.

Fabian, D., M. Sabol, K. Domaracké and D. Bujnékovâ. 2006. Essential oils, their antimicrobial activity against *Escherichia coli* and effect on intestinal cell viability. Toxicol. *In Vitro* 20: 1435–1445.

Farag, R.S., Z.Y. Daw, F.M. Hewedi and G.S.A. El-Baroty. 1989. Antimicrobial activity of some Egyptian spice essential oils. J. Food Prot. 52: 665–667.

Gill, A.O., P. Delaquis, P. Russo and R.A. Holley. 2002. Evaluation of antilisterial action of cilantro oil on vacuum packed ham. Int. J. Food Microbiol. 73: 83–92.

Guenther, E. 1948. The Essential Oils; D. Van Nostrand Company Inc.: New York, NY, USA, p. 427.

Gustafson, J.E., Y.C. Liew, S. Chew, J.L. Markham, H.C. Bell, S.G. Wyllie, J.R. Warmington. 1998. Effects of tea tree oil on *Escherichia coli*. Lett. Appl. Microbiol. 26: 194–198.

Hart, P.H., C. Brand, C.F. Carson, T.V. Riley, R.H. Prager and J.J. Finlay-Jones. 2000. Terpinen-4-ol, the main component of the essential oil of Malaleuca altemifolia (tea tree oil), suppresses inflammatory mediator production by activated human monocytes. Inflamm. Res. 49: 619–626.

Hong, E.J., K.J. Na, I.G. Choi, K.C. Choi and E.B. Jeung. 2004. Antibacterial and antifungal effects of essential oils from coniferous trees. Biol. Pharm. Bull. 27: 863–866.

Kamatou, G.P.P. and A.M. Viljoen. 2010. A review of the application and pharmacological properties of α-Bisabolol and α-Bisabolol-rich oils. J. Am. Oil Chem. Soc. 87: 1–7.

Kilani, S., A. Abdelwahed and R. Ben Ammar. 2008. Chemical Composition of the essential oil of Juniperus phoenicea L. from Algeria. J. Essent. Oil 20: 695–700.

Knobloch, K., H. Weigand, N. Weis, H.M. Schwarm and H. Vigenschow. 1986. Action of terpenoids on energy metabolism. pp. 429–445. *In*: E.J. Brunke (ed.). Progress in Essential Oil Research: 16th International Symposium on Essential Oils; De Walter de Gruyter: Berlin, Germany.

Koh, K.J., A.L. Pearce, G. Marshman, J.J. Finlay-Jones and P.H. Hart. 2002. Tea tree oil reduces histamine-induced skin inflammation. Br. J. Dermatol. 147: 1212–1217.

Lambert, R.J.W., P.N. Skandamis, P. Coote and G.J.E. Nychas. 2001. A study of the minimum inhibitory concentration and mode of action of oregano essential oil, thymol and carvacrol. J. Appl. Microbiol. 91: 453–462.

Macías, F.A., N. Chinchilla, R.M. Varela and J.M. Molinillo. 2006. Bioactive steroids from *Oryza sativa* L. Steroids 71: 603–608.

Manjamalai, A., G.J. Jiflin and V.M. Grace. 2012. Study on the effect of essential oil of Wedelia chinensis (Osbeck) against microbes and inflammation. Asian J. Pharm. Clin. Res. 5: 155–163.

Marino, M., C. Bersani and G. Comi. 1999. Antimicrobial activity of the essential oils of *Thymus vulgaris* L. measured using a bioimpedometric method. J. Food Prot. 62: 1017–1023.

Maruyama, N., N. Sekimoto and H. Ishibashi. 2005. Suppression of neutrophil accumulation in mice by cutaneous application of geranium essential oil. J. Inflamm. 2: 1–11.

Milner, J.A. 2001. A historical perspective on garlic and cancer. Recent advances on the nutritional effects associated with the use of garlic as a supplement. J. Nutr. 131: 1027–1031.

Milner, J.A. 2006. Preclinical perspectives on garlic and cancer. Signifiance of garlic and its constituents in cancer and cardiovascular disease. J. Nutr. 136: 827–831.

Moon, T., J.M. Wilkinson and H.M. Cavanagh. 2006. Antiparasitic activity of two Lavandula essential oils against Giardia duodenalis, Trichomonas vaginalis and Hexamitainflata. Parasitol. Res. 99: 722–728.

Nickerson, G. and S. Likens. 1996. Gas chromatographic evidence for the occurrence of hop oil components in beer. J. Chromatogr. 21: 1–5.

Nychas, G.J.E. 1995. Natural antimicrobials from plants. pp. 58–89. *In*: G.W. Gould (ed.). New Methods of Food Preservation, 1st ed.; Blackie Academic & Professional: London, UK.

Parveen, M., M.K. Hasan, J. Takahashi, Y. Murata, E. Kitagawa, O. Kodama and H. Iwahashi. 2004. Response of Saccharomyces cerevisiae to a monoterpene: Evaluation of antifungal potential by DNA microarray analysis. J. Antimicrob. Chemother. 54: 46–55.

Pauli, A. 2001. Antimicrobial properties of essential oil constituents. Int. J. Aromather. 11: 126–133.

Pophof, B., G. Stange and L. Abrell. 2005. Volatile organic compounds as signals in a plant-herbivore system: Electrophysiological responses in olfactory sensilla of the Moth Cactoblastis cactorum. Chem. Senses 30: 51–68.

Priestley, C.M., I.F. Burgess and F.M. Williamson. 2006. Lethality of essential oil constituents towards the human louse, Pediculus humanus, and its eggs. Fitoterapia 77: 303–309.

Pyun, M.S. and S. Shin. 2006. Antifungal effects of the volatile oils from Asiium plants against Trichophyton species and synergism of the oils wih ketoconazole. Phytomedicine 13: 394–400.

Ratledge, C. and S.G. Wilkinson. 1988. An overview of microbial lipids. pp. 3–22. *In*: C. Ratledge and S.G. Wilkinson (eds.). Microbial Lipids; Academic Press Limited: London, UK, 1988; Volume 1.

Ravizza, R., M.B. Gariboldi, R. Molteni and E. Monti. 2008. Linalool, a plant-derived monoterpene alcohol, reverses doxorubicin resistance in human breast adenocarcinoma cells. Oncol. Rep. 20: 625–630.

Reichling, J., P. Schnitzler, U. Suschke and R. Saller. 2009. Essential oils of aromatic plants with antibacterial, antifungal, antiviral, and cytotocic properties-an overview. Forsch. Komplement. 16: 79–90.

Rim, I.S. and C.H. Jee. 2006. Acaricidal effects of herb essential oils against Dermatophagoides farinae and D. pteronyssinus (Acari: Pyroglyphidae) and qualitative analysis of a herb Mentha pulegium (pennyroyal). Korean J. Parasitol. 44: 133–138.

Rota, C., J.J. Carraminana, J. Burillo and A. Herrera. 2004. *In vitro* antimicrobial activity of essential oils from aromatic plants against selected foodborne pathogens. J. Food Prot. 67: 1252–1256.

Sacchetti, G., S. Maietti, M. Muzzoli, M. Scaglianti, S. Manfredini, M. Radice and R. Bruni. 2005. Comparative evaluation of 11 essential oils of different origin as functional antioxidants, antiradicals and antimicrobials in food. Food Chem. 91: 621–632.

Santos, S., M.L.L. Moraes, M.O.O. Rezende and A.P.S. Souza-Filho. 2011. Potencial alelopático e identificação de compostos secundários em extratos de calopogônio (Calopogonium mucunoides) utilizando eletroforese capilar. Eclética Quím. 36: 51–68.

Schnitzler, P., C. Koch and J. Reichling. 2007. Susceptibility of drugresistant clinical HSV-1 strains to essential oils of Ginger, Thyme, Hyssop and Sandalwood. Antimicrob. Agents Chemother. 51: 1859–1862.

Sell, C.S. 2006. The Chemistry of Fragrance. From Perfumer to Consumer, 2nd ed; The Royal Society of Chemistry: Cambridge, UK, p. 329.

Senatore, F., F. Napolitano and M. Ozcan. 2000. Composition and antibacterial activity of the essential oil from Crithmum maritimum L. (Apiaceae) growing wild in Turkey. Flav. Frag. J. 15: 186–189.

Sharififar, F., M. Mirtajadini, M.J. Azampour and E. Zamani. 2012. Essential oil and methanolic extract of Zataria multiflora Boiss with anticholinesterase effect. Pak. J. Biol. Sci. 15: 49–53.

Shelef, L.A. 1983. Antimicrobial effects of spices. J. Food Saf. 6: 29–44.

Si, W., J. Gong, R. Tsao, T. Zhou, H. Yu, C. Poppe, R. Johnson and Z. Du. 2006. Antimicrobial activity of essential oils and structurally related synthetic food additives towards selected pathogenic and beneficial gut bacteria. J. Appl. Microbiol. 100: 296–305.

Sikkema, J., J.A.M. de Bont and B. Poolman. 1994. Interactions of cyclic hydrocarbons with biological membranes. J. Biol. Chem. 269: 8022–8028.

Singh, H.P., S. Kaur, S. Mittal, D.R. Batish and R.K. Kohli. 2009. Essential oil of Artemisia scoparia inhibits plant growth by generating reactive oxygen species and causing oxidative damage. J. Chem. Ecol. 35: 154–162.

Sonboli, A., B. Babakhani and A.R. Mehrabian. 2006. Antimicrobial activity of six constituents of essential oil from Salvia. Z. Naturforschung 61: 160–164.

Tellez, M.R., M. Kobaisy, S.O. Duke, K.K. Schrader, F.E. Dayan and J. Romagni. 2002. Terpenoid based defense in plants and other organisms. *In*: T.M. Kuo and H.W. Gardner (eds.). Lipid Technology; Marcel Dekker: New York, NY, USA, p. 354.

Ultee, A., M.H.J. Bennink and R. Moezelaar. 2002. The phenolic hydroxyl group of carvacrol is essential for action against the food-borne pathogen *Bacillus cereus*. Appl. Environ. Microbiol. 68: 1561–1568.

Vainstein, A., E. Lewinsohn, E. Pichersky and D. Weiss. 2001. Floral fragrance. New inroads into an old commodity. Plant Physiol. 27: 1383–1389.

Wu, C.C., L.Y. Sheen, H.W. Chen, W.W. Kuo, S.J. Tsai and C.K. Lii. 2002. Differential effects of garlic oil and its three major organosulfur components on the hepatic detoxification system in rats. J. Agric. Food Chem. 50: 378–383.

Yoon, H.S., S.C. Moon, N.D. Kim, B.S. Park, M.H. Jeong and Y.H. Yoo. 2000. Genistein induces apoptosis of RPE-J cells by opening mitochondrial PTP. Biochem. Biophys. Res. Commun. 276: 151–156.

Yu, H.S., S.Y. Lee and C.G. Jang. 2007. Involvement of 5-HT1A and GABAA receptors in the anxiolytic-like effects of Cinnamomum cassia in mice. Pharmacol. Biochem. Behav. 87: 164–170.

Zeytinoglu, H., Z. Incesu and K.H. Baser. 2003. Inhibition of DNA synthesis by carvacrol in mouse myoblast cells bearing a human NRAS oncogene. Phytomedicine 10: 292–299.

5

Essential Oils of Medicinal Plants from Northern Peru
Traditional and Scientific Knowledge

Mayar L. Ganoza-Yupanqui,[1,*] Luz A. Suárez-Rebaza,[2]
Edmundo A. Venegas-Casanova,[2] Segundo G. Ruiz-Reyes,[2]
Ricardo D.D.G. de Albuquerque[3] and Mayer M. Ganoza-Suárez[4]

Introduction

Aromatic medicinal plants in Northern Peru are widely used in a popular way in traditional medicine. Its knowledge comes from ancient generations through its expert connoisseurs, who had a broad notion of its properties, applying them properly for the prevention and treatment of various diseases (Torres-Guevara et al. 2020). In the north of Peru, there are three natural regions that vary in their climate, altitude, radiation and edaphoclimatic factors and these factors affect the physiological development of plants. In the Costa region, *Lantana camara* L. is the representative; in the Sierra region, *Minthostachys mollis* (Kunth) Griseb and *Myrcianthes rhopaloides* Mc Vaugh are the representatives; for the Selva region, *Ocotea aciphylla* (Nees & Mart.) Mez. is the representative; other species that grow in different geographical areas include *Schinus molle* L., *Chenopodium ambrosioides* L., *Aloysia citriodora* Palau and *Justicia pectoralis* Jacq.

[1] Departamento de Farmacología, Facultad de Farmacia y Bioquímica, Universidad Nacional de Trujillo, Av. Juan Pablo II S/N, Ciudad Universitaria, Trujillo, 13011, Perú.
[2] Departamento de Farmacotecnia, Facultad de Farmacia y Bioquímica, Universidad Nacional de Trujillo, Av. Juan Pablo II S/N, Ciudad Universitaria, Trujillo, 13011, Perú.
[3] Laboratório de Tecnologia em Productos Naturais, Universidade Federal Fluminense, Niterói, Brazil
[4] Escuela de Medicina, Facultad de Medicina, Universidad Nacional de Trujillo, Av. Roma 338, Trujillo, 13011, Perú.
* Corresponding author: mganoza@unitru.edu.pe

Essential oils differ from the other metabolites of medicinal plants due to their physical properties, such as density, viscosity, refractive index and optical activity (Castro-Alayo et al. 2019). Chemically, they are composed of monoterpenes and sesquiterpenes. In biological tests, they have been shown to have antibacterial, antiviral, antifungal, antiparasitic, antiseptic and repellent activity (Benites et al. 2016, Malca-García et al. 2017). Aromatic medicinal plants produce essential oils with higher concentration and quality in higher altitude geographic areas (Passos et al. 2022), such as *Schinus molle* L. There is an innovative experience in the communities of the Paramos of the Piura region, which consists of obtaining the essential oils from native species using an artisanal prototype of extraction by hydrodistillation (Torres-Guevara et al. 2021).

Traditional Uses, Chemical Composition and Biological Activities of Essential Oils from Medicinal Plants in Northern Peru

Aloysia citriodora Palau

Aloysia citriodora Palau is a spontaneous plant from South America, originally from Peru and is popularly known as *cedrón, cidrón, lemon verbena, verbena, yerba luisa, wari pankara* or *princess grass*, depending on the country or the region (Rojas-Armas et al. 2015). Ethnopharmacological information refers to various uses of *A. citriodora* in popular medicine, mainly including the intake orally infusion or decoction of the aerial parts as antispasmodic, tranquilizer, nervous calming, expectorant and stomachic. In addition to these uses, Bolivia is also used for high blood pressure, whereas in Ecuador the plant is used to treat fever and headache and as a diuretic. On the other hand, it is also traditionally used in Asia to treat gastrointestinal spasms, the common cold and as a sedative (Rojas et al. 2012, Girault 1987). Regarding the chemical composition of essential oils of the species collected from Northern Peru, the work of Ruiz-Reyes (2020) showed two different chemotypes according to the site of collection. The essential oil of aerial parts collected in the provinces of La Libertad and Ancash presented geranial (60%) as the major component, which was followed by carvone (30%), while in the province of Cajamarca (Figure 1), limonene was predominant with 75% of the total composition. Geranial and carvone presented concentrations in the range between 10% and 8% (Ruiz-Reyes 2020). Another work conducted by Huerta-León (2021) showed that α-citral was found at the concentration of 25.13% with β-citral (18.15%) and limonene (15.62%) as other main compounds of the oil (Huerta-León 2021). In other chemical studies carried out in different regions of Peru and Latin American countries, geranial and limonene were considered the major substances of the oil and often accompanied by neral and, in some cases, also by citronellal (Bardales-Huamán and Farfán-Chaupis 2018, De Figueiredo et al. 2004, Lira et al. 2008, Rudas-González 2017, León 2020). On the other hand, Elechosa et al. (2017) demonstrated that these chemotypes are globally divided according to the following major components: thujones, citronellal, carvone and neral/geranial (Elechosa et al. 2017).

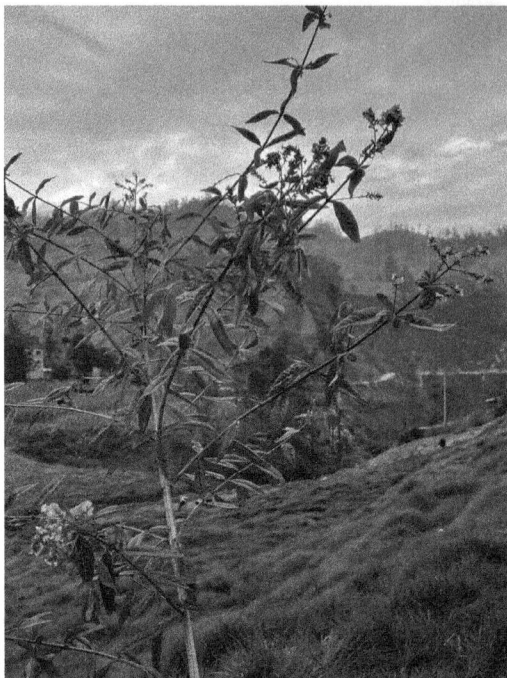

Figure 1: *Aloysia citriodora* Palau in its natural habitat from Cajamarca, Peru (Photo by Segundo G. Ruiz-Reyes in May, 2022).

The essential oils of *A. citriodora* from Northern Peru showed antibacterial, anti-trypanosoma and antispasmodic activity. Antibacterial effects were observed against *S. aureus* ATCC 25923 (100% concentration with an average inhibition halo of 35.14 mm) (Anaya-Huánuco 2018), *Streptococcus mutans* ATCC 25175 (100% concentration with average inhibition halo of 21.3 mm) and *Porphyromonas gingivalis* ATCC 33277 (100% concentration with average inhibition halo of 35.5 mm) (Huerta-León 2021), in addition to the activity against *Escherichia coli* ATCC 25922 and *Staphylococcus aureus* ATCC 25923 in the study of Chicoma-Gutiérrez and Malca-Alvarado (2015). For the activity against *Trypanosoma cruzi*, the essential oil produced a significant reduction of 85.4% of peak parasitaemia in rats with the 250 mg/kg dose and also produced a reduction in the number of amastigotes and inflammatory infiltrates in the heart (Rojas et al. 2010). These results were corroborated by the study of Rojas-Armas (2015), which demonstrated similar inhibition of the parasitemia peak through the administration of citral, which is the main component of the studied oil. This study also reported the negative result of the essential oil in a toxicity test in albino rats (Rojas-Armas et al. 2015). In the study of Rojas et al. (2012), the essential oil from *A. citriodora* inhibited the growth of *T. cruzi* epimastigote form with IC_{50} of 96.49 mg/mL (Rojas et al. 2012). Finally, the essential oil also presented antispasmodic activity on ileum of *Cavia porcellus*, so that the 0.2 mL dose (0.3% essential oil) decreased spontaneously and acetylcholine and potassium chloride-induced motility in similar patterns to atropine and nifedipine (Pérez-Robles and Vela-Aguiar 2014).

Regarding the biological activities found in oils from other regions and countries, the EOs from *A. citriodora* presented larvicidal, antimicrobial and anesthetic activities. The larvicidal effect was observed against *Culex quinquefasciatus* in a binary mixture 1:1 of leaves EO from *A. citriodora* (geranial/neral chemotype) and flowering aerial parts EO from *Satureja montana*, presenting a LC_{50} = 18.3 μL/L (Benelli et al. 2017). This same chemotype presented antimicrobial activity against six urinary and vaginal pathogens (*Escherichia coli*, *Klebsiella ozaenae*, *Enterobacter aerogenes*, *Proteus mirabilis*, *Staphylococcus aureus* and *Enterococcus* sp.) with MIC values of 10–50 mg/mL (Rojas et al. 2010), besides also presented activity against *Salmonella typhirium* with a MIC of 1.6% (Rudas-González 2017). The anesthetic activity was demonstrated by the work of Parodi et al. (2014), so the efficacy of the aerial parts EO of *A. citriodora* acted as an anesthetic for albino and gray strains of silver catfish, *Rhamdia quelen*, in a 100–800 μL/L (11.1–1.24 minutes) range (Parodi et al. 2014).

Chenopodium ambrosioides L.

This species is found in Northern Peru, besides other regions of America, including the Caribbean countries and belongs to the Amaranthaceae family, which is known by several popular names such as *paico, paico macho, pichin, pazote* and *té de los jesuítas* (Berrios-Acevedo 2019, Ocampo and Valverde 2000). Different parts of the plant are used in folk medicine as stomachic, carminative and anthelmintic due to its paralyzing and narcotic action on roundworms, pinworms and hookworms (Jaramillo et al. 2012). Chemical studies of the essential oils collected in Peru showed the majority presence of ascaridole; the main substance related to the antiparasitic activity as well as bornilene, *m*-cymene, δ-carene, pinenes, α-phellandrene, α-terpinene, limonene, eucalyptol, *cis*-anetol, carvone, α-terpineol, timol, carvacrol and linalool (Puma-Mamani 2009, León-Romaní 2009).

Several studies about the biological activity of the essential oils from *C. ambrosioides* showed antifungal, insecticide, antimicrobial, antiparasitic, antileishmanial and cytotoxic activity. The antifungal activity was observed for the leaves EO against *Aspergillus*, *Colletotrichum* and *Fusarium* species at a concentration of 0.1% (Jardim et al. 2008) and also against the dermatophytes *Trychophyton mentagrophytes* and *Microsporum audouinii* (50 ppm) (Kishore et al. 1996) as well as against the phytopathogen *Rhizoctonia solani* with a 100% inhibition using a concentration of 2% (Berrios-Acevedo 2019). Meanwhile, the EO from aerial parts was active against *Candida* species at concentrations in a range of 0.25 to 2 mg/mL for *in vitro* analysis and presented *in vivo* efficacy in rats at a concentration of 0.1% (Chekem et al. 2010). The leaves EO presented also an insecticidal effect against *Sitophilus zeamais*, one of the main cereal pests, through fumigant activity and repellent and feeding deterrent effects in a concentration of 0.5% (Aros et al. 2019). The antibacterial activity of EOs from *C. ambrosioides* was observed on several bacteria species and strains but with a more pronounced effect against *E. coli*, *S. aureus* and *Erwinia* spp. (Aquino-Apaza 2017, Lezama-Paredes 2019, Condori-Espinoza et al. 2018). The effect on *S. aureus* can be mainly mediated by the inhibition of NorA efflux-pump caused by α-terpinene (Morais Oliveira-

Tintino et al. 2018). The ethnomedicinal use of *C. ambrosioides* in the treatment of parasitosis is based on the observed antiparasitic activities of the essential oil of aerial parts against the nematode *Meloidogyne incognita* as well as the effect of leaves EO on *Ascaris lumbricoides* (Bai et al. 2011, Marín-Diaz 2019). The antileishmanial activity of the essential oil extracted from the leaves of *C. ambrosioides* and its main components was observed against *L. amazonensis*. Isolated ascaridole showed the highest activity against the parasite; however, the essential oil showed higher specificity (cytotoxicity 15 times higher). Furthermore, the essential oil controlled the development of the infection when administered orally and intralesionally in infected mice. On the other hand, the formation of a free radical from the essential oil and ascaridole was demonstrated. The products inhibited Complex III of the respiratory chain and it was possible to observe interactions of the essential oil with the DNA of the parasite (Monzote-Fidalgo 2010). Finally, the EO from leaves also presented considerable cytotoxic activity against the tumor cells RAJI and MCF-7 lines (IC_{50} = 1.0 and 9.45 µg/mL, respectively) (Jia-liang et al. 2013).

Justicia pectoralis Jacq.

Justicia pectoralis Jacq. is a medicinal plant with large traditional use in Central and South America. The plant is popularly known as *chambá, anador, trevo-cumaru, trevo-do-Pará, cachamba* in Brazil, *curia* in México, Venezuela, Trinidad and Panama, *tilo* in Cuba and Costa Rica, *chapantye* in Caribbean and Haiti, *zèb chapantye* in Dominica and Martinica, *fresh cut* in Jamaica and as *amansa guapo* in Colombia. Regarding its ethnomedicinal use, *Justicia pectoralis* is traditionally used for prostate problems, treatment of menopause symptoms, menstrual pains and dysmenorrhea, as anxiolytics and sedatives, and in the treatment of respiratory tract disorders, such as cough, colds, bronchitis and asthma as well as for the treatment of diabetes and infections in Colombia and compose the therapeutics arsenal of the National Health System in Cuba and Brazil (Venegas-Casanova et al. 2018). Its EO is rich in coumarin, alcohols, cetones and aldehydes; for example, nonanal, *trans*-(E)-2-hexenal, 2-hexanone, 2-hexanol and 1-octen-3-ol (Pino et al. 2011, Venegas-Casanova et al. 2018). In addition, the EO from aerial parts of *J. pectoralis* showed anthelmintic activity against the larvae of the nematode *Strongyloides stercoralis*, a human parasite that causes strongyloidiasis (Mijares-Palacios 2017).

Lantana camara L.

Lantana camara L. (Verbenaceae) is a hairy shrub native to South and Central America and has been introduced in many countries as a hedge or ornamental plant and is a highly invasive weed in many places around the world. In Peru, this species is known as *siete colores* or *yerba de la maestranza* and has been completely naturalized along the coast. Different parts of *L. camara* are used in the local traditional medicine systems for the treatment of health problems, including rheumatism, malaria, tumors, tetanus as well as bilious fever, dysentery, catarrh, itches, dermatitis, ulcers, eczema and swellings (Brick 1999, Ghisalberti et al. 2000, Sundufu and Shoushan 2004). According to the work by Benites et al. (2009), the chemical composition of the essential oil of aerial parts from *Lantana camara* collected in Northern Peru (Trujillo, La Libertad) (Figure 2) has a chemotype quite different from those found in other parts of the world since carvone (75.9%) and limonene (16.9%) were the major substances in the essential oil in this location, while in the species found on the African continent β-caryophyllene and ar-curcumene were the main markers with davanone also being the majority in some species collected in Madagascar. In addition, β-caryophyllene, sabinene, germacrene D, curcumenes and alpha-zingiberene were also recurrent major compounds in the species collected in Asia and other countries of America, so that the chemical profile found in the specimens collected in Trujillo is considered quite particular (Benitez et al. 2009, Valdez-Tenezaca 2018, Nea et al. 2020). Moreover, another chemical study carried out with essential oil from *Lantana camara* collected in Peru was reported by Inga-Chavelón (2016) in which the majority presence of caryophyllenes, α-pinene and limonene was demonstrated; however, the species were collected in the department of Lima, central coast of the country (Inga-Chavelón 2016).

The essential oils collected in the province of Trujillo showed antibacterial, antioxidant and anti-inflammatory activity. The work of Benites et al. informed that *L. camara* essential oil presented moderate antibacterial activity against the human pathogen *Staphylococcus aureus* (MIC 200 µg/ml) and high antioxidant activity evaluated by the Trolox equivalent antioxidant capacity assay (TEAC) (29.0 mmol TE/kg), besides relative moderated anti-inflammatory activity due to its ability for inhibiting lipoxygenase (IC$_{50}$ = 81.5 µg/mL) (Benites et al. 2009). In another study, Casanova also demonstrated the antibacterial activity of the oil through the disk diffusion method with inhibition of *S. aureus* and *E. coli* colonies (33% and 12%, respectively, using 100% of the oil) (Venegas del Castillo and Vásquez-Valles 2016).

Figure 2: *Lantana camara* L. in its natural habitat from La Libertad, Peru (Photo by Mayar L. Ganoza-Yupanqui in May, 2022)

Minthostachys mollis (Kunth) Griseb

M. mollis, known as *muña* in Peru (Figure 3) and Bolivia, has its traditional use in the conservation of potatoes and as a medicinal plant (Linares-Otoya 2020). The placement of branches of *Minthostachys* spp. *Muña* among potatoes harvested acts as a preservative since it prevents insect attack and inhibits germination (Ormachea 1979) and the infusions of its leaves and stems are used to treat abdominal infections and inflammations, colic of gas and breathing problems, while stems, roots and flowers are used diseases of the genital organs. Other preparations such as decoction and rubbing are used for cardiovascular, neurological and kidney diseases and rheumatism (Linares-Otoya 2020). In a recent chemical study with EO from its aerial parts, it was identified that menthone (13.2%), pulegone (12.4%), *cis*-dihydrocarvone (9.8%) and carvacrol acetate (8.8%) were the main essential oil components. The same study demonstrated the cytotoxic activity of this EO, which in turn showed an IC_{50} value of around 0.2 mg/mL against three human cancer cell lines (T24, DU-145 and MCF-7) (Benites et al. 2018). Furthermore, another study reported the antibacterial and antifungal properties of EO from leaves and branches so that this oil presented activity against several oral pathogen microorganisms, which included *Streptococcus mutans*, *Lactobacillus acidophilus*, *Enterococcus faecalis*, *Porphyromonas gingivalis* y *Candida albicans* as well as against Erwinia spp. (Sánchez-Tito and Collantes-Díaz 2021, Condori-Espinoza et al. 2017). Still, a repellent cream containing 10% of EO from leaves presented activity against adult insects of the Culicidae family, vectors of Zika and Chikungunya virus with a repellency grade I (Dávila-Guerra 2016).

Figure 3: *Minthostachys mollis* (Kunth) Griseb in its natural habitat from Cajamarca and Peru (Photo by Mayar L. Ganoza-Yupanqui in April, 2019).

Myrcianthes rhopaloides Mc Vaugh

This species is mostly known as *arrayan* and is used by folk medicine in the control of diabetes (Lizcano Ramón and Vergara González 2008). Chemical studies with EOs from leaves of *M. rhopaloides* collected in the American continent demonstrated great variation in the quality of their compositions, depending on the collection site. Citronelal, *β*-myrcene, rodinol (Silva-Carrero et al. 2016), linalool (17.7%), *α*-cadinol and spathulenol (Cole et al. 2008), linalool (19.7%), eucalyptol, limonene and terpineol (Maldonado-Rodríguez 2006) as well as geranial, neral and *α*-pinene were reported as main constituents of leaves EOs (Malagón et al. 2003). Regarding the biological activity, the EO reported in the work of Cole et al. (2008) presented high cytotoxic activity against melanoma cancer line SK-Mel-28 at the concentration of 100 µg/mL, causing the lethality of 9% of the cells (Werka et al. 2007). The antibacterial effect of leaves EO was also observed by Maldonado-Rodríguez (2006), specifically against *Streptococcus mutans*, *S. pyogenes* and, in a less extension, against *Staphylococcus aureus* and *S. epidermidis* (Maldonado-Rodriguez 2006).

Schinus molle L.

S. molle, popularly known as *Peruvian peppertree* or *Molle del Peru* (Figure 4), has been used as a medicinal resource by indigenous people in the Americas so that its sap is used as a diuretic and purgative, and the entire plant is used externally as a topical antiseptic and for fractures. The oleoresin has wound healing properties and is also used to treat toothaches, rheumatism and *susto*. Moreover, berries are used in syrups, vinegar and beverages (Huaman et al. 2004). Regarding the chemical composition of essential oils from species collected in the Americas, the main constituents in leaves oil are *α*- and *β*-pinene, sabinene, limonene and bicyclogermacrene (Duarte et al. 2018, Santos et al. 2009, Rocha et al. 2012, Díaz et al. 2008), whereas myrcene is the major compound in fruits (Huaman et al. 2004, Pérez-López et al. 2011).

Regarding the biological activities of these oils, Morillo-Horna (2015) investigated the acaricide effect on *Varroa destructor*, an ectoparasite of the bee colony. The EO from leaves caused 89% mortality with a concentration of 2 mL/L and is considered an alternative in the maintenance of beekeeping activity (Morillo-Horna 2015). In addition, the leaves EO presented antibacterial activity against *Staphylococcus aureus* with an inhibition halo of 31.0 mm in a concentration of 50% (Quispe 2018). Still concerning the antibacterial property, the study of Rocha et al. (2012) demonstrated the strong/moderate effect on *Escherichia coli* and moderate/weak effect on *Pseudomonas aeruginosa* of the EO from leaves and fruits (Rocha et al. 2012) and Pérez-López et al. (2011) reported the activity of fruits EO on *S. pneumoniae* (MIC = 125 µg/mL) (Pérez-López et al. 2011). The EO from leaves also presented cytotoxic activity against several cell lines, which is more effective on breast carcinoma and leukemic cell lines (Diaz et al. 2008). Still, regarding the biological effect of leaves EO, Peña-Caiza (2018) evaluated its inhibitory effect against different stages of *Premnotrypes vorax*, the white potato worm, so that within 24 hours, the 8% EO concentration caused larval and adult mortality

Figure 4: *Schinus molle* L. in its natural habitat from La Libertad, Peru (Photo by Mayar L. Ganoza-Yupanqui in May 2022).

(36.67%), inhibition of larval hatching (80%), and food inhibition (94.78%) (Peña-Caiza 2018). Furthermore, the antifungal activity was also observed with EO from leaves, such as this oil showed 91% inhibition against *Botrytis cinerea*, the agent of plant gray rot in a concentration of 15% (Espinoza-Pantigozo 2016) besides to presenting antifungal activity against *Candida albicans* that reaches 18% of halo inhibition in a concentration of 10% (Zegarra-Carrera 2019). The oil from leaves was still active against the fungus *Colletotrichum* spp., *Fusarium* spp. and *Alternaria* spp., showing total inhibition of colonies in 24 hours at the concentrations of 25%, 50% and 50%, respectively (Santos et al. 2010). Another study demonstrated the antifungal effect of fruit EO on *Colletotrichum* spp., an important phytopathogen. At a concentration of 750 mg/mL, the oil showed 37% fungal inhibition, using the good diffusion method (Toro and Piedra 2019). Finally, technological products based on EOs from *S. molle* were developed, which included a wound healing ointment to treat cattle and mice (Alba-Gonzalez et al. 2009) and an active nanoemulsion against *T. cruzi*, the etiological agent of Chagas Disease (Baldissera et al. 2013).

Ocotea aciphylla (Nees and Mart.) Mez.

The species *O. aciphylla*, popularly known as *alcanfor, tinchi-takak, canela moena, roble amarillo, moena amarilla, mantayá, kaikua, eáyua, palta moena* or *canelón*, is used in the Peruvian Amazon as an anti-rheumatic, depurative and the treatment of dental caries, bloody diarrhea, abdominal disorders and vaginal cyst (Marques 2001, Roumy et al. 2020). In a chemical study of its EO bark, the most abundant constituents were δ-cadinene (19.5%), β-selinene (16.3%), γ-muurolene (14.1%), benzyl benzoate (9.9%), linalol (6.0%), eucalyptol (6.2%) and 3-carene (4.7%). Moreover, this EO presented class IV repellency with 73.3% repellency against *Aedes aegypti* in a 20% solution (Guzmán-Ramírez and López-Tuesta 2018).

Conclusions

This chapter presented various medicinal species found in northern Peru that are characterized by the production of essential oils, which in turn are related to the pharmacological activity of each species. The ethnomedicinal use for all plants was presented, followed by the chemical data of the essential oil, which often proved to be peculiar when compared with the chemical characterization of the same oils collected in different regions or countries. These differences are sometimes related to the diversity of the ethnomedicinal use of each plant in different regions of the planet in addition to influencing the pharmacological profile, either qualitatively or quantitatively. Finally, the pharmacological studies carried out with the essential oils of each species described in this chapter were presented, which scientifically validates the contribution of these oils to the medicinal utility of these plants.

References

Alba-González, A., P. Bonilla-Rivera and R. Arroyo-Acevedo. 2009. Healing activity of ointment the essential oil of the *Schinus molle* L. "molle" in ront to the wounds infected in the cattle and in mice. Rev. UNMSM. OAI: ojs.csi.unmsm:article/3384.

Anaya-Huánuco, E.R. 2018. Efecto antibacteriano del aceite esencial de *Aloysia triphylla* "Cedrón" sobre *Staphylococcus aureus* ATCC 25923 comparado con oxacilina. Graduation Thesis. Universidad Nacional de Trujillo, Trujillo, Perú.

Aquino-Apaza, E. 2017. Actividad antimicrobiana de aceites esenciales de *Chenopodium ambrosioides, Artemisia absinthium, Caiophora cirsiifolia* sobre bacterias gram negativas *Staphylococcus aureus* y su toxicidad en *Artemia salina*. Graduation Thesis. Universidad Nacional del Altiplano, Puno, Perú.

Aros, J., G. Silva-Aguayo, S. Fischer, I. Figueroa, J.C. Rodríguez-Maciel, A. Lagunes-Tejeda, A.S. Castañeda-Ramírez and L. Aguilar-Marcelino. 2019. Actividad insecticida del aceite esencial del paico *Chenopodium ambrosioides* L. sobre *Sitophilus zeamais* Motschulsky. Chil. J. Agric. Anim. Sci. 35: 282–292.

Bai, C.Q., Z.L. Liu and Q.Z. Liu. 2011. Nematicidal constituents from the Essential Oil of *Chenopodium Ambrosioides* aerial parts. E-Journal Chem. 8(S1): S143–S148.

Baldissera, M.D., A.S. Da Silva, C.B. Oliveira, C.E.P. Zimmermann, R.A. Vaucher, R.C.V. Santos, V.C. Rech, A.A. Tonin, J.L. Giongo and C.B. Mattos. 2013. Trypanocidal activity of the essential oils in their conventional and nanoemulsion forms: *In vitro* tests. Exp. Parasitol. 134: 356–361.

Bardales-Huamán, M. and M.R. Farfán-Chaupis. 2018. Determinación de los componentes mayoritarios del aceite esencial del cedrón (*Aloysia triphylla*) mediante destilación por arrastre de vapor. Graduation Thesis. Universidad Nacional de Callao, Callao, Perú.

Benelli, G., R. Pavela, A. Canale, K. Cianfaglione, G. Ciaschetti, F. Conti, M. Nicoletti, S. Senthil-Nathan, H. Mehlhorn and F. Maggi. 2017. Acute larvicidal toxicity of five essential oils (*Pinus nigra, Hyssopus officinalis, Satureja montana, Aloysia citrodora* and *Pelargonium graveolens*) against the filariasis vector *Culex quinquefasciatus*: Synergistic and antagonistic effects. Parasitol. Int. 66(2): 166–171.

Benites, J., A. Guerrero-Castilla, F. Salas, J.L. Martinez, R. Jara-Aguilar, E.A. Venegas-Casanova, L. Suarez-Rebaza, J. Guerrero-Hurtado and P.B. Calderón. 2018. Chemical composition, *in vitro* cytotoxic and antioxidant activities of the essential oil of Peruvian *Minthostachys mollis* Griseb. Bol. Latinoam. Caribe Plantas Med. Aromat. 17(6): 566–574.

Benites, J., C. Moiteiro, A.C. Figueiredo, P. Rijo, P. Buc-Calderon, F. Bravo, S. Gajardo, I. Sánchez, I. Torres and M. Ganoza. 2016. Chemical composition and antimicrobial activity of essential oil of Peruvian *Dalea strobilacea* barneby. Bol. Latinoam. Caribe Plantas Med. Aromat. 15(6): 429–435.

Benites, J., C. Moiteiro, G. Miguel, L. Rojo, J. López, F. Venâncio, L. Ramalho, S. Feio, S. Dandlen, H. Casanova and I. Torres. 2009. Composition and biological activity of the essential oil of peruvian *Lantana Camara*. J. Chil. Chem. Soc. [Online] 54(4): 379–384.

Berrios-Acevedo, C.R. 2019. Efecto antifúngico comparativo *in vitro* del aceite esencial y extracto etanólico de hojas de *Chenopodium ambrosioides* L. "paico" sobre *Rhizoctonia solani* K. Graduation Thesis, Universidad Nacional de Trujillo, Trujillo, Perú.

Brick, A. 1999. Diccionario Enciclopédico de Plantas útiles del Perú. Programa de las Naciones Unidas para el Desarrollo. Centro de Estudios Regionales Andinos "Bartolomé de Las Casas". Cusco, Peru, 541–550.

Castro-Alayo, E.M., S.G. Chávez-Quintana, E.A. Auquiñivín-Silva, A.B. Fernández-Jeri, O. Acha-De la Cruz, N. Rodríguez-Hamamura, G.I. Olivas-Orozco and D.R. Sepúlveda-Ahumada. 2019. Aceites esenciales de plantas nativas del Perú: Efecto del lugar de cultivo en las características fisicoquímicas y actividad antioxidante. Scientia Agropecuaria 10(4): 479–487.

Chekem, M.S.G., P.K. Lunga, J.D. Tamokou, J.R. Kuiate, P. Tane, G. Vilarem and M. Cerny. 2010. Antifungal properties of *Chenopodium ambrosioides* Essential Oil against *Candida* species. Pharmaceuticals 3: 2900–2909.

Chicoma-Gutiérrez, R. and A.Y. Malca-Alvarado. 2015. Efecto antibacteriano *in vitro* del aceite esencial de las hojas de *Aloysia triphylla* P. "cedrón" de la Región Cajamarca, frente a las bacterias patógenas *Escherichia coli* ATCC 25922 y *Staphylococcus aureus* ATCC 25923. Graduation Thesis. Universidad Privada Antonio Guillermo Urrelo, Trujillo, Perú.

Cole, R.A., W.A. Haber, R.O. Lawton and W.N. Setzer. 2008. Leaf essential oil composition of three species of *Myrcianthes* from Monteverde, Costa Rica. Chem. Biodivers. 5(7): 1327–1334.

Condori-Espinoza, P.E. 2018. Evaluación del control antibacteriano de *Erwinia* spp. en papa con dos aceites esenciales: muña (*Minthostachys mollis*) y paico (*Chenopodium ambrosioides*). Graduation Thesis. Universidad Nacional del Altiplano, Puno, Perú.

Dávila-Guerra, C.E. 2016. Actividad repelente del aceite esencial de *Minthostachys mollis* Grisebach; y elaboración de una crema repelente contra insectos adultos de la familia Culicidae. Graduation Thesis, Universidad Mayor de San Marcos, Lima, Perú.

De Figueiredo, R.O., M.B. Stefanini, L.C. Ming, M.O.M. Marques and R. Facanali. 2004. Essential Oil composition of *Aloysia Triphylla* (L'herit) Britton Leaves Cultivated in Botucatu, São Paulo, Brazil. Acta Hortic. 629: 131–134.

Díaz, C., S. Quesada, O. Brenes, G. Aguilar and J.F. Cicció. 2008. Chemical composition of *Schinus molle* essential oil and its cytotoxic activity on tumour cell lines. Nat. Prod. Res. 22: 1521–1534.

Duarte, J.A., L.A.B. Zambrano, L.D. Quintana, M.B. Rocha, E.C. Schimitt, A.A. Boligon, M.M.A. Campos, L.F.S. Oliveira and M.M. Machado. 2018. Immunotoxicological evaluation of *Schinus molle* L. (Anacardiaceae) Essential Oil in lymphocytes and macrophages. Evid. Based Compl. Alt. Med. 2018: Article ID 6541583.

Elechosa, M.A., P.L. Lira, M.A. Juárez, C.I. Viturro, C.I. Heit, A.C. Molina, A.J. Martínez, S. López, A.M. Molina, C.M. van Barren and A.L. Bandoni. 2017. Essential oil chemotypes of *Aloysia citrodora* (Verbenaceae) in Northwestern Argentina. Biochem. System. Ecol. 74: 19–29.

Espinoza-Pantigozo, I.K. 2016. Efecto comparativo de aceite esencial y extracto acuoso de hojas de *Schinus molle* L. "molle" sobre el crecimiento de *Botrytis cinerea*. Graduation Thesis. Universidad Nacional de Trujillo, Trujillo, Perú.

Ghizalberti, E.L. 2000. *Lantana camara* L. (Verbenaceae). Fitoterapia 71: 467–86.

Girault, L. 1987. *In*: UNICEF - OPS e OMS (Ed.), Kallawaya, curanderos itinerantes de los Andes. La Paz.

Guzmán-Ramírez, K.R. and E. López-Tuesta. 2018. Determinación de los constituyentes químicos del aceite esencial de *Ocotea aciphylla* y evaluación de la actividad repelente frente a *Aedes aegypti*. Graduation Thesis, Universidad Nacional de la Amazonía Peruana, Iquitos, Perú.

Huamán, Y., O.A. De La Cruz, A. Boliscov and I. Batiu. 2004. Essential oil from the fruits of *Schinus molle* L. from Peru. J. Essent. Oil Res. 7: 223–227.

Huerta-León, J.L. 2021. Efecto antimicrobiano del aceite esencial de *Aloysia triphylla* "Cedrón" y su aplicación para el control de la halitosis. Graduation Thesis. Universidad Nacional Mayor de San Marcos, Lima, Perú.

Inga-Chavelón, L. 2016. Identificación de los componentes del aceite esencial de *Lantana camara* L. Formulación y elaboración de una forma farmacéutica repelente de insectos. Graduation Thesis, Universidad Mayor de San Marcos, Lima, Perú.

Jaramillo, B.E.C., E.R. Duarte and W. Delgado. 2012. Bioactividad del aceite esencial de *Chenopodium ambrosioides* colombiano. Rev. Cubana Plant. Med. 17: 54–64.

Jardim, C.M., G.N. Jham, O.D. Dhingra and M.M. Freire. 2008. Composition and antifungal activity of the Essential Oil of the Brazilian *Chenopodium ambrosioides* L. J. Chem. Ecol. 34: 1213–1218.

Jia-liang., W., M. Dan-wei, W. Ya-nam, Z. Hong, H. Bing, L. Qun, Z. Zhi-yan and F. Jing. 2013. Cytotoxicity of Essential Oil of *Chenopodium ambrosioides* L. against human breast cancer MCF-7 cells. Trop. J. Pharm. Res. 12: 929–933.

Kishore, N., J.P.N. Chansouria and N.K. Dubey. 1996. Antidermatophytic action of the Essential Oil of *Chenopodium ambrosioides* and an ointment prepared from it. Phytother. Res. 10: 453–455.

León, J.R.H., J.W.S. Joaquin and C.F. Ruiton. 2020. Composición Química del Aceite Esencial de *Aloysia Triphylla* "Cedrón" como Insumo para la Elaboración de un Enjuague Bucal. Rev. Ágora. 7(2): 70–74.

León-Romaní, C.Z. 2009. Estudio de la extracción y determinación de la composición química del aceite esencial de paico (*Chenopodium ambrosioides* L.). Universidad Nacional del Callao, Vicerrectorado de Investigación: Editorial Universitaria 12: 6–12.

Lezama-Paredes, M.J. 2019. Efecto antibacteriano *in vitro* del aceite esencial de hojas de *Chenopodium ambrosioides* (L.) (PAICO) sobre *Staphylococcus aureus*. Graduation Thesis. Universidad Católica Los Ángeles de Chimbote, Chimbote, Perú.

Linares-Otoya, V. 2020. Considerations for the use and study of the Peruvian "muña" *Minthostachys mollis* (Benth.) Griseb and *Minthostachys setosa* (Briq.) Epling. Ethnobot. Res. Appl. 19: 1–9.

Lira, P.L., C.M. van Barren, D. Retta, A.L. Bandoni, A. Gil, M. Gattuso and S. Gattuso. 2008. Characterization of lemon verbena (*Aloysia citriodora* Palau) from Argentina by the Essential Oil. J. Essent. Oil Res. 20(4): 350–353.

Lizcano Ramón, A.J. and J.L. Vergara González. 2008. Evaluación de la actividad antimicrobiana de los extractos etanolicos y aceites esenciales de las especies vegetales *Valeriana pilosa*, *Hesperomeles ferruginea*, *Myrcianthes rhopaloides*, y *Passiflora manicata*, frente a microorganismo patógenos. Graduation Thesis, Pontificia Universidad Javeriana, Bogotá, Colombia.

Malagón, O., R. Vila, J. Iglesias, T. Zaragoza and S. Cañigueral. 2003. Composition of the essential oils of four medicinal plants from Ecuador. Flav. Fragr. J. 18: 527–531.

Malca-García, G.R., L. Hennig, M.L. Ganoza-Yupanqui, A. Piña-Iturbe and R.W. Bussmann. 2017. Constituents from the bark resin of *Schinus molle*. Rev. Bras. Farmacogn. 27: 67–69.

Maldonado-Rodríguez, M.E. 2006. Estudio químico y evaluación de la actividad antimicrobiana del aceite esencial de *Myrcianthes rhopaloides* McVaugh. Master Thesis, Universidad Politécnica Salesiana, Cuenca, Ecuador.

Marín-Diaz, V.M. 2019. Eficacia *in vitro* del aceite esencial de *Chenopodium ambrosioides* (paico), sobre el *Ascaris lumbricoides* comparado con Albendazol. Graduation Thesis. Universidad César Vallejo, Trujillo, Perú.

Marques, C. 2001. Importância econômica da família Lauraceae Lindl. Floresta e Ambiente 8: 195–206.

Mijares-Palacios, G.H. 2017. Caracterización de principios bioactivos de la especie botánica de uso tradicional *Justicia pectoralis* Jacq. con potencial efecto antihelmíntico. Graduate Thesis, Universidad Central de Venezuela, Caracas, Venezuela.

Monzote-Fidalgo, L. 2010. Potencial terapéutico del aceite esencial de *Chenopodium ambrosioides* y algunos de sus componentes frente a *Leishmania*. Ph.D. Thesis. Centro Nacional de Información de Ciencias Médicas, Habana, Cuba.

Morais Oliveira-Tintino, C.D., S.R. Tintino, P.W. Limaverde, F.G. Figueiredo, F.F. Campina, F.A.B. Cunha, R.H.S. Costa, P.S. Pereira, L.F. Lima, Y.M.L.S. Matos, H.D.M. Coutinho, J.P. Siqueira-Júnior, V.Q. Balbino and T.G. Silva. 2018. Inhibition of the essential oil from *Chenopodium ambrosioides* L. and α-terpinene on the NorA efflux-pump of *Staphylococcus aureus*. Food Chem. 262: 72–77.

Morillo-Horna, M.A. 2015. Efecto del aceite esencial de *Schinus molle* L. "molle", sobre *Varroa destructor* en colonias de *Apis mellifera* L. Graduation Thesis, Universidad Nacional de Trujillo, Trujillo, Perú.

Nea, F., D.A. Kambiré, M. Genva, E.A. Tanoh, E.L. Wognin, H. Martin, Y. Brostaux, F. Tomi, G.C. Lognay, Z.F. Tonzibo and M.-L. Fauconnier. 2020. Composition, seasonal variation, and biological activities of *Lantana camara* Essential Oils from Côte d'Ivoire. Molecules 25: 2400.

Ocampo, R. and R. Valverde. 2000. Manual de cultivo y conservación de plantas medicinales. 1° Ed. Edit. Tramil, Costa Rica. Avaliable at: http://www.manioc.org/gsdl/collect/recherch/import/tramil/manualdecu1.pdf.

Ormachea, E.C. 1979. Usos tradicionales de la "muña" (*Minthostachys* spp., Labiatae) en aspectos fitosanitarios de Cusco y Puno. Rev. Per. Entomol. 22(1).

Parodi, T.V., M.A. Cunha, A.G. Becker, C.C. Zeppenfeld, D.I. Martins, G. Koakoski, L.G. Barcellos, B.M. Heinzmann and B. Baldisserotto. 2014. Anesthetic activity of the essential oil of *Aloysia triphylla* and effectiveness in reducing stress during transport of albino and gray strains of silver catfish, *Rhamdia quelen*. Fish Physiol. Biochem. 40(2): 323–334.

Passos, B.G., R.D.D.G. de Albuquerque, A. Muñoz-Acevedo, J. Echeverria, A.M. Llaure-Mora, M.L. Ganoza-Yupanqui and L. Rocha. 2022. Essential oils from Ocotea species: Chemical variety, biological activities and geographic availability. Fitoterapia 156: 105065.

Peña-Caiza, J.M. 2018. Evaluación del efecto antialimentario y actividad insecticida del aceite esencial de molle (*Schinus molle* L.) frente al gusano blanco de la papa (*Premnotrypes vorax* Hustache). Graduation Thesis, Universidad Técnica de Ambato, Ambato, Ecuador.

Pérez-López, A., A.T. Cirio, V.M. Rivas-Galindo, R.S. Aranda and N.K. Torres. 2011. Activity against Streptococcus pneumoniae of the Essential Oil and δ-cadinene isolated from *Schinus molle* fruit. J. Essent. Oil Res. 23: 25–28.

Pérez-Robles, V.N. and A. Vera-Aguilar. 2014. Efecto del aceite esencial de *Aloysia tryphylla* en íleon aislado de *Cavia porcellus*. Graduation Thesis. Universidad Nacional de Trujillo, Trujillo, Perú.

Pino, J.A. 2011. Volatile constituents from leaves of *Justicia pectoralis* Jacq. var. tipo. J. Essent. Oil Bear Plants 14: 161–163.

Puma-Mamani, R.Y. 2019. Extracción y caracterización de aceite esencial de paico (*Chenopodium ambrosioides*) mediante arrastre de vapor. Graduation Thesis, Universidad Nacional del Altiplano, Puno, Perú.

Quispe, E.E.R. 2018. Actividad antibacteriana *in vitro* del aceite esencial de las hojas de *Schinus molle* L. (Molle) frente a cultivos de *Staphylococcus aureus*. Gradaute Thesis, Universidad Católica Los Ángeles de Chimbote, Trujillo, Perú.

Rocha, P.M.D.M., J.M. Rodilla, D. Díez, H. Elder, M.S. Guala, L.A. Silva and E.B. Pombo. 2012. Synergistic antibacterial activity of the Essential Oil of Aguaribay (*Schinus molle* L.). Molecules 17: 12023–12036.

Rojas, J., O. Palacios and S. Ronceros. 2012. The effect of the essential oil from *Aloysia triphylla* Britton (lemon verbena) on *Trypanosoma cruzi* in mice. Rev. Peru. Med. Exp. Salud Publica 29(1): 61–68.

Rojas, L.B., J. Velasco, T. Díaz, R.G. Otaiza, J. Carmona and A. Usubillaga. 2010. Chemical composition and antibacterial effects of the essential oil of *Aloysia triphylla* against genito-urinary pathogens. Bol. Latinoam. Caribe Plant. Med. Arom. 9(1): 56–62.

Rojas-Armas, J., O. Palacios-Aguero, J.M. Ortiz-Sánchez and L. Lopez de la Peña. 2015. *In vivo* evaluation of *Aloysia triphylla* britton (lemon verbena) essential oil toxicity and citral anti-*Trypanosma cruzi* activity. An. Fac. Med. 76(2): 129–134.

Roumy, V., J.C.R. Macedo, M. Bonneau, J. Samaille, N. Azaroual, L.A. Encinas, C. Riviere, T. Hennebelle, S. Saphaz, S. Antherieu, C. Pinçon, C. Neut, A. Siah, A.L. Gutiérrez-Choquevilca and L. Ruíz. 2020. Plant therapy in the Peruvian Amazon (Loreto) in case of infectious diseases and its antimicrobial evaluation. J. Ethnopharmacol. 249: 112411.

Rudas-González, D.D. 2017. Composición química, fraccionamiento y actividad *in vitro* del aceite esencial de *Aloysia citriodora* Palau ("Cedrón") sobre las bacterias *Escherichia coli* y *Salmonella typhimurium*. Gradaute Thesis, Universidad Peruana Cayetano Heredia, Lima, Perú.

Ruiz-Reyes, S.G. 2020. Variabilidad química del aceite esencial de *Aloysia triphylla* (l'her) Britton (cedrón) procedente de tres regiones del Perú. Profesor Promotion Thesis. Universidad Nacional de Trujillo, Trujillo, Perú.

Sanchez-Tito, M.A. and I. Collantes-Diaz. 2021. Actividad antimicrobiana de fracciones obtenidas del aceite esencial de *Minthostachys mollis* frente a patógenos orales. Rev. Habanera Cien. Med. 20(4): e3971.

Santos, A.C.A., M. Rossato, F. Agostini, L.A. Serafini, P.A. Santos, R. Molon and E. Dellacassa. 2009. Chemical composition of the Essential Oils from leaves and fruits of *Schinus molle* L. and *Schinus terebinthifolius* Raddi from Southern Brazil. J. Essent. Oil Bear. Plant 12: 16–25.

Santos, A.C.A., M. Rossato, L.A. Serafini, M. Bueno, L.B. Crippa, V.C. Sartori, E. Dellacassa and P. Moyna. 2010. Efeito fungicida dos óleos essenciais de *Schinus molle* L. e *Schinus terebinthifolius* Raddi, Anacardiaceae, do Rio Grande do Sul. Braz. J. Pharmacog. 20(2): 154–159.

Silva-Carrero, D.A., J.A. Matulevich-Pelaez and B.O. Devia-Castillo. 2016. Composición química del aceite esencial de hojas de *Myrcianthes rhopaloides* (Kunt) McVaugh (Myrtaceae). Rev. Facul. Cien. Basic. 12(1): 84–91.

Sundufu, A.J. and H. Shoushan. 2004. Chemical composition of the essential oils of *Lantana camara* L. occurring in south China. Flavour Fragr. J. 19(3): 229–232.

Toro, A.M.B. and J.L.L. Piedra. 2019. Obtención de aceite esencial de molle (*Schinus molle* l.) Y su evaluación antifungica sobre *Colletotrichum* spp. *in vitro*. Rev. Cien. Inst. 11(4): 101–109.

Torres-Guevara, F.A., M.L. Ganoza-Yupanqui, L.A. Suárez-Rebaza, G.R. Malca-García and R.W. Bussmann. 2020. Wild plants of Northern Peru: Traditions, scientific knowledge and innovation. *In*: M. Rai, S. Bhattarai and C.M. Feitosa (eds.). Wild Plants: The Treasure of Natural Healers CRC Press, Boca Raton.

Torres-Guevara, F.A., M.L. Ganoza-Yupanqui, L.A. Suárez-Rebaza, G.R. Malca-García and R.W. Bussmann. 2021. Ethnopharmacology of wild plants from the tropical mountains of Northern Peru. *In*: M. Rai, S. Bhattarai and C.M. Feitosa (eds.). Ethnopharmacology of Wild Plants. CRC Press, Boca Raton.

Valdez-Tenezaca, A.V. 2018. Evaluación de índices de toxicidad del aceite esencial de *Lantana camara* en *Drosophila Melanogaster*. Master Thesis, Universidad de Cuenca, Cuenca, Ecuador.

Venegas del Castillo, A. and M.N. Vásquez-Valles. 2016. Efecto del aceite esencial de *Lantana camara* sobre el crecimiento de *Staphylococcus aureus* y *Escherichia coli*. Rev. Cien. Facultad Cien. Biol. 36(1): 29–37.

Venegas-Casanova, E.A., S.G. Ruiz-Reyes, J.G. Gavidia-Valencia, R. Jara-Aguilar, J.C. Uribe-Villarreal, Y.F. Curo-Vallejos, R.A. Rengifo-Penadillos, J.L. Martinez and A. Cuéllar. 2018. Variability in the chemical composition of *Justicia pectoralis* jacq. (two varieties): Essential oils in over several months. Pharmacology Online 3: 402–411.

Werka, J.S., A.K. Boehme and W.N. Setzer. 2007. Biological activities of Essential Oils from Monteverde, costa rica. Nat. Prod. Commun. 2(12): 1215–1219.

Zegarra-Carrera, P.M. 2019. Efecto antimicótico *in vitro* del aceite esencial de hojas de *Schinus molle* L. Frente a cepas de *Candida albicans*. Gradaution Thesis, Universidad Católica Los Ángeles de Chimbote, Trujillo, Perú.

6

Traditional Use of Medicinal Plants and Essential Oils

*Golshan Zare, N. Yağmur Diker and I. Irem Tatlı Çankaya**

Introduction

Essential Oils

Essential oils are complex mixtures of volatile secondary metabolites, which are usually liquid at room temperature, sometimes solid, that are obtained from plants or herbal drugs by water/steam distillation and contain terpenic, aromatic substances, nitrogen, and sulphurous heterosides. Essential oils have their natural scent and their relaxing and healing effect has made them popular in current supportive treatments (Table 1).

Known as "Essential oil" in English, *Aetherische Öle* in German, and *Huile essentielle* in French, this group of substances is called *Uçucu yağlar* in Turkish. They are sometimes erroneously referred to as "essential fatty acids". However, these are in the class of fixed oils. Essential oils are part of the plant's immune system, a defence mechanism against environmental threats. Studies have shown that odours are effective on neurophysiological and autonomic functions, which in turn affect our physical and mental states (Tayfun 2019).

The use of essential oils in the body changes body chemistry supports body systems and improves mental and emotional states. Humans can distinguish more than ten thousand odours. The molecules in inhaled essential oils reach the olfactory receptors in the nose. Different molecules bind to different sites of these receptors. The receptors convert odours into electrical impulses and these electrochemical messages, which are formed by the binding of molecules to the receptors, are transmitted to the limbic system via the olfactory bulb and olfactory pathway. These

Hacettepe University, Faculty of Pharmacy, Department of Pharmaceutical Botany, Sıhhiye, Ankara, Turkey, 06100.

Emails: golshanzare@gmail.com; yagmurkumser@gmail.com

* Corresponding author: itatli@hacettepe.edu.tr

Table 1: The list of common essential oils, their nomenclature, family, part used, and organoleptic properties (essential oils that are generally obtained from parts of the plants by water-steam distillation).

Essential Oil	Plant Name	Family	Part Used	Organoleptic Properties
Angelicae radix aetheroleum	*Angelica archangelica* L.	Apiaceae	Roots	Pale yellow, smell fresh, woody, and peppery aroma.
Boswellia carterii aetheroleum/ Frankincense	*Boswellia carteri* Birdw.	Burseraceae	Resin	Colourless to light yellow, viscous, and characteristic scented.
Boswellia serratae aetheroleum/Indian Frankincense (Olibanum indicum) Essential Oil	*Boswellia serrata* Roxb	Burseraceae	Resin	Amber colour, viscous, fresh limon, and terpen scented.
Cananga aetheroleum/ Ylang Ylang Essential Oil	*Cananga odorata* (Lam.) Hook. f. and Thomson	Annonaceae	Fresh flower	Light amber to yellow-brown colour, fluid, sweet, and floral scented.
Carvi aetheroleum/ Caraway Essential Oil	*Carum carvi* L.	Apiaceae	Fruits (crushed)	Colourless or yellow, clear, and fluid.
Cedrus aetheroleum/ Cedarwood Essential Oil	*Cedrus atlantica* (Endl.) Manetti ex Carrière	Pinaceae	Woods	Golden yellow-brown colour, fluid, and characteristic scented.
Chamomillae romanae aetheroleum/ Roman Chamomillae oil	*Anthemis nobilis* L.	Asteraceae	Capitulums/ Flowers	Pale yellow, clear, fluid, bright, crispy, sweet, and fruity scented.
Cinnamomi camphorae aetheroleum/ Ravintsara Essential Oil	*Cinnamomum camphora* var. *linaloolifera* Y. Fujita	Lauraceae	Leaves and twigs	Colourless or pale-yellow colour, clear, fluid, and kafur scented.
Cinnamomi zeylanici corticis aetheroleum/ Ceylon, Cinnamon Bark Oil	*Cinnamomum zeylanicum* Blume	Lauraceae	Branch barks	Yellow colour that turns red over time, clear, fluid, and characteristic scented reminiscent of cinnamic aldehyde.
Cisti cretici aetheroleum/Cistus Essential Oil and Labdanum Oil	*Cistus creticus* L.	Rutaceae	Fresh leafy branch tips and Resin (Labdanum)	Pale-yellow, from yellow to deep red and brown colour, clear, fluid, balsamic, and amber scented.
Citri bergamiae aetheroleum/ Bergamot Essential Oil	*Citrus aurantium* L. var. *bergamia* (Wight et Arnott) Engler	Rutaceae	Fresh peels of fruits	Pale-yellow, yellow, yellowish-brown colour, clear, fluid, and characteristic scented.

Table 1 contd. ...

...Table 1 contd.

Essential Oil	Plant Name	Family	Part Used	Organoleptic Properties
Limonis aetheroleum/ Lemon Essential Oil	*Citrus limon* (L.) Burman fil.	Rutaceae	Fresh peels of fruits	Pale yellow or greenish yellow colour, clear that be blurry at low temperatures, and strong lemon scent.
Citri paradisi aetheroleum/ Grapefruit Essential Oil	*Citrus* x *paradisi* (L.) Macfady	Rutaceae	Fresh peels of fruits	Yellow or reddish-yellow colour, clear, fluid, and characteristic scented.
Citri reticulatae aetheroleum/ Mandarin Essential Oil	*Citrus reticulata* Blanco.	Rutaceae	Fresh peels of fruits	Greenish-yellow or red, orange colour, fluid, and characteristic scented.
Aurantii dulcis aetheroleum/Sweet Orange Essential Oil	*Citrus* x *sinensis* (L.) Osbeck.	Rutaceae	Fresh peels of fruits	Pale yellow to orange colour, fluid, and clear that may become blurry when shaken.
Cumini aetheroleum/ Cumin Essential Oil	*Cuminum cyminum* L.	Apiaceae	Fruits (lightly pounded)	Dark brown to dark amber colour, clear, fluid, and energising spice scented.
Cupressi aetheroleum/ Cypress Essential Oil	*Cupressus sempervirens* L.	Cupressaceae	Fresh leaves and twigs	Colourless, very light yellow colour, fluid, woody, and slightly spicy scented.
Cymbopogon martinii motiae aetheroleum/ Palmarosa Essential Oil (Turkish geranium oil)	*Cymbopogon martinii* L. var. *motia*	Poaceae	Aerial parts	Pale yellow to olive oil colour, fluid, and rose-like scented.
Cymbopogon martinii sofiae aetheroleum/ Gingergrass	*Cymbopogon martinii* L. var. *sofia*	Poaceae	Aerial parts	Pale yellow colour, fluid, and characteristic scented.
Cymbopogon nardi aetheroleum/ Citronella Essential Oil	*Cymbopogon nardus* L.	Poaceae	Freshly cut, or dried leaves	From yellow to olive oil, fluid, characteristic, lemon-like scent (Super quality Java-derived essential oil is colourless or light yellow colour).
Eucalypti citriodorae aetheroleum/Lemon-Scented (Gum) Eucalypt Essential Oil	*Eucalyptus citriodora* Hook.	Myrtaceae	Young leaves and branch tips	Light yellow to greenish-yellow, fluid, and citronellal scented.

Table 1 contd. ...

...Table 1 contd.

Essential Oil	Plant Name	Family	Part Used	Organoleptic Properties
Eucalypti aetheroleum/ Eucalyptus Essential Oil	*Eucalyptus globulus* Labill.	Myrtaceae	Young leaves	Colourless or pale yellow colour and eucalyptol scented.
Eucalypti radiatae aetheroleum/Narrow-leaved peppermint Essential Oil	*Eucalyptus radiata* Sieber ex DC	Myrtaceae	Leaf and branch tips	Colourless to light yellow colour, clear, fluid, and eucalyptol scented.
Eucalypti smithii aetheroleum/ Eucalyptus smithii Essential Oil (Gully Gum Oil)	*Eucalyptus smithii* F. Muell. ex R.T. Baker	Myrtaceae	Young leaves and branch tips	Colourless or up to pale yellow, fluid, sweetish, woody, and eucalyptol scented.
Foeniculi dulcis aetheroleum/Sweet Fennel Essential Oil	*Foeniculum vulgare* Mill. ssp. *vulgare* var. *dulce* (Miller) Thellung	Apiaceae	Fruits (lightly pounded)	Light yellow colour, semi-fluid, and anise scented.
Helichrysi italici aetheroleum/ Immortelle Essential Oil	*Helichrysum italicum* (Roth) G. Don.	Asteraceae	Inflorescence	Light yellow colour, semi-fluid, and characteristic scented.
Helichrysi splendidi aetheroleum/ Everlasting Oil	*Helichrysum splendidum* (Thunb.) Less	Asteraceae	Inflorescence	Light yellow colour, semi-fluid, and characteristic scented.
Juniperi aetheroleum/ Juniper Essential Oil	*Juniperus communis* L.	Cupressaceae	Ripe and unfermented cones	Colourless or yellowish colour, fluid, and characteristic scented.
Juniperi virginianae aetheroleum/Eastern Red Cedar (Virginian Cedarwood) Essential Oil	*Juniperus virginiana* L.	Cupressaceae	Woods	Colourless or light yellow colour, fluid, and characteristic scented.
Lauri folii aetheroleum/Bay Laurel Leaf Essential Oil	*Laurus nobilis* L.	Lauraceae	Fresh or dried leaves	Colourless, light green or pale yellow colour, clear, fluid, and characteristic scented.
Lavandulae aetheroleum/Fine Lavender Essential Oil	*Lavandula angustifolia* Mill.	Lamiaceae	Flowers	Colourless to pale yellow colour, clear, fluid, complex, fresh, floral, and intense lavender scented.

Table 1 contd. ...

...Table 1 contd.

Essential Oil	Plant Name	Family	Part Used	Organoleptic Properties
Lavandulae aetheroleum/Lavandin Essential Oil	*Lavandula* x *intermedia* Emeric ex Loisel.	Lamiaceae	Spica inflorescence flowers and flowering twigs	Light yellow colour, clear, fluid, and less lavender scent compared to medicinal lavender essential oil.
Spicae aetheroleum/ Spike Lavender Essential Oil	*Lavandula latifolia* Medik.	Lamiaceae	Flowers and flowering twigs	Pale-yellow or greenish-yellow colour, clear, fluid, eucalyptol and camphor scented.
Leptospermi scoparii aetheroleum/Manuka Essential Oil	*Leptospermum scoparium* J.R. Forst. and G. Forst.	Myrtaceae	Leaves and twigs	Yellow to brown colour, clear, fluid, and characteristic scented.
Matricariae aetheroleum/ Chamomile Essential Oil	*Matricaria recutita* L.	Asteraceae	Fresh or dried flowering or flowering upper stem	Navy blue colour, clear, viscous, and dense characteristic scented.
Melaleucae alternifolii aetheroleum/Tea Tree Essential Oil	*Melaleuca alternifolia* L.	Myrtaceae	Leaves and thin twig	Colourless or pale yellow colour, clear, fluid, and characteristic scented.
Melaleucae aetheroleum/Cajeput Essential Oil	*Melaleuca cajuputi* Powell	Myrtaceae	Leaves and thin twig	Colourless or pale-yellow colour, clear, fluid, and characteristic scented.
Melissae aetheroleum/ Lemon Balm Essential Oil	*Melissa officinalis* L.	Lamiaceae	Fresh and dried leaves	Colourless to light yellow colour, clear, fluid, and scented reminiscent of lemon.
Menthae arvensis aetheroleum partim mentholum depletum/ Mint Essential Oil, Partly Dementholised	*Mentha canadensis* L.	Lamiaceae	Fresh flowering aerial parts	Colourless, pale yellow or greenish-yellow colour, fluid, and characteristic scented.
Menthae piperitae aetheroleum/ Peppermint Essential Oil	Menthae x piperita L.	Lamiaceae	Fresh flowering aerial parts	Colourless, pale yellow or greenish-yellow colour, fluid, characteristic scented, and refreshing.
Myrrh Aetheroleum/ Myrrh Essential Oil	*Commiphora myrrha* (Nees) Engl.	Burseraceae	Resin	–

Table 1 contd. ...

...Table 1 contd.

Essential Oil	Plant Name	Family	Part Used	Organoleptic Properties
Basilici aetheroleum/ Sweet Basil Essential Oil	*Ocimum basilicum* L.	Lamiaceae	Leaves and upper branches with buds	Colourless to pale yellow colour, fluid, and characteristic scented.
Pelargonii aetheroleum/ Geranium Essential Oil	*Pelargonium graveolens* L'Hér.	Geraniaceae	Leaves	Yellow colour, clear, fluid, and rose-like scented.
Anisi aetheroleum/ Anise Essential Oil	*Pimpinella anisum* L.	Apiaceae	Dried ripe fruits (lightly pounded)	Colourless or pale yellow colour, and clear.
Pini pumilionis aetheroleum/Dwarf Pine Essential Oil	*Pinus mugo* Turra	Pinaceae	Coniferous leaves and shoots	Colourless or pale yellow colour, clear, fluid, and characteristic scented.
Pini sylvestri aetheroleum/Scotch Pine Essential Oil	*Pinus sylvestris* L.	Pinaceae	Fresh leaves and branches	Colourless or pale yellow colour, clear, fluid, and characteristic scented.
Pogostemon cablin aetheroleum/Patchouli Essential Oil	*Pogostemon cablin* (Blanco) Benth.	Lamiaceae	Leaves	Light yellow to reddish-brown colour, clear, fluid, and woody scented.
Rosae damascenae aetheroleum/Damask Rose Essential Oil	*Rosa* x *damascena* Mill.	Rosaceae	Freshly collected petals	Pale yellow colour, fluid, and characteristic scented.
Rosmarini Aetheroleum/ Rosemary Essential Oil	*Rosmarinus officinalis* L.	Lamiaceae	Aerial parts with flowers	Colourless or pale yellow, fluid, and characteristic scented.
Salviae trilobae aetheroleum/ Sage Triloba Oil, and Turkish Sage Essential Oil	*Salvia fruticosa* Mill.	Lamiaceae	Leaves	Colourless or pale yellow colour, clear, fluid, and characteristic scented.
Salviae lavandulifoliae aetheroleum/Spanish Sage Essential Oil	*Salvia lavandulifolia* Vahl.	Lamiaceae	Aerial parts of the plant collected at the flowering stage	Colourless or pale yellow colour, fluid, and characteristic scented.
Salviae sclareae aetheroleum/Clary Sage Essential Oil	*Salvia sclarea* L.	Lamiaceae	Flowering branch tips, spike inflorescence	Colourless or slightly yellowish colour, fluid, and characteristic scented.

Table 1 contd. ...

...Table 1 contd.

Essential Oil	Plant Name	Family	Part Used	Organoleptic Properties
Santali ligni aetheroleum/ Santalwood Essential Oil	*Santalum album* L.	Santalaceae	Woods	Pale yellow to light yellow, and light viscose with characteristic scented.
Caryophylli flos aetheroleum/Clove Bud Essential Oil	*Syzygium aromaticum* (L.) Merr. et L.M. Perry	Myrtaceae	Dried flower buds that have turned reddish-brown in colour	Yellow in colour that turns red-brown over time, clear, and eugenol scented.
Thymi serpylli aetheroleum/Turkish Thyme Essential Oil	*Thymus serpyllum* L.	Lamiaceae	Fresh flowering aerial parts	Pale yellow colour and fluid.
Thymi aetheroleum typo geraniolo/Thyme Essential Oil C.T. Geraniol	*Thymus vulgaris* L., *T. zygis* L. or a mix of both types	Lamiaceae	Fresh flowering aerial parts	Yellow or reddish-brown colour, clear, fluid, and geranium scented.
Thymi aetheroleum typo linalool/Thyme Essential Oil C.T. Linalool	*Thymus vulgaris* L., *T. zygis* L. or a mix of both types	Lamiaceae	Fresh flowering aerial parts	Yellow or reddish-brown colour, clear, fluid, and slight lavender scented.
Thymi typo thymolo aetheroleum/Thyme Essential Oil	*Thymus vulgaris* L., *T. zygis* L. or a mix of both types	Lamiaceae	Fresh flowering aerial parts	Yellow or reddish-brown colour, clear, fluid, and thymol scented.
Valerianae aetheroleum/ ValerianaeEssential Oil	*Valeriana officinalis* L.	Valerianaceae	Roots	Yellowish-brown colour, clear, viscose, and characteristic scented.
Vetiverae aetheroleum/Vetiver Essential Oil	*Vetiveria zizanioides* (L.) Nash.	Poaceae	Roots	Yellowish-brown colour, clear, viscose, and characteristic scented.

messages activate memory and emotional responses through the hypothalamus, allowing the resulting response to be sent to other parts of the brain and body. These messages provide euphoric, relaxation, sedation, and stimulating actions. For example, it is known that some components with analgesic effects in essential oils affect the release of substances, such as dopamine, endorphins, norepinephrine, and serotonin in the brain stem and thus exert an analgesic effect. Briefly, inhaling essential oils transmits signals from the olfactory system to the brain, which releases neurotransmitters, such as serotonin and dopamine, regulating anxiety, depression, and mood disorders, and exerts an analgesic effect (Babar et al. 2015).

When applied to the skin, essential oils begin to act immediately in body tissues. Essential oil molecules are absorbed from the pores in the skin in topical applications and mixed with the bloodstream and mixed with all parts of the body. It is known that the manipulation of the soft tissues of the body relieves mental and physical tension, relieves pain, stimulates healthy circulation, and restores the balance of one's health. General information and usage properties of commonly used essential oils are given separately in Tables 1 and 2.

Obtaining Methods of Essential Oils

Essential oils used in aromatherapy are obtained by distillation (water distillation, steam distillation, and water-steam distillation) and mechanical methods used for *Citrus* peels (Kaya and Ergönül 2015).

In the production of essential oil, criteria such as which part of the plant the oil is located, whether it is obtained from fresh or dry material, sensitivity to heat, volatility, boiling point, pH, solubility, and oxidation is very important.

Distillation

In distillation, the most widely used method, steam distillation is suitable for plants that have decomposed by heat or that carry essential oil on their surface, and water-steam distillation is preferred for plants that carry essential oils in deep tissues. In water-steam distillation, the plant is boiled in water that is added for 2–8 hours and the essential oil is stored by the disintegration of the plant cell walls, which are separated from the cells and passed into the cooling tanks with the steam. The essential oil condensed in the cooler is then collected. The collected essential oils are packaged, and the underlying liquid is sometimes used as aromatic water.

In the steam distillation method, the steam applied to the fresh plant material placed in the glass container with the help of pressure is brought to the collection container by dragging the oil droplets along with it, and the oil is condensed and separated there. It is suitable for plants that are sensitive to heat, such as cinnamon and thyme.

Mechanical Method

The preferred squeezing method for obtaining essential oils from fresh fruit peels of *Citrus* is mechanical. It is based on the principle of obtaining essential oils by bursting the secretory pockets in the fruit peels. Essential oils are obtained by placing the peels of these fruits in a cloth bag and squeezing them in cold hydraulic presses. The absence of temperature during application provides the essential oils obtained with an intense aroma feature (Tayfun 2019, Kaya et al. 2015, Trease and Evans 2002).

After the essential oils are obtained, quantification, quality control, and pharmacopoeia analyses must be performed. The chemical profile and purity of essential oils are determined by gas chromatography/mass spectrometer analysis. In this way, the quality of their therapeutic activity can also be evaluated (Tatlı 2012).

Table 2: Traditional and evidence-based use information of common essential oils.

Essential Oil	(Traditional) Use in Folk Medicine	Evidence-Based Use
Angelicae radix aetheroleum	It is used for coughs, sinus infections, arthritis, gout, fatigue, psoriasis, stress and quitting smoking, and nicotine addiction.	It is used in the treatment of anxiety and stress-related depression, as an invigorating and strengthening body, respiratory ailments, and digestive problems.
Boswellia carterii aetheroleum	In respiratory diseases and externally as a wound healer, hemostatic in uterine haemorrhages.	Externally an epithelial, cicatrizant, broad-spectrum antimicrobial, antiseptic, astringent, and sedative. In asthma and shortness of breath.
Boswellia serratae aetheroleum	Externally as an anti-inflammatory, pain reliever in skin diseases, as a pain reliever in muscle and rheumatic pains, as well as in depression, and stress situations.	Externally an antimicrobial, anti-inflammatory, analgesic in muscle and rheumatic pains, as an adjuvant in radiotherapy, as a skin protector, epithelializing, preventive and healing of pregnancy cracks, and also for sedative purposes.
Cananga aetheroleum	Against depression, stress, lowering blood pressure, increasing sexual power, and also externally healing the skin.	Pain reliever and as a sedative effect, as well as improving cognitive performance and calming. Alone or in combination with other essential oils (mainly *Lavandula angustifolia*, *Matricaria recutita*, *Rosa damascena*, *Salvia sclarea*, and *Melissa officinalis*) for sedative and pain relief at the time of birth.
Carvi aetheroleum	Externally in scabies and fungal diseases; support the digestive function, and relieve colic and bloating.	Externally in relief of irritable bowel syndrome symptoms. In meteorism, dyspepsia, and in children or infants as an anti-colic and carminative. It has antimicrobial and antiviral effects.
Cedrus aetheroleum	As a sedative and sexual enhancer.	The respiratory regulator in cough and bronchitis, relieving physical stress, externally antiseborrheic, antifungal, as an aid in the treatment of alopecia (*Alopecia partiale* and *A. areata*), osteoarthritis, and oily skin acne (*Acne vulgaris*) for epithelial purposes and cellulite.
Chamomillae romanae aetheroleum/ Roman Chamomillae oil	It has anti-inflammatory, antiseptic and bactericidal properties. It is used in infected wounds, inflammations, cuts, burns, allergies and skin problems, such as insect bites. It is applied as a compress or bath.	It has anti-inflammatory, antiseptic and bactericidal properties. It is used in infected wounds, inflammations, cuts, burns, allergies, and skin problems, such as insect bites.
Cinnamomi camphorae aetheroleum	As a breath freshener in viral enteritis and flu infections.	For expectorant and antiviral purposes, in flu infections, neuromuscular disorders, and externally in the treatment of shingles.

Table 2 contd. ...

...Table 2 contd.

Essential Oil	(Traditional) Use in Folk Medicine	Evidence-Based Use
Cinnamomi zeylanici corticis aetheroleum	In colds, rheumatism, stress, and as a circulatory regulator.	As antimicrobial, antirheumatic, and anti-inflammatory.
Cisti cretici aetheroleum	As an anti-inflammatory, antiseptic, external antiviral, and wound healing.	Anti-inflammatory, antimicrobial, antiviral, cytotoxic, vasodilator, and in upper respiratory tract infections and externally as a wound healer.
Citri bergamiae aetheroleum	As an antipyretic, against parasites, as therapeutic for tonsillitis, sore throat, and externally minor wounds.	Externally as an antimicrobial, reducing the loss of pigmentation on the skin, as well as a sleep regulator, anxiolytic, sedative and pain reliever.
Limonis aetheroleum	For memory-enhancing purposes, and as an antiseptic.	Immunostimulant, skin whitening, antimicrobial, anxiolytic, antidepressant and to increase cognitive performance.
Citri paradisi aetheroleum	As an antiseptic, in the treatment of migraine, and as a pain reliever.	Antimicrobial, anxiolytic, antidepressant, cognitive performance enhancer, appetite suppressant and helping to reduce the appearance of cellulite.
Citri reticulatae aetheroleum	Against insomnia and stress.	For refreshing, calming and antiseptic purposes and especially as a respiratory enhancer in children.
Aurantii dulcis aetheroleum	As a pain reliever, sedative and respiratory relaxant.	Anxiolytic, externally antimicrobial purposes, and in the removal of skin blemishes, acne and acne treatment.
Citri bergamiae aetheroleum	As an antipyretic, against parasites, tonsillitis, sore throat, and externally therapeutic for minor wounds.	Externally as an antimicrobial, reducing the loss of pigmentation on the skin, as well as a sleep regulator, anxiolytic, sedative, and pain reliever.
Cumini aetheroleum	As carminative.	In dyspepsia, epigastralgia, mild analgesic, as a concentration enhancer, in flu infections in children, and for antimicrobial purposes.
Cupressi aetheroleum	As a sedative, depressant and for respiratory tract disorders.	Antiseptic, wound healing, astringent, cleanser for oily skin, antispasmodic, sedative, relief of venous circulation disorders and haemorrhoid problems, hemostatic in nose bleeding, foot sweating, cough, bronchitis, asthma, pertussis, emphysema, influenza, menstrual disorders, as relieving arthritis, and rheumatic pain.
Cymbopogon martinii motiae aetheroleum	For sore throat, tension reliever, and mosquito repellent.	As a broad spectrum antibacterial, strong antimycotic, purifier for bronchitis, rhinopharyngitis, urethritis, vaginitis, dermatitis, acne, and oily skin.
Cymbopogon martinii sofiae aetheroleum	For sore throat, tension reliever and mosquito repellent.	As an antimicrobial, purifying in rhinopharyngitis, and oily skin.

Table 2 contd. ...

...Table 2 contd.

Essential Oil	(Traditional) Use in Folk Medicine	Evidence-Based Use
Cymbopogon nardi aetheroleum	As a mosquito repellent and for sore throat.	For antimicrobial, anti-inflammatory, spasmolytic, arthritis, deodorant, insect repellent, air disinfectant, and epithelializing purposes.
Eucalypti citriodorae aetheroleum	In upper respiratory tract infections.	As an antiviral, antimicrobial, anti-inflammatory, antirheumatic, analgesic, sedative, and external wound healer in upper respiratory tract infections.
Eucalypti aetheroleum	In upper respiratory tract infections.	In upper respiratory tract infections and in essential oil mixtures used to heal foul-smelling necrotic wounds (cineole chemotype), in rheumatoid arthritis (camphor chemotype), hair loss and against head lice (verbenon chemotype).
Eucalypti radiatae aetheroleum	In sinusitis, bronchitis, gingival and middle ear inflammation, externally wound treatment, fungal infections, abdominal pain, and rheumatism.	In antimicrobial, antiviral, expectorant, anti-inflammatory, throat infections, toothache, acute and chronic sinusitis. It has a stimulating effect on the sympathetic nervous system.
Eucalypti smithii aetheroleum	Against upper respiratory tract infections, joint and muscle pain, microbial, and viral infections.	As an adjunct to the treatment of recurrent upper respiratory tract infections for antimicrobial and antiviral purposes.
Foeniculi dulcis aetheroleum	In massage to treat various digestive disorders, and to support weight loss programs.	As an antispasmodic, respiratory system stimulant (bronchodilator), for relieving arthritis, rheumatism, cellulite, oedema and anti-inflammatory properties.
Helichrysi italici aetheroleum	Externally for the improvement of blood accumulation in subcutaneous tissues, bruises and scars, stopping bleeding in injuries, and relieving inflammation, and pain in joint rheumatism.	Externally for the healing of cuperosis, hematoma, scars, traumas, decubitus, keloid, epithelializing, radiotherapy as an adjuvant to protect and heal the skin, in rheumatoid arthritis as an anti-inflammatory, pain reliever and give vitality.
Helichrysi splendidi aetheroleum	As a curative and calming agent, for pain, as an anti-inflammatory, for relief of muscle spasms, for colds, pneumonia, and against infections.	Externally as an antimicrobial and epithelializing and mucolytic.
Juniperi aetheroleum	Externally to help relieve muscle and joint pain and to relieve pain in urinary tract infections.	Externally as an antimicrobial, antiviral, in rheumatic diseases, joint pain, helping to reduce the appearance of first and second stage cellulite, and also in bronchitis and as a cognitive performance enhancer.

Table 2 contd. ...

...Table 2 contd.

Essential Oil	(Traditional) Use in Folk Medicine	Evidence-Based Use
Juniperi virginianae aetheroleum	Externally as an antispasmodic, wound healer and externally in eczema, to help relieve muscle and joint pain and to relieve pain in urinary tract infections.	As an antimicrobial, antispasmodic, and anxiolytic, and in psoriasis and venous circulation disorders.
Lauri folii aetheroleum	Against haemorrhoids and relieving rheumatic pains, throat and nose infections, fungal infections on the skin, and digestive disorders.	Against upper respiratory tract infections (bronchitis), as a broad spectrum antibacterial, anticandidal, antiviral (HSV-1), expectorant, insect repellent, in the treatment of hair loss, greasy and dandruff hair, to increase attention, muscle aches, neuralgia, arthritis, flatulence, dental infections, stomatitis, aphthosis, circulatory disorders, and neurovegetative dystonia.
Lavandulae aetheroleum/ Fine Lavender Essential Oil	Externally for abscesses, acne, bedsores, burns, itching, skin allergies, haemorrhoids, hair loss, insect bites, colds, mental and tension problems, and concentration disorders.	Anxiety, bipolar disorder, borderline status, primary and secondary level depression, sleep disorder, dementia, arterial hypertonia and as an immunomodulator, against insect bites and head lice in pain, dysmenorrhea, as well as fungal and bacterial infections (especially *Staphylococcus aureus*), eczema, dermatitis, erythema, sunburn and radiotherapy. As physical and mental support for postpartum mothers, alone or together with other essential oils (mainly *Matricaria recutita, Cananga odorata, Rosa damascena, Salvia sclarea, Melissa officinalis*) for sedative and pain relief at the time of delivery. As a sedative and pain reliever through massage from the 34th week as preparation for the birth and at the time of birth.
Lavandulae aetheroleum/ Lavandin Essential Oil	Externally for abscesses, acne, bedsores, burns, itching, skin allergies, haemorrhoids, hair loss, insect bites, colds, as well as mental and tension problems, and concentration disorders.	Against anxiety, bipolar disorder, borderline condition, primary and secondary depression, sleep disorders, pain, fungal and bacterial infections, eczema, dermatitis, erythema, and insect bites.
Spicae aetheroleum	Externally for abscesses, acne, bedsores, itching, skin allergies, mild burns, haemorrhoids, hair loss, insect bites, as well as relieving pain, colds, and concentration disorders.	As an anti-inflammatory, fungal and bacterial infections (especially *Staphylococcus aureus*), eczema, dermatitis, decubitus, first-degree burns, pain, insect bites, as well as tension state, sleep disorder, bronchitis, sinusitis.

Table 2 contd. ...

...Table 2 contd.

Essential Oil	(Traditional) Use in Folk Medicine	Evidence-Based Use
Leptospermi scoparii aetheroleum	As relieve cough and colds, and pain in arthritis, as well as sprains, and bruises.	Antibacterial, antiviral, strong antimycotic, anti-inflammatory, mucolytic, wound healing, candidosis, psoriasis, decubitus, stomatitis. Combination with Kanuka (*Kunzea ericoides*) essential oil in radiotherapy-induced mucous inflammation and Immortelle (*Helichrysum italicum*) oil in first and second-degree burns.
Matricariae aetheroleum	In gastrointestinal disorders, inflammations, antiseptic, antispasmodic, sedative, and topically in a suitable carrier oil against migraine.	In skin and mucosal irritations in the anal and genital areas, dermatitis, acne, *Herpes simplex*, and oral mucositis, as anti-inflammatory, antispasmodic, antimicrobial and antiseptic in gastrointestinal system disorders. Along with other essential oils (mainly *Lavandula angustifolia, Cananga odorata, Rosa damascena, Salvia sclarea, Melissa officinalis*) for sedative and pain relief at the time of birth. As a sedative and pain reliever from the 34th week as preparation for the birth and at the time of birth.
Melaleucae alternifolii aetheroleum	Externally for wounds, burns, and insect bites.	In upper respiratory tract infections, antimicrobial, acne, eczema, *Herpes simplex*, recurrent *Herpes labialis* and other viral skin diseases (*Molluscum contagiosum*), psoriasis, gingivitis, cuts, 1st and second-degree burns, insect bites, head lice (*Pediculus humanus capitis*) and against athlete's foot (*Tinea pedis*) and nail fungus (*Tinea ungium*).
Melaleucae aetheroleum	In acne, boils, bronchitis, viral upper respiratory tract infections, sinusitis, and calcification.	As an antimicrobial and in upper respiratory tract infections, as well as reducing radiotherapy-induced skin damage and wound healing, in the treatment of psoriasis, fructose, insect bites, athlete's foot (*Tinea pedis*) and nail fungus (*Tinea ungium*) and *Herpes genitalis*.
Melissae aetheroleum	Relieve sedative, sleep-inducing, nervous sleep disorders, and functional disorders of the gastrointestinal tract.	as antiviral (*Herpes labialis* and *Herpes genitalis*), cystitis, hypotensive, immunomodulatory, pregnancy nausea, amenorrhoea, drowsiness, sedative, cardiac arrhythmia, anxiety, irritability, varicose veins, birth stress reducer and pre-radiotherapy skin care (Citral chemotype). The antiviral effect is stronger. In antibacterial, rhinitis, sinusitis, laryngitis, cognitive performance enhancer, tension-type headache and migraine, analgesic and dyspepsia (Citronellal chemotype).

Table 2 contd. ...

...Table 2 contd.

Essential Oil	(Traditional) Use in Folk Medicine	Evidence-Based Use
Menthae arvensis aetheroleum partim mentholum depletum	In cold, cough and bronchitis, pharynx and mouth inflammation, pain and infection conditions, muscle aches, and neuralgia.	In upper respiratory tracts due to its antimicrobial, antiviral, cooling and mild analgesic effect in muscle pains, neuralgia and migraine, sports injuries, and joint rheumatism.
Menthae piperitae aetheroleum	In cough and cold, muscle contractions and cramps, neuralgia and migraine, and also as a cognitive performance enhancer.	As an antimicrobial, antispasmodic in the upper respiratory tract, as an aid in the treatment of irritable bowel syndrome, as an analgesic, mild antiemetic, in rheumatoid arthritis, muscle pain, tension-type headache, migraine, cognitive performance enhancer, and as a wound healer, against cold sores and gingivitis.
Myrrh aetheroleum	In cold, flu, sinusitis, cough, sore throat, gum problems, and skin disorders.	As an analgesic, antiseptic, antioxidant, anti-inflammatory, astringent, antispasmodic, and carminative.
Basilici aetheroleum	Against headaches and migraine, against depression, stomach and respiratory tract disorders, fever, menstrual cramps, colds, and insect bites.	Antimicrobial, mild analgesic, used in headache, sinusitis, common cold, arthritis, muscle aches and oily scalp. Relieves insecurity, indecision, negative thoughts, stress, depression, fear, anger, and mental depression.
Origani dubii aetheroleum	In tooth and headaches, cramps, sprains and crushes, arthritis, cellulite treatment, antiseptic and toenail fungus.	In arthritis, muscle pain, antimicrobial, toenail fungus (*Tinea pedis – T. ungium*), anxiolytic, and dysmenorrhea.
Pelargonii aetheroleum	In external haemorrhoids and heavy menstrual bleeding.	In neurodegenerative diseases such as Parkinson's, Alzheimer's, ALS and Multiple sclerosis. It has antimicrobial effects, and insecticidal and insect repellent activity.
Anisi aetheroleum	As a milk enhancer and as an expectorant in upper respiratory tract infections.	As an analgesic, spasmolytic, and estrogenic effects, in dental infections, PMS, amenorrhea and oligomenorrhea, meteorism, dyspepsia and for expectorant purposes. Mixtures are used in migraine attacks and against head lice (*Pediculus humanus capitis*).
Pini pumilionis aetheroleum	Against nasal congestion and skin diseases and as a pain reliever.	As a decongestant, cold, dry cough, against skin diseases, as an antirheumatic and neurological pain reliever.
Pini sylvestri aetheroleum	Against nasal congestion and skin diseases and as a pain reliever.	As an anti-inflammatory, antimicrobial, decongestant, anti-rheumatic and neurological pain reliever in colds, dry cough, and skin diseases.

Table 2 contd. ...

...Table 2 contd.

Essential Oil	(Traditional) Use in Folk Medicine	Evidence-Based Use
Pogostemon cablin aetheroleum	As a wound healer, skin and hair care, mood stabilizer, aphrodisiac, and insect repellent.	Antiphlogistic, cell regenerative, acne, eczema, antiseptic, foot fungus, insecticide, appetite control, antidepressant, sedative, concentration enhancer, and anxiolytic purposes. The stimulant in high doses and sedative in low doses.
Rosae damascenae aetheroleum	The improvement of sadness, nervous stress and blood pressure problems, persistent cough, cognitive performance, and gynaecological diseases.	As a mild local anaesthetic, anxiolytic, bacteriostatic, epithelializing, to improve the cognitive status, and improve post-dressing pain in patients with burns, increase concentration, for first and second-degree depression, sexual dysfunction, in PMS, reduce menopausal symptoms, and migraine pain. Alone or in combination with other essential oils for sedative and pain relief at the time of delivery.
Rosmarini aetheroleum	As an antiseptic, wound healer, improving cognitive performance, in minor peripheral circulatory disorders, relieving minor muscle, and joint pain.	Relief of antirheumatic, minor muscle and joint pains and as an adjunct to cellulite treatment (camphor chemotype), female and male type alopecia and epithelializing purpose (verbenone chemotype), improving cognitive performance, anxiolytic purpose and removal of hematomas (1,8-cineol chemotype).
Salviae trilobae aetheroleum	As mouth rinse and mouthwash for mouth and throat infections, as a rub on the abdomen and soles of the infants, carminative, colic pain, wound healing, and antiseptic.	In respiratory tract infections, bacterial infections in the genital area and gastrointestinal system disorders as a strong expectorant, mucolytic, antimicrobial, and antiviral.
Salviae lavandulifoliae aetheroleum	As an antiseptic, carminative and cognitive performance enhancer.	As an expectorant, antimicrobial, antiviral effective in respiratory tract infections, mild analgesic, spasmolytic effects in gastrointestinal system disorders, Alzheimer's type dementia, and cognitive performance enhancer.
Salviae sclareae aetheroleum	As a tension reliever and in the treatment of acne.	in bacterial infections, anxiety, to facilitate birth from the 34th week of pregnancy, PMS, amenorrhea, dysmenorrhea, asthenia and migraine. Alone or in combination with other essential oils (mainly *Lavandula angustifolia*, *Cananga odorata*, *Rosa damascena* and *Matricaria recutita*) for sedative, and pain relief at the time of birth.

Table 2 contd. ...

...Table 2 contd.

Essential Oil	(Traditional) Use in Folk Medicine	Evidence-Based Use
Santali ligni aetheroleum	As a cognitive performance enhancer.	As a calming, anti-inflammatory, antimicrobial, antiproliferative, epithelializing effect, in the treatment of acne, psoriasis, eczema, warts, HPV (*Human papilloma virus*), herpes (*Herpes simplex*), *Molluscum contagiosum* and atopic dermatitis, as well as in haemorrhoids, additionally, in dermatitis caused by radiotherapy in patients with head, neck, and breast cancer.
Caryophylli flos aetheroleum	Toothaches, minor infections of the mouth and skin, minor wounds, insect bites, nausea, gas pains, mouth-throat inflammations and sore throats are associated with the common cold.	Dentistry is a mild analgesic and anaesthetic, and for antimicrobial purposes.
Thymi serpylli aetheroleum	As a strong antiseptic for cough relief and cold relief, rheumatism, sciatica, menstrual pain, and skin diseases.	As a broad spectrum antimicrobial, in acute bronchitis and as a reliever of joint and muscle pain.
Thymi aetheroleum typo geraniolo	Mild spasmolytic purposes and urethritis.	Sore throat, stomatitis, vaginitis and urethritis with strong antimicrobial purpose.
Thymi aetheroleum typo linalool	As an immune system stimulant and in bronchitis in children.	In spastic bronchitis (suitable for children), bronchopneumonia, stomatitis, enterocolitis, muscle pain, urethritis, and vaginitis.
Thymi typo thymolo aetheroleum	Relieve cough and colds, by massage to relieve pain in arthritis, as well as in sprains and bruises by applying in the form of a bath.	Cough and cold symptoms, fungal and bacterial skin diseases, antimicrobial in ringworm disease (*Alopecia areata*), and analgesic in rheumatic pains.
Valerianae aetheroleum	Traditionally, it is used for the relief of mild symptoms of mental stress and to aid sleep.	It is used as a sedative, and for sleep disorders/problems.
Vetiverae aetheroleum	Against bacteria and fungi, as an anti-inflammatory, corrective of cognitive functions, insecticide, and insect repellent.	As an antiseptic, aphrodisiac, anxiolytic, sedative effect, calming in stress situations, insomnia problems, modulating brain functions, wrinkle remover and moisturizer, and especially in the treatment of acne.

Chemical Composition of Essential Oils

Plants carrying essential oils (usually taxa belonging to Lamiaceae, Rutaceae, Apiaceae, and, Rosaceae families) are aromatic and generally grow in hot regions such as the Mediterranean Region. Many essential oils are obtained from these plants, each of which has its unique smell. There are about 450,000 plant species in the world and about 1/3 of them are aromatic. Flora of Turkey is also rich in aromatic plants and contains about 3,000 taxa of essential oil.

Essential oils can be found in flowers, leaves, fruits, seeds, bark, rhizomes, or roots of aromatic plants. Examples of where essential oils are found are leaves (eucalyptus, medicinal mint, patchouli, and laurel), leafy branches (pine), flowers (rose and jasmine), aerial parts (thyme), stem bark (cinnamon), wood (sandalwood), roots (vetiver), rhizome (ginger), fruits (anise and juniper), fruit peel (bergamot, orange, and lemon), seeds (cardamom and coriander), and resin (frankincense) can be given (Babar et al. 2015).

Essential oils are found in both superficial and subsurface tissues, glandular hairs, secretory ducts, secretory cells, and parenchyma cells in plants. They protect the plant from microorganisms, facilitate pollination, reduce water loss, act as an insect repellent or attractant, and remove unwanted products in biosynthesis (Dhifi et al. 2016, Sell 2006). The chemical contents of essential oils are complex. They contain terpenic (mono-, sesqui-, and even diterpenes, such as hydrocarbons and oxygenated derivatives, alcohols, acids, esters, ethers, aldehydes, and ketones), aromatic substances, amines, and sulphur compounds in their structure. In addition, phenylpropanoids, fatty acids, and esters are also found in some essential oils. Most of these essential oils contain between 20 and 60 different bioactive components. Bioactive essential oils consist of approximately 90% monoterpenes. Terpenic structures consist of isoprene units and are synthesised in the cytoplasm of the cell via mevalonic acid (Degenhardt et al. 2009, Lang and Buchbauer 2012, Nazzaro et al. 2013).

Some of the introduced chemical compositions of essential oils: monoterpene hydrocarbons (p-cymene, limonene, α-pinene, and α-terpinene), oxygenated monoterpenes (camphor, carvacrol, eugenol, and thymol), diterpenes (caurene and camphorene), sesquiterpene hydrocarbons (β-caryophyllene and germacrene D and humulene), monoterpene alcohols (geraniol, linalool, and nerol), sesquiterpene alcohol (patulol), aldehydes (citral and cumin), acids (geranic acid and benzoic acid), ketones (acetophenone and benzophenone), lactones (bergapten), phenols (eugenol, thymol, carvacrol, and catechol) and esters (bornyl acetate) can be given as examples (Böhme et al. 2014, Swamy et al. 2016, Sell 2006).

Application Ways

Inhalation

Due to the volatilisation feature of essential oils, applications are usually done by inhalation. This process is carried out directly or indirectly through the respiratory tract. It is a preferred method because of the respiratory antiseptic and spasmolytic effects of essential oils. Application forms include:

(a) Mist: The essential oil added to the boiling water is entrained with water vapour, and the airways are opened by inhaling the steam while breathing.

(b) Essential oil preparations prepared in appropriate amounts are dripped onto the chest and neck parts of children's pillows and clothes, and the evaporated essential oil is inhaled.

(c) Inhalation pomades prepared by mixing essential oil into vaseline are applied to the chest, neck and nostrils, and the essential oil then evaporates with body temperature when inhaled.

Vaporizers/Diffusers

These are ceramic teapots heated by a small candle or diffusers used with electricity, mostly carrying essential oils, used in inhalation applications. They emit an aromatic odour and provide spiritual and mental relaxation to the person and at the same time disinfect the environment.

Massage

With the rubbing action in the massage application, the essential oils are better absorbed by the skin and their scent is revealed. Essential oils are not applied in pure form, but together with carrier oils. In the application of essential oils through massage, the nervous system is calmed by creating a state of relaxation or energising in the body, and blood and lymphatic circulations are stimulated to restore physical functions. It also reduces tension and pain in overworked and tired muscles.

Compress

Compressing is effective in relieving pain and inflammation. It is applied cold or hot. Warm compresses are suitable for muscle aches, arthritis, rheumatism, toothaches, earaches, and abscesses, while cold compresses are suitable for headaches, sprains, and swellings.

Baths

Aromatic bath therapy using essential oils, called "Balneotherapy" or "Phytobalneotherapy", is an aromatherapy technique used to improve mood, insomnia and stress, rheumatic disorders, and upper respiratory tract infections, such as colds and skin diseases. It is an easy and multi-purpose method to apply at home. A bath is prepared by dripping 20 drops of one of the recommended essential oils into a tub full of hot water. A mixture of up to two essential oils is used. Essential oils are insoluble in water but form a thin film on the surface. The heat of the water evaporates them and helps their absorption through the skin. The duration of the bath should not exceed 10 minutes. A post-bath massage is recommended. After the bath and massage therapy, the body should not be activated immediately, the body should be rested for a while. This can also be supported with herbal teas.

Topical Applications

Preparations of essential oils in cream, lotion, shampoo, and gel form facilitate the application of skin wounds, cuts, bruises, redness, and itching.

Mouthwashes

Mouth ulcers, gum diseases, throat infections, and mouthwashes containing antiseptic essential oils are excellent treatments for bad breath. Not suitable for children (Tayfun 2019). Excretion of aromatherapeutic agents applied by inhalation and massage occurs through respiration, sweat, and urine (Bilgiç 2017).

Information related to the application of commonly used essential oils is given in Table 3.

Considerations for Essential Oils

Interactions

For the essential oils described in the chapter, no interaction is known when used externally and at recommended doses. However, it may be inconvenient to use it for a long time and in high doses of bergamot essential oil. It should not be used with homoeopathic products.

Fine lavender, lavandin, spike lavender, and lemon balm essential oil may interact with barbiturates, benzodiazepines, and other sedatives/antidepressants in inhalation/massage applications. Fine lavender and spike lavender essential oils should not be used together with other drugs in topical applications. Chamomile essential oil can interact with all drugs and fats metabolized in the liver by the cytochrome P450 enzyme pathway (CYP2D6, CYP1A2, CYP2C9, and CYP3A4). Therefore, it may potentiate the effect of conventional antidepressants. Partly dementholised mint essential oil should not be administered simultaneously with other drugs. Sweet Basil essential oil should not be used on sensitive skin due to its irritant properties.

Contraindications

For the essential oils described in the chapter, people who have an allergic reaction to its ingredients should not use them. It should not be used in epileptic patients. Due to its strong irritant effect, Ceylon, and cinnamon bark oil should not be used on sensitive skin and mucous membranes. Owing to its bergapten content, bergamot essential oil should not be used in children under 12 years of age. Essential oils, such as citronella and lemon balm essential oils, with high monoterpene aldehyde content, such as citral, etc., are contraindicated in patients with glaucoma. *Trans*-anethole has an estrogen-like effect, so sweet fennel oil should never be used in pregnant women, patients with endometriosis, prostatic hyperplasia, or estrogen-dependent cancer. Eucalypti essential oil, known as gully gum oil, should not be used in children under eight years of age. Narrow-leaved peppermint and gully gum oil should not be applied to the head and neck area of children under the age of 10. It may cause respiratory problems due to its effects on the central nervous system due to 1,8-cineole.

People with bronchial asthma and whooping cough should not use dwarf pine essential oil. Peppermint and partly dementholised mint essential oil should not

Table 3: Application method related to common essential oils.

Essential Oil	Application Methods
Angelicae radix aetheroleum	Massage oil and inhalation.
Boswellia carterii aetheroleum	Massage oil, bath, and in the diffuser, and mist inhalation.
Boswellia serratae aetheroleum	Massage oil, bath, diffuser, and mist inhalation.
Cananga aetheroleum	Massage oil, bath, diffuser and mist inhalation.
Carvi aetheroleum	To support digestive function, relieve colic and for bloating; the essential oil is used externally, prepared preferably in rosehip seed oil from the 12 months to its elderly stage.
Cedrus aetheroleum	Massage oil, bath, diffuser and mist inhalation. The mixture of essential oil in carrier oil is dripped onto a standard cotton swab and applied directly to the acne area as a point until the problem disappears. Since the essential oil is irritating, a hydrolate (lavender, rose, medicinal chamomile, or immortelle) is applied as a spray after 30 minutes and the excess is removed so that the skin remains moist without pressing with a paper towel.
Chamomillae romanae aetheroleum	Compress or bath.
Cinnamomi camphorae aetheroleum	Massage oil, bath, diffuser and mist inhalation.
Cinnamomi zeylanici corticis aetheroleum	Massage oil, diffuser, and mist inhalation.
Cisti cretici aetheroleum	Massage oil, diffuser, and mist inhalation.
Citri bergamiae aetheroleum	Massage oil and the diffuser. In order to reduce the loss of pigmentation on the skin and be applied to the relevant area with the advice of a physician and pharmacist.
Limonis aetheroleum	Massage oil, diffuser, and mist inhalation. As a mouthwash, for three to four times a day.
Citri paradisi aetheroleum	Massage oil, diffuser and mist inhalation. As a mouthwash, for three to four times a day.
Citri reticulatae aetheroleum	Massage oil, bath, diffuser and mist inhalation.

Table 3 contd. ...

...Table 3 contd.

Essential Oil	Application Methods
Aurantii dulcis aetheroleum	Massage oil, diffuser and mist inhalation. As a mouthwash, for three to four times a day. In the treatment of acne, applied form a mixture with *Ocimum basilicum* essential oil until the problem disappears.
Cumini aetheroleum	In cases of gas and bloating problems (flatulence), in adults, in the form of massage by diluting the lower abdomen with carrier oil in the direction of the solar plexusor used in a diffuser. 0.5–1 mL/L is used for 10–20 minutes in daily practice for sitz bath. Sitz bath water temperature should be 38–40°C for the effect to be seen.
Cupressi aetheroleum	As a foot bath and in diffuser and mist inhalation to help control astringent and foot sweating. In the form of steam therapy for coughs, bronchitis, asthma, whooping cough, emphysema, and influenza and it helps to calm the mind and relieve anger. Mixed massage oil or cypress oil in the bath as a massage oil or diluted in the bath for arthritis, asthma, cellulite, cramps, sweaty feet, rheumatism, varicose veins, heavy menstrual bleeding, and menopause. As a cold compress diluted in a suitable carrier oil for nosebleeds. In the form of lotions and creams to clean oily and clogged skin in venous circulation disorders.
Cymbopogon martinii motiae aetheroleum	In the diffuser, in cases of bronchitis and rhinopharyngitis, three or four times a day. As a mouthwash for antibacterial purposes in sore throat. In urethritis and vaginitis, daily for 10–20 minutes and sitz bath water temperature should be 38–40°C for the effect to be seen. In dermatitis, acne and oily skin, a suitable carrier oil (preferably jojoba or rosehip seed oil) is applied two times a day or as a steam bath until the problem disappears.
Cymbopogon martinii sofiae aetheroleum	In the diffuser, in cases of bronchitis and rhinopharyngitis, three-four times a day. As a mouthwash for antibacterial purposes in sore throat. In urethritis and vaginitis, daily for 10–20 minutes, sitz bath water temperature should be 38–40°C for the effect to be seen. In dermatitis, acne and oily skin, a suitable carrier oil (preferably jojoba or rosehip seed oil) is applied two times a day or as a steam bath until the problem disappears.
Cymbopogon nardi aetheroleum	For antimicrobial, anti-inflammatory, spasmolytic and epithelialising purposes, and twice a day by diluting in a suitable carrier oil (preferably jojoba/rosehip seed oil) In arthritis, coconut oil, cherry seed oil, and flaxseed oil are diluted in the form of massage two to three times a day. As a bath for antimicrobial purposes. As a deodorant, insect repellent, or air disinfectant in a diffuser or preferably diluted in jojoba oil three times a day as spraying. It is used four times.
Eucalypti citriodorae aetheroleum	Especially in children's upper respiratory tract infections, massage oil to the chest, in the form of a bath, in the diffuser, and in mist inhalation. Applications on the wound are prepared in carrier oil and applied by spraying. Locally in rheumatoid arthritis in the form of semi-solid preparations for periods not exceeding three weeks.

Table 3 contd. ...

...Table 3 contd.

Essential Oil	Application Methods
Eucalypti aetheroleum	Cineole chemotype. In upper respiratory tract infections of adults, massage oil to the chest, in the form of a bath, in a diffuser, and as a mist inhalation. Especially on bad-smelling wounds, prepared in carrier oil, and applied by spraying. Camphor chemotype. In rheumatoid arthritis, locally for periods not exceeding three weeks, preferably in the form of semi-solid preparations prepared in coconut oil. Verbenone chemotype. In case of hair loss, mixtures prepared preferably in sesame or coconut oil are applied by spraying. Since coconut oil becomes liquid at a temperature exceeding 24°C, it should be prepared in a way that it will be h/h. Combination with essential oil of borne on chemotype *Rosmarinus officinalis* in head lice (*Pediculus humanus capitis*).
Eucalypti radiatae aetheroleum	For toothache, dilute in a cold-pressed carrier oil and apply one to two drops to the gingiva or around the tooth, which can be repeated when necessary. To stimulate the sympathetic nervous system, inhalation one or two times a day. As an expectorant in bronchitis, two drops of essential oil are freshly dripped onto a tissue paper and sniffed directly four/five times a day, or as a 15-minute inhalation four/five times a day in a diffuser and as a mist inhalation. Influenza, as direct inhalation two to four times a day. In upper respiratory tract infections and applying massage oil to the thorax.
Eucalypti smithii aetheroleum	Massage oil to the chest, in the form of a bath, in a diffuser, and in mist inhalation.
Foeniculi dulcis aetheroleum	In the form of massage diluting in a suitable carrier oil in arthritis, rheumatism, cellulite and for antispasmodic purposes. In a diffuser and as a mist inhalation for bronchodilator purposes.
Helichrysi italici aetheroleum	As massage when prepared in a carrier oil, to heal hematomas, wounds, and locally rheumatoid arthritis several times a day. In a diffuser and as a mist type inhalation to give mucolytic effect and vitality.
Helichrysi splendidi aetheroleum	Massage oil, diffuser and mist inhalation.
Juniperi aetheroleum	Topically by adding it to a carrier oil as a massage oil in rheumatic diseases. Since coconut oil becomes liquid at a temperature exceeding 24°C, it should be prepared in a way that it will be h/h. In bronchitis, as mist type inhalation or as a massage oil, added to a carrier oil (preferably jojoba or cold-pressed coconut oil), to the chest. In the diffuser as a cognitive performance enhancer. For its antifungal effect in adolescents, adults, and the elderly, it is added to the bathwater at 35–38°C as a foot bath and applied for 10–20 minutes three or four times a week.

Table 3 contd. ...

...Table 3 contd.

Essential Oil	Application Methods
Juniperi virginianae aetheroleum	For venous circulation disorders for adults and the elderly, for psoriasis, for adolescents, adults and the elderly, a cold compress three or four times a week for 10–20 minutes. For antispasmodic purposes, preferably half an hour after as a massage oil in the evening. In diffuser for anxiolytic purpose. For foot fungus (*Tinea pedis*), in the form of a foot bath with water at 35–38°C for 10–15 minutes.
Lauri folii aetheroleum	In upper respiratory tract infections and as expectorant only in adults, in diffuser and mist type inhalation. In mouthwashes for stomatitis and aphthosis antimicrobial purposes. As an insect repellent, preferably diluted in sesame or argan oil, by spraying and then massaging. In muscle pains, neuralgia, and arthritis, preferably diluted in coconut oil, by spraying and then massaging. In the treatment of hair loss, oily and dandruff hair is preferably diluted in sesame oil and by spraying and then massaging.
Lavandulae aetheroleum/ Fine Lavender Essential Oil	Locally as a mixture with Cananga odorata essential oil for pain As a bath for not more than 15 minutes in case of eczema, tension, and sleep disorders. In dysmenorrhea, suitable mixtures containing essential oils of *Lavandula angustifolia, Salvia sclarea* and *Origanum dubium*, by massaging the lower abdomen from the last menstrual bleeding to the beginning of the next menstrual bleeding. In eczema, first and second-level depression, sleep disorder and dementia in the form of a bath, for no more than 15 minutes. Topically against eczema, dermatitis, erythema, sunburn and radiotherapy, insect sting and head lice the problem area in the form of spraying. After the application of a bath in eczema, to the problem area in the form of spraying.
Lavandulae aetheroleum/ Lavandin Essential Oil	As a diffuser or mist inhalation in sleep disorders. In adults, directly to the problem area (acne) in the form of dots on the face until the problem disappears. To the soles of the feet in hyperactive children aged two to three years, and to the temples and earlobes of children older than three years. For bronchitis, it is applied topically to the rib cage alone or in a suitable carrier oil. Locally as a mixture with *Cananga odorata* essential oil for pain As a bath for not more than 15 minutes in case of eczema, tension, and sleep disorders. Topically against eczema, dermatitis, insect sting to the problem area in the form of spraying. After the application of a bath in eczema, to the problem area in the form of spraying.
Spicae aetheroleum	Topically on the skin in pure form in adults. In the diffuser or form of mist inhalation in sinusitis. For bronchitis, topically to the rib cage alone or in a suitable carrier oil. Locally as a mixture with *Cananga odorata* essential oil for pain. As a bath for not more than 15 minutes in case of eczema, tension, and sleep disorders. Topically against eczema, dermatitis, and insect sting to the problem area in the form of spraying. After the application of a bath in eczema, to the problem area in the form of spraying.

Table 3 contd. ...

...Table 3 contd.

Essential Oil	Application Methods
Leptospermi scoparii aetheroleum	Massage oil and in the diffuser and mist inhalation.
Matricariae aetheroleum	Daily for 10–20 minutes as a sitz bath for skin and mucous irritations in the anal and genital area. Sitz bath water temperature should be 38–40°C for the effect to be seen. Local application against dermatitis, acne, and Herpes simplex until the problem disappears in the form of semi-solid preparations containing medicinal chamomile essential oil. As a mouthwash in oral mucositis. As massage oil is a suitable carrier oil, anti-inflammatory, antispasmodic, and in gastrointestinal system disorders. In the form of mist for anti-inflammatory purposes.
Melaleucae alternifolii aetheroleum	Locally by diluting in a suitable carrier oil in eczema, psoriasis, viral skin diseases and burns. In the treatment of acne, once a day until the problem disappears, apply aromatic water or hydrolate (lavender, rose, medicinal chamomile, or immortelle) to the acne-prone area by spraying, after removing the excess without pressing with a paper towel so that the skin remains moist. 1–2 drops of essential oil preferably by diluting with 10–12 drops of jojoba oil. In the form of mouthwash for gingivitis. In herpes simplex, 1–2 drops of essential oil are applied directly on a standard cotton swab and applied directly as a point. For upper respiratory tract infections and in the diffuser and the form of mist inhalation; repeated two-four times a day. In the form of a spray for cuts, head lice and foot fungus. In nail fungus, the solution prepared with essential oil is applied with the help of a thin brush.
Melaleucae aetheroleum	As an antimicrobial and in upper respiratory tract infections, in a diffuser and the form of mist inhalation, repeated two to four times a day or as a massage oil on the chest. Diluting in a suitable carrier oil to reduce skin damage caused by radiotherapy. In the form of a spray for psoriasis, fructose, insect bites, as a wound healer, and foot fungus, while the solution is prepared with essential oil two to three times a day with the help of a fine brush. In the treatment of *Herpes genitalis*, by spraying directly or by spraying on an organic pad and placing it on the relevant area.
Melissae aetheroleum	Citral Chemotype: In the form of sitz bath, in a diffuser for hypotensive, immunomodulatory purposes, cardiac arrhythmia and pregnancy nausea. In stomach cramps, diluted in a suitable carrier oil and massaged clockwise. In the form of a bottom-up massage, diluted in a suitable carrier oil on varicose veins. As a birth stress reducer, to the perineal area, preferably in argan oil, together with *Lavandula angustifolia, Cananga odorata, Matricaria recutita, Salvia sclarea* essential oil, by massage and in a diffuser, starting from the 34th week. Citronellal Chemotype: In fixed oil suitable for earlobe and/or temples in tension-type headache and migraine. In dyspepsia, applied clockwise to the upper abdomen in suitable fixed oil. In rhinitis, sinusitis and laryngitis mist, diffuser, or mouthwash are in the form of inhalation. In the form of spraying on the area to be in skincare before radiotherapy.

Table 3 contd. ...

...Table 3 contd.

Essential Oil	Application Methods
Menthae arvensis aetheroleum partim mentholum depletum	In the diffuser or as a mist inhalation or bath for antimicrobial and antiviral purposes in the upper respiratory tract. In muscle pains, neuralgia, sports injuries and joint rheumatism, essential oil, locally in a carrier oil (preferably coconut oil) as semi-solid or liquid preparations containing essential oil. Since coconut oil becomes liquid at a temperature exceeding 24°C, it should be prepared in a way that it will be h/h. As massage oil, to the temples and earlobes in migraine.
Menthae piperitae aetheroleum	In the upper respiratory tract, for mild antiemetic and cognitive performance, over the age of 12, three times a day in a diffuser and as a mist inhalation. In the upper respiratory tract in the form of semi-solid or liquid preparations containing essential oil in a fixed oil containing essential oil (preferably coconut oil) by massaging the chest. For antispasmodic, analgesic purposes, in muscle pains and rheumatoid arthritis, the essential oil is applied locally in a carrier oil (preferably coconut oil) as semi-solid or liquid preparations containing essential oil. Since coconut oil becomes liquid at a temperature exceeding 24°C, it should be prepared. As a massage oil to the solar plexus area as an aid in the treatment of irritable bowel syndrome. In tension-type headaches and migraines, massage oil is applied to the temples and earlobes. It is applied by spraying by mixing with borage fixed oil or immortelle essential oil as a wound healer. The mixture prepared in coconut or argan oil, containing essential oil against herpes, was applied to herpes by dropping two drops on a cotton swab. Over the age of 12, in gingivitis, as a mouthwash three or four times a day or by spraying the essential oil diluted in coconut oil.
Myrrh Aetheroleum	Applied topically by inhalation, massage and bath as well as for skin/wound care.
Basilici aetheroleum	For mild analgesic purposes and against headache, diluted in a suitable carrier oil and applied to the temples or neck as a massage two or three times a day. Diluted in a suitable carrier oil to relieve insecurity, indecision, negative thoughts, stress, depression, fear, anger and mental depression and as a massage to the earlobe two or three times a day. In the form of massage, preferably diluted in coconut oil, cherry seed oil or linseed oil on arthritis, muscle pains and oily scalp. In sinusitis, insecurity, indecision, negative thoughts, stress, depression, fear, anger and mental depression, for antimicrobial purposes, and in colds, up to four times a day, preferably before going to bed, in a diffuser or as a mist type inhalation for over 12 years of age.
Origani dubii aetheroleum	As massage oil in arthritis and muscle pain, and in the treatment of athlete's foot (*Tinea pedis*), preferably diluted in a carrier oil rich in vitamin E. In the diffuser in cases of cold, flu and anxiety. In dysmenorrhea, suitable mixtures containing essential oils of *Lavandula angustifolia, Salvia sclarea* and *Origanum dubium,* by massaging the lower abdomen from the last menstrual bleeding to the beginning of the next menstrual bleeding.

Table 3 contd. ...

...Table 3 contd.

Essential Oil	Application Methods
Pelargonii aetheroleum	In external haemorrhoids, and as an insecticide and insect repellent, preferably by spraying in coconut oil directly. By massage in muscle stiffness in Parkinson's and multiple sclerosis, and by diluting with coconut oil in Alzheimer's by applying to the temples or in a diffuser.
Anisi aetheroleum	In dental infections, by dropping two drops on a cotton swab and press on the aching tooth for 8–10 seconds. In the diffuser and the form of mist inhalation in upper respiratory tract infections. The inhalation time should not exceed ten minutes. In the form of semi-solid or liquid preparations containing essential oil, applied to the temples and earlobes, in mixtures used in migraine attacks. PMS, amenorrhea and oligomenorrhea to the lower abdomen (in the direction of the solar plexus), in meteorism and dyspepsia, as a light massage application to the upper abdomen (clockwise).
Pini pumilionis aetheroleum	Massage oil and in the diffuser and mist inhalation and mouthwash. For children under the age of 6, only in the diffuser. In bath therapy, the essential oil is transferred to the bath water in a half-filled tub with the water temperature not exceeding 39°C. Bathing time should not exceed 15 minutes.
Pini sylvestri aetheroleum	In adults, massage oil, and in the diffuser and mist inhalation and mouthwash. For children under the age of 4–6, only in the diffuser. As bath therapy, the essential oil is prepared in a half-filled tub, with the water temperature not exceeding 39°C. Bathing time should not exceed 15 minutes.
Pogostemon cablin aetheroleum	Sedative, appetite control, antidepressant and anxiolytic in a diffuser and the form of mist inhalation. For concentration enhancing and stimulating purposes, used in diffuser and mist type inhalation. Diluted in a suitable carrier oil for insecticide purposes and sprayed, renewed every two hours. In the form of massage, preferably diluted in jojoba oil for antiphlogistic, cell regenerative, acne, eczema, antiseptic, and antifungal. In the form of direct application to the problem area with the help of a cotton ear swabs in acne until the problem disappears.
Rosae damascenae aetheroleum	Diluted in a suitable carrier oil (preferably rosehip seed oil) and applied to the temples and earlobes as a mild local anesthetic, to improve the cognitive status, as a concentration enhancer, in the first and second-degree depressions. In the diffuser as a concentration enhancer and anxiolytic. At the time of birth, diluted in a suitable carrier oil (preferably rosehip seed oil) for sedative and pain relief and applied by massage. In order to facilitate childbirth, together with *Lavandula angustifolia, Cananga odorata, Matricaria recutita, Salvia sclarea* essential oil, preferably applied in argan oil to the perineum from the 34th week onwards, by massage and in a diffuser. In PMS, preferably applied in borage oil or primrose oil by massaging the temples and earlobes one to two days before the menstruation period and until the end of menstruation. For burns is applied locally as a spray by dripping into distilled water.

Table 3 contd. ...

...Table 3 contd.

Essential Oil	Application Methods
Rosmarini aetheroleum	Camphor chemotype. As massage oil in antirheumatic and cellulite treatment. As a bath for antirheumatic purposes. In semi-solid and liquid dosage forms or as a bath for the relief of minor muscle and joint pains. Verbenon chemotype. Diluting it in a suitable carrier oil (preferably argan or sesame fixed oil) as an epithelial and in alopecia. 1,8-cineole chemotype. In a diffuser for anxiolytic purposes and to improve cognitive performance. To remove hematomas together with *Helichrysum* essential oil.
Salviae trilobae aetheroleum	As an antimicrobial, antiviral, strong expectorant, and mucolytic purpose in respiratory tract infections in the form of massage to the chest and bath, in a diffuser and as a mist type inhalation. As a massage oil to be applied in the direction of the solar plexus in gastrointestinal system disorders. In bacterial infections in the genital area, as a sitz bath alone or in a mixture with *Salvia sclarea* or *Melaleuca alternifolia* by adding it to water.
Salviae lavandulifoliae aetheroleum	In respiratory tract infections, as an expectorant, antimicrobial and antiviral massage to the chest, in the form of a bath, in a diffuser and in the form of mist inhalation. In gastrointestinal system disorders by massaging in the direction of the solar plexus. For mild analgesic and antibacterial purposes, as a spray in tonsillitis alone or a mixture with *Melaleuca alternifolia*, preferably diluted in coconut oil. In Alzheimer's type dementia and as a cognitive performance enhancer in a diffuser.
Salviae sclareae aetheroleum	In the form of massage diluting it in a suitable carrier oil for sedative and pain relief at the time of delivery to facilitate delivery, as well as in the case of PMS, amenorrhea, and dysmenorrhea. In asthenia and migraine by massaging the earlobe and neck muscles. As a sitz bath in bacterial infections in the genital area alone or combination with *Salvia fruticosa* or *Melaleuca alternifolia*. In the diffuser for sedative purposes in anxiety and at the time of delivery.
Santali ligni aetheroleum	As a massage after being diluted in a suitable carrier oil in anti-inflammatory, antimicrobial, antiproliferative, epithelial, psoriasis and eczema. In dermatitis, it is diluted in a suitable carrier oil and used as a spray or bath. It is used directly with semi-solid preparations against warts, HPV, and *Herpes simplex*, preferably in rosehip seed oil or jojoba oil in acne. As a sitz bath in haemorrhoids. In a diffuser and by mist inhalation for soothing and anti-inflammatory purposes.
Caryophylli flos aetheroleum	In adults aged 12 years and over, the essential oil can be dripped onto a piece of cotton or as a solution and gel for toothache relief. As a mouth rinse in cases of mouth-throat inflammation, as a mouthwash for sore throats, and also applied in insect bites.
Thymi serpylli aetheroleum	In adults, in liquid or semi-solid dosage form, applied to the skin (chest or back) three times a day for the relief of cough and cold symptoms. It has applications as a bathroom additive. In the diffuser and the form of mist-type inhalation. Locally as a massage in a suitable carrier oil (preferably coconut or sesame oil) as a pain reliever.

Table 3 contd. ...

...Table 3 contd.

Essential Oil	Application Methods
Thymi aetheroleum typo geraniolo	In diffuser for antimicrobial purposes and the form of mist type inhalation. In adults, in liquid or semi-solid dosage form, applied to the skin (chest or back) three times a day. As a bath additive, as well as a sitz bath in urethritis and vaginitis, once a day every day until the symptoms disappear. It is applied once every other day, with a bath time not exceeding 15 minutes for adults and 10 minutes for children. The recommended bath temperature is 36–38°C. As massage oil, it is applied to the chest in children aged 3–6, preferably not in jojoba oil in children aged 6 years and older. It can be applied as a mouthwash. The last application should preferably be done half an hour before going to bed.
Thymi aetheroleum typo linalool	As inhalation and mist type inhalation. In a liquid or semi-solid dosage form, it is applied to the skin (chest or back) three times a day. As a bath additive, as well as a sitz bath in urethritis and vaginitis, once a day every day until the symptoms disappear. It is applied once every other day, with a bath time not exceeding 15 minutes for adults and 10 minutes for children. The recommended bath temperature is 36–38°C. As massage oil, it is applied to the chest in children aged 3–6, preferably not in jojoba oil in children aged 6 years and older. Can be applied as a mouthwash. The last application should preferably be done half an hour before going to bed.
Thymi typo thymolo aetheroleum	In the diffuser for the relief of cough and cold symptoms. Thymol chemotype is not recommended for mist inhalation. As a bath additive in some skin diseases. Applied once every other day with a bath time not exceeding 15 minutes for adults and 10 minutes for children. The recommended bath temperature is 36–38°C. Diluting in a suitable carrier oil (preferably grape seed, jojoba, argan, or sesame fixed oil) in *Alopecia*. Locally as a massage in a suitable carrier oil (preferably coconut or sesame oil) as a pain reliever.
Valerianae aetheroleum	Bath application. It can be applied as a single dose (240–400 mg) for a full bath up to three to four times per week.
Vetiverae aetheroleum	A diffuser as an aphrodisiac, anxiolytic, sedative, calming in stress situations, insomnia problems, and modulating brain functions. Diluted in borage oil, as a wrinkle remover, and moisturiser. The mixture, which is preferably prepared with one-two drop of essential oil/5 mL of carrier oil in rosehip seed oil or jojoba oil, is dripped onto a standard cotton swab and applied directly to the acne-prone area as a point until the problem disappears. Since the essential oil is irritating, a hydrolate (lavender, rose, medicinal chamomile, or evergreen flower) is applied by spraying after 30 minutes and the excess is removed so that the skin remains moist without pressing with a paper towel.

be used in children under four years of age. Additionally, collapse and apnea may develop in partly dementholised mint essential oil. Chamomile essential oil should not be used in people who are sensitive to the ingredients and plants of the Asteraceae (Compositae) family. Those with widespread skin damage, open wounds, acute skin disease, high fever, chronic diseases, chronic heart, and circulatory disorders should not use rosemary essential oil. Similarly, thyme essential oils and chemotypes should not apply during bathing. Clove bud essential oil should not be used in cancer patients in the presence of oral ulcers.

Warnings

Essential oils are secondary metabolites that are generally considered safe to use with minimal side effects. Some of these are approved as food additives and are included in the category recognised as safe at certain rates by the American Food and Drug Administration (FDA) (Bilsland and Strong 1990). However, essential oil applications are usually performed externally, and essential oils are not used internally in aromatherapy applications.

Since it may cause dermatitis or irritation on the skin, it should not be applied directly to the skin without diluting it with a fixed/carrier oil. Lavender essential oil is one of the rare essential oils that can be applied directly to the skin. However, it is necessary to be careful about the development of an allergic reaction (Tatlı 2012). If it is applied to the skin in pure form and above the recommended dose, skin sensitivity may develop and it should not come into contact with the eyes and mucous membranes (Babar et al. 2015). A case of reversible prepubertal gynecomastia due to long-term topical use of lavender and tea tree essential oils has been reported in the literature (Henley et al. 2007).

It should not be used in higher doses than the specified dose, it may trigger asthma attacks. It should not be used in children under three years old. In the case of hypertension, bathing should not be applied. In case of use as a bath additive, dyspnea, fever, and purulent sputum, the physician should be consulted. Hypersensitivity to its components may develop.

Essential oils should be kept away from children and must be tightly closed. An oxidised essential oil should not be used. Since essential oils obtained from *Citrus* species carry furanocoumarins, they are photosensitive and should not be exposed to sunlight or UV rays after applying them to the skin. They are not recommended for use during pregnancy and lactation due to insufficient data.

Consultation with a specialist physician should be sought before the use of essential oils in infants and children, particularly in cases of allergies, chronic diseases such as epilepsy and hypertension, in the presence of continuous medication, psychiatric treatment, and homoeopathic treatments (Tatlı 2012). Warnings regarding commonly used essential oils are presented in Table 4.

Use of Essential Oils During Pregnancy and Lactation

In large, essential oils are not recommended for use in women during their pregnancy and lactation due to insufficient data. Since citral has been reported to be dose-dependent teratogenic because it inhibits the retinoic acid synthesis and inhibits fetal development, lemon essential oil should not be used in women during pregnancy and lactation. Palmarosa essential oil, known as Turkish Geranium oil, citronella,

Table 4: Warnings regarding commonly used essential oils.

Essential Oil	Warning
Angelicae radix aetheroleum	It is responsible for phototoxicity. It is not used for diabetic ailments.
Boswellia carterii aetheroleum	Hypertension patients should use it with caution. An oxidised essential oil should not be used. It may cause skin sensitivity. In inhalation applications, high doses and long-term use may cause dizziness. It should not be applied to the skin in pure form and above the recommended dose and contact with the mucous membranes and eyes should be avoided.
Boswellia serratae aetheroleum	Hypertension patients should use it with caution. An oxidised essential oil should not be used. It may cause skin sensitivity. It should not be applied to the skin in pure form and above the recommended dose, and contact with the mucous membranes and eyes should be avoided. In inhalation applications, high doses and long-term use may cause dizziness.
Cananga aetheroleum	Hypertension patients should use it with caution. It should not be applied to the skin in pure form and above the recommended dose, and contact with the mucous membranes and eyes should be avoided. An oxidised essential oil should not be used. It may cause sensitization on the skin. It should not be used on damaged skin and in children under two years of age. It can cause headaches in high doses and long-term use. Care should be taken as it may cause short-term circulatory disorders in high-temperature baths.
Carvi aetheroleum	Hypertension patients should use it with caution. An oxidised essential oil should not be used. It should not be applied to the skin in pure form and above the recommended dose, and contact with the mucous membranes and eyes should be avoided.
Cedrus aetheroleum	Hypertension patients should use it with caution. An oxidised essential oil should not be used. Allergy and skin irritation to its components may develop. It should not be applied to the skin in pure form and above the recommended dose, and contact with the mucous membranes and eyes should be avoided.
Chamomillae romanae aetheroleum	In the Asteraceae family, it can cause dermatitis in susceptible individuals. Care should be taken not to get the essential oil into the eyes.
Cinnamomi camphorae aetheroleum	Hypertension patients should use it with caution. Due to the high 1,8-cineol content, care should be taken with dose adjustment in children. An oxidised essential oil should not be used. Allergy and skin irritation to its components may develop. It should not be applied to the skin in pure form and above the recommended dose, and contact with the mucous membranes and eyes should be avoided. The source from which the essential oil is obtained is the Ravintsera Madagascar camphor tree, not the *Ravensara aromatica* plant. Attention should be paid to label information and analysis certificates. It should not contain methyl eugenol, safrole and methyl chavicol compounds.

Table 4 contd. ...

...Table 4 contd.

Essential Oil	Warning
Cinnamomi zeylanici corticis aetheroleum	Hypertension patients should use it with caution. An oxidised essential oil should not be used. Allergy and skin irritation to its components may develop. It should not be applied to the skin in pure form and above the recommended dose, and contact with the eyes should be avoided. Mouthwashes containing essential oils should not be used in children under 12 years of age.
Cisti cretici aetheroleum	Hypertension patients should use it with caution. An oxidised essential oil should not be used. Hypersensitivity to its components may develop. It should not be applied to the skin in pure form and above the recommended dose, and contact with the mucous membranes and eyes should be avoided. Cistus ladanifer should not be confused as it is a different species. However, the differences in their minor components are observed and evaluated under this monograph.
Citri bergamiae aetheroleum	Hypertension patients should use it with caution. Since photosensitivity may develop due to furanocoumarins, the application area should not be exposed to sunlight or solarium UV rays. It should not be applied to the skin in pure form and above the recommended dose, and contact with the mucous membranes and eyes should be avoided. Allergy to its components, skin irritation and hyperpigmentation may develop. An oxidised essential oil should not be used. It should not be used during radiotherapy.
Limonis aetheroleum	Hypertension patients should use it with caution. Since photosensitivity may develop due to furanocoumarins, the application area should not be exposed to sunlight or solarium UV rays. It should not be applied to the skin in pure form and above the recommended dose, and contact with the mucous membranes and eyes should be avoided. Allergy and skin irritation to its components may develop. It should not be used for a long time. An oxidised essential oil should not be used. Cross-allergy may develop when used with Peruvian balm.
Citri paradisi aetheroleum	Hypertension patients should use it with caution. Since photosensitivity may develop due to furanocoumarins, the application area should not be exposed to sunlight or solarium UV rays. Allergy and skin irritation to its components may develop. Oxidised and essential oils containing l-limonene should not be used. It should not be applied to the skin in pure form and above the recommended dose, and contact with the mucous membranes and eyes should be avoided. Adding it to hot bath water may cause skin irritation.
Citri reticulatae aetheroleum	Hypertension patients should use it with caution. Depending on the method of obtaining, allergies to furanocoumarin and coumarin may develop. Since photosensitivity may develop, direct sunlight and solarium UV application should be avoided during use. It should not be applied to the skin in pure form and above the recommended dose, and contact with the mucous membranes and eyes should be avoided. Oxidised and essential oils containing l-limonene should not be used.

Table 4 contd. ...

...Table 4 contd.

Essential Oil	Warning
Aurantii dulcis aetheroleum	Hypertension patients should use it with caution. Oxidised and essential oils containing l-limonene should not be used. Skin sensitivity may develop at a concentration of 1.2%. Since photosensitivity may develop due to furanocoumarins, the application area should not be exposed to sunlight or solarium UV rays. Allergy to its components may develop. It should not be applied to the skin in pure form and above the recommended dose, and contact with the mucous membranes and eyes should be avoided.
Cumini aetheroleum	Hypertension patients should use it with caution. An oxidised essential oil should not be used. Allergy to its components may develop. It should not be applied to the skin in pure form and above the recommended dose, and contact with the mucous membranes and eyes should be avoided. To prevent photosensitivity, it should not be applied to areas of the skin exposed to sunlight or solarium UV rays; a high dose of essential oil can be an irritant.
Cupressi aetheroleum	Oxidised oil can cause skin sensitivity. It should not be used in children under age.
Cymbopogon martinii motiae aetheroleum	An oxidised essential oil should not be used. It should not be applied on the skin in its pure form, and contact with the mucous membranes and eyes should be avoided.
Cymbopogon martinii sofiae aetheroleum	An oxidised essential oil should not be used. It should not be applied on the skin in its pure form, and contact with the mucous membranes and eyes should be avoided.
Cymbopogon nardi aetheroleum	Since it is well tolerated due to its low content of monoterpene ketone components, especially Java-derived essential oil should be preferred. An oxidised essential oil should not be used. It should not be applied on the skin in its pure form, and contact with the mucous membranes and eyes should be avoided.
Eucalypti citriodorae aetheroleum	Hypertension patients should use it with caution. An oxidised essential oil should not be used. Allergy to its components may develop. It should not be applied to the skin in pure form and above the recommended dose, and contact with the mucous membranes and eyes should be avoided.
Eucalypti aetheroleum	Hypertension patients should use it with caution. An oxidised essential oil should not be used. Allergy to its components may develop. Due to its effects on the central nervous system due to 1,8-cineole, it may cause respiratory problems. Other chemotypes should be used with caution due to similar risks. It should not be applied to the skin in pure form and above the recommended dose, and contact with the mucous membranes and eyes should be avoided.
Eucalypti radiatae aetheroleum	Hypertension patients should use it with caution. An oxidised essential oil should not be used. It should not be applied to the skin in pure form and above the recommended dose, and contact with the mucous membranes and eyes should be avoided. It may cause skin irritation. It should not be applied to children under 1-year-old.

Table 4 contd. ...

...Table 4 contd.

Essential Oil	Warning
Eucalypti smithii aetheroleum	Hypertension patients should use it with caution. An oxidised essential oil should not be used. Hypersensitivity to its components may develop. It should not be applied to the skin in pure form and above the recommended dose, and contact with the mucous membranes and eyes should be avoided.
Foeniculi dulcis aetheroleum	Since the chemical compositions of bitter fennel essential oil are very different, they should not be used interchangeably. Bitter fennel essential oil may have a carcinogenic effect due to its high methyl cavicol content. The oxidised essential oil can cause skin sensitivity.
Helichrysi italici aetheroleum	Hypertension patients should use it with caution. An oxidised essential oil should not be used. It may cause skin sensitivity. It should not be applied in pure form and above the recommended dose, and contact with the mucous membranes and eyes should be avoided. It should not be used on hypersensitive and damaged skin. It should be used in children under 4 years of age and in women in the third trimester of pregnancy, in consultation with a physician or pharmacist.
Helichrysi splendidi aetheroleum	Hypertension patients should use it with caution. An oxidised essential oil should not be used. It may cause skin sensitivity. It should not be applied in pure form and above the recommended dose, and contact with the mucous membranes and eyes should be avoided. It should not be used on hypersensitive, damaged skin and in children under six years old. It should be used in children older than 6 years and women during pregnancy, in consultation with a physician or pharmacist.
Juniperi aetheroleum	Hypertension patients should use it with caution. An oxidised essential oil should not be used. It should not be applied to the skin in pure form and above the recommended dose, and contact with the mucous membranes and eyes should be avoided. The amount of sabinen should be at most 20% because of the risk of epilepsy-like side effects and low and renal dysfunction. Therefore, Juniperus sabina essential oil should never be used. If symptoms persist for more than 2 weeks during the use of the aromatherapeutic product, a physician or pharmacist should be consulted. Not recommended for use in children under 12 years of age.
Juniperi virginianae aetheroleum	Hypertension patients should use it with caution. An oxidised essential oil should not be used. Hypersensitivity to its components may develop. It should not be applied in pure form and above the recommended dose, and contact with the mucous membranes and eyes should be avoided.

Table 4 contd. ...

...Table 4 contd.

Essential Oil	Warning
Lauri folii aetheroleum	Due to its potential carcinogenicity, methyl eugenol should be at most 3%. Due to the risk of sensitization, it is recommended to be careful in dermal use on sensitive skin. There is a risk of skin sensitization and mucous membrane irritation. It should not come into contact with the eyes. It can cause problems in the central nervous system and breathe in children due to 1,8-cineole. It should not be used in children under 2 years of age as it may cause hypersensitivity. The source from which the essential oil is obtained is the *Laurus nobilis* plant, not the *Pimenta racemosa* known as 'West Indian bay leaf', *Cinnamomum tamala* known as 'Indian bay leaf', *Syzygium polyanthum* known as 'Indonesian bay leaf', *Umbellularia californica* known as 'Californian bay leaf' and *Litsea glaucescens* species known as 'Mexican bay leaf'. Attention should be paid to the label information and analysis certificates.
Lavandulae aetheroleum	For use against anxiety, linalool, a monoterpenol, and lavandulyl acetate, a monoterpene ester, are responsible for the effect and should be used in a standardised form. The lavender essential oil can be applied topically on the skin in pure form in adults. However, when used by dripping on the pillow, it should be dripped onto the corners of the pillow so that it does not come into contact with the skin. In its pure form, it can cause sensitivity and nausea if it is smelled for a long time. An oxidised essential oil should not be used. Allergy to its components may develop. It may cause dermatitis or skin irritation. Caution should be exercised in the use of vehicles and machinery and in other works that require attention. Alcohol should not be consumed during and after inhalation.
Lavandulae aetheroleum	Linalool, a monoterpenol, and lavandulyl acetate, a monoterpene ester, are responsible for the effect and should be used in a standardised form for anxiety-related uses. In its pure form, it can cause sensitivity and nausea if it is smelled for a long time. An oxidised essential oil should not be used. Since it is one of the rare essential oils used directly, skin sensitivity may develop in case of using oxidised oil. Allergy to its components may develop. Caution should be exercised in the use of vehicles and machinery and in other works that require attention. Alcohol should not be consumed during and after inhalation.
Spicae aetheroleum	In its pure form, it can cause sensitivity and nausea if it is smelled for a long time. An oxidised essential oil should not be used. Since it is one of the rare essential oils used directly, skin sensitivity may develop in case of using oxidised oil. Allergy to its components may develop. Caution should be exercised in the use of vehicles and machinery and in other works that require attention. Alcohol should not be consumed during and after inhalation.

Table 4 contd. ...

...Table 4 contd.

Essential Oil	Warning
Leptospermi scoparii aetheroleum	Hypertension patients should be used with caution. An oxidised essential oil should not be used. Hypersensitivity to its components may develop. It should not be used as a substitute for Melaleuca alternifolia, known as the tea tree. Attention should be paid to label information and analysis certificates. It should not be applied in pure form and above the recommended dose, and contact with the mucous membranes and eyes should be avoided.
Matricariae aetheroleum	Hypertension patients should use it with caution. An oxidised essential oil should not be used. It should not be applied to the skin in pure form and above the recommended dose, and contact with the mucous membranes and eyes should be avoided. It should not be used in children under 12 years of age as there is insufficient data. Allergic reactions (dyspnea, Quincke's disease, vascular collapse or anaphylactic shock) may occur after mucosal contact in hypersensitive individuals. Patients with chronic heart and circulatory disorders should not take a full bath with chamomile essential oil without consulting a physician. Those with open wounds, widespread skin damage or acute (short-term) skin disease, high fever and chronic infection should not take a full or partial bath with medicinal chamomile essential oil in any way. Caution should be exercised in the use of vehicles and machinery and in other works that require attention.
Melaleucae alternifolii aetheroleum	Hypertension patients should use it with caution. An oxidised essential oil should not be used. It should not be used as a mouthwash in children under six years of age, in case of swallowing. If the essential oil is applied directly, it should be used with caution due to its methyl eugenol content. It should not be used directly in cases of contact dermatitis and eczema. It can cause excessive dryness on the skin and cause changes in pH and humidity. May increase sensitivity in photosensitivity.
Melaleucae aetheroleum	Hypertension patients should use it with caution. An oxidised essential oil should not be used. It should not be applied to the skin in pure form and above the recommended dose, and contact with the mucous membranes and eyes should be avoided. A skin test should be done. Due to its effects on the central nervous system due to 1,8-cineole, it may cause respiratory problems. It should not be used in children under six years of age.
Melissae aetheroleum	In children under 2 years of age, dermal use may cause hypersensitivity on problematic skin. For sensitive skin, dermal use should be a maximum of 0.9%. Essential oils of *Cymbopogon* and *Lippia* species should not be used instead of *Melissa officinalis* essential oil. Attention should be paid to label information and analysis certificates

Table 4 contd. ...

...Table 4 contd.

Essential Oil	Warning
Menthae arvensis aetheroleum partim mentholum depletum	Hypertension patients should use it with caution. An oxidised essential oil should not be used. It should not be applied to the skin in pure form and above the recommended dose, and contact with the mucous membranes and eyes should be avoided. Due to the menthol and its derivatives in its content, it can cause bronchial spasms and breathing and swallowing difficulties, especially in children and COPD patients, depending on the dose. When applied externally, skin sensitivity and contact dermatitis may occur.
Menthae piperitae aetheroleum	Hypertension patients should use it with caution. An oxidised essential oil should not be used. It should not be applied to the skin in pure form and above the recommended dose, and contact with the mucous membranes and eyes should be avoided. When applied externally, skin sensitivity and contact dermatitis may occur. Due to the menthol and its derivatives in its content, it can cause bronchial spasms and breathing and swallowing difficulties, especially in children and COPD patients, depending on the dose. It should not be used in epileptic patients. It should not be used for longer than 3 months in massage applications. It should not be used for more than two weeks in inhalation and mouthwash applications.
Myrrh Aetheroleum	An oxidised essential oil should not be used. Should not come into contact with mucous membranes and eyes.
Basilici aetheroleum	Attention should be paid to the methyl chavicol content of basil essential oil. Hypertension patients should use it with caution. An oxidised essential oil should not be used. It may cause skin sensitivity. It should not be applied in pure form and above the recommended dose, and contact with the mucous membranes and eyes should be avoided.
Origani dubii aetheroleum	Hypertension patients should use it with caution. An oxidised essential oil should not be used. It should not be applied in pure form and above the recommended dose, and contact with the mucous membranes and eyes should be avoided.
Pelargonii aetheroleum	Hypertension patients should use it with caution. An oxidised essential oil should not be used. It may cause skin sensitivity. Some people may experience a rash or burning sensation when applied to the skin. In inhalation applications, high doses and long-term use may cause dizziness. It should not be applied in pure form and above the recommended dose, and contact with the mucous membranes and eyes should be avoided.
Anisi aetheroleum	Hypertension patients should use it with caution. An oxidised essential oil should not be used. Hypersensitivity to its active ingredients (*trans*-anethole) and other plants of the Apiaceae (Umbelliferae) family (usually due to methyl cavicol and umbelliferon) may develop. It should not be used in dermatitis, inflammatory or allergic skin conditions. It should not be applied to the skin in pure form and above the recommended dose, and contact with the mucous membranes and eyes should be avoided. Tisserand recommended a dermal max. of 2.4%.

Table 4 contd. ...

...Table 4 contd.

Essential Oil	Warning
Pini pumilionis aetheroleum	In bath therapy applications, care should be taken not to slip when getting out of the tub. Not recommended for those with hypertension and heart disease. It should be used with caution as it may cause bronchospasm. An oxidised essential oil should not be used. In case of oxidation, skin sensitivity may develop. Therefore, the amount of peroxide is important. Allergy and skin irritation to its components may develop. It should not be applied in pure form and above the recommended dose, and contact with the mucous membranes and eyes should be avoided.
Pini sylvestri aetheroleum	In bath therapy applications, care should be taken not to slip when getting out of the tub. Not recommended for those with hypertension and heart disease. It should be used with caution as it may cause bronchospasm. Allergy and skin irritation to its components may develop. An oxidised essential oil should not be used. It should not be applied to the skin in pure form and above the recommended dose, and contact with the mucous membranes and eyes should be avoided.
Pogostemon cablin aetheroleum	An oxidised essential oil should not be used.
Rosae damascenae aetheroleum	An oxidised essential oil should not be used. Hypersensitivity to its components may develop. It should not be applied to the skin in pure form and above the recommended dose, and contact with the mucous membranes and eyes should be avoided.
Rosmarini aetheroleum	An oxidised essential oil should not be used. It should not be applied to the skin in pure form and above the recommended dose, and contact with the mucous membranes and eyes should be avoided. The camphor chemotype should be used with caution. It should not be used in children under 12 years of age. Sudden cough, bronchial and laryngeal spasm may be seen in the use of 1,8-cineole chemotype at high doses. The use of 1,8-cineole chemotype is not recommended under 8 years of age due to insufficient data. Contact dermatitis and asthma may be pure in hypersensitive individuals. It should not be applied to the face area when used in a hematoma. In case of undesirable effects, the doctor or pharmacist should be consulted. It should not be applied as a hot bath in people with hypertension. For use on the skin, it should not be applied close to the eyes and mucous membranes.
Salviae trilobae aetheroleum	Hypertension patients should use it with caution. An oxidised essential oil should not be used. It should not be applied to the skin in pure form and above the recommended dose, and contact with the mucous membranes and eyes should be avoided. It should not be used in infants and children near the face and eyes. Due to its effects on the central nervous system due to 1,8-cineole, it may cause respiratory problems.

Table 4 contd. ...

...Table 4 contd.

Essential Oil	Warning
Salviae lavandulifoliae aetheroleum	Hypertension patients should use it with caution. An oxidised essential oil should not be used. It should not be applied to the skin in pure form and above the recommended dose, and contact with the mucous membranes and eyes should be avoided. It should not be used in children under 12 years of age. Due to its effects on the central nervous system due to 1,8-cineole, it may cause respiratory problems. It should not be used in epileptic patients. The amount of geraniol should be at most 1%. Toxicity due to sabinyl acetate should be noted.
Salviae sclareae aetheroleum	An oxidised essential oil should not be used. It should not be applied to the skin in pure form and above the recommended dose, and contact with the mucous membranes and eyes should be avoided.
Santali ligni aetheroleum	Hypertension patients should use it with caution. Since it is a highly adulterated oil, there are synthetics in the market. Because of this reason, care should be taken to use well-analyzed certified products. An oxidised essential oil should not be used. Allergy to its components may develop. It should not be applied to the skin in pure form and above the recommended dose, and contact with the mucous membranes and eyes should be avoided. Do not go out in the sun after use, it should be used in the evening.
Caryophylli 49üre aetheroleum	Since data outside these values will indicate oxidation, hypersensitivity to its components may develop. Hypertension patients should use it with caution. It should not be applied to the skin in pure form and above the recommended dose, and contact with the eyes should be avoided. Care should be taken not to swallow it. Depending on the phenolic components it contains, it irritates the skin even at a concentration of 0.03%.
Thymi serpylli aetheroleum	Use in children under 3 years of age is not recommended. Hypertension patients should use it with caution. In the case of hypertension, bathing should not be applied. In use as a bath additive; in case of dyspnea, fever, or purulent sputum, the physician should be consulted. An oxidised essential oil should not be used. Hypersensitivity to its components may develop. It is used by diluting in a carrier oil. Otherwise, allergic reactions due to irritation may occur. It should not be applied to the skin in pure form and above the recommended dose, and contact with the mucous membranes and eyes should be avoided. The efficacy observed against dry cough at low concentration is not seen at high concentration. For this reason, it should not be used in higher doses than the specified dose, it may trigger asthma attacks.

Table 4 contd. ...

...Table 4 contd.

Essential Oil	Warning
Thymi aetheroleum typo geraniolo	It should not be used in children under 3 years old. In the case of hypertension, bathing should not be applied. In use as a bath additive; In case of dyspnea, fever, or purulent sputum, the physician should be consulted. An oxidised essential oil should not be used. Hypersensitivity to its components may develop. It should not be applied in pure form and above the recommended dose, and contact with the mucous membranes and eyes should be avoided.
Thymi aetheroleum typo linalool	It should not be used in children under 3 years old. In the case of hypertension, bathing should not be applied. In use as a bath additive; In case of dyspnea, fever, or purulent sputum, the physician should be consulted. An oxidised essential oil should not be used. Hypersensitivity to its components may develop. It should not be applied in pure form and above the recommended dose, and contact with the mucous membranes and eyes should be avoided.
Thymi typo thymolo aetheroleum	It should not be used in higher doses than the specified dose, it may trigger asthma attacks. Crystalline synthetic thymol should not be used as it is not an aromatherapeutic product. It should not be used in children under 3 years old. In the case of hypertension, bathing should not be applied. In use as a bath additive; In case of dyspnea, fever, or purulent sputum, the physician should be consulted. An oxidised essential oil should not be used. Hypersensitivity to its components may develop. It should not be applied in pure form and above the recommended dose, and contact with the mucous membranes and eyes should be avoided.
Valerianae aetheroleum	Hypersensitivity reactions can be developed to active substances. Full baths are contraindicated in cases of open wounds, large skin injuries, acute skin diseases, high fever, severe infections, severe circulatory disturbances, and cardiac insufficiency.
Vetiverae aetheroleum	Hypertension patients should use it with caution. An oxidised essential oil should not be used. It May cause skin sensitivity. In inhalation applications, high doses and long-term use may cause dizziness. It should not be applied to the skin in pure form and above the recommended dose, and contact with the mucous membranes and eyes should be avoided.

and clary sage essential oil is not recommended to be used in the first four months of pregnancy due to insufficient data. Due to its spasmolytic effect, manuka essential oil should not be used in women during pregnancy and lactation. Sweet basil essential oil is recommended not to be used by pregnant and lactating women because it contains methyl cavicol and isoeugenol. Due to the monoterpene ketone type components in its content, pregnant and lactating women should not use partly dementholised mint essential oil. Cajeput essential oil should not be used in the first trimester, especially since it contains hormone-like components.

Storage Conditions of Essential Oils

Essential oils should be stored in tightly closed dark bottles, defined in the legislation, at room temperature. It has fast oxidising properties and the brownish colour of the oil indicates that it has been oxidised. Essential oils prepared in small quantities for use should be stored in a cool and dark environment. It can be stored in the refrigerator for three to six months. Essential oils obtained from *Citrus* species should be consumed within three months, and other oils within six months. In the literature, there are studies reporting that the ratio of essential oil components in essential oils decreases when kept waiting (especially essential oils opened for use) (Kumar et al. 2013, Arabhosseini et al. 2007).

In addition to the general storage conditions, it is recommended to store Palmarosa essential oil at temperatures below 25°C. Grapefruit essential oil can be stored for one year at temperatures below 21°C and must be stored in the dark. Manuka essential oil can be stored for three years without opening and for two years after opening. Sandalwood essential oil can be stored for three years without opening. The shelf life of the Ceylon, cinnamon bark oil should not exceed one year. Pregnant women should not use juniper essential oil because of sabinene content. Damask rose essential oil should be stored at 25–28°C and be stored in the dark. If it is used frequently after the cover is opened, bay laurel leaf, tea tree, and patchouli essential oil should not be used for more than one year. Therefore, small packages should be preferred.

Essential Oils in Plants

Essential oils and their constituents are a group of bioactive components in gymnosperm and angiosperm that are often associated with a broad-spectrum of biological activity in plants (Table 1).

Gymnosperm

All members of conifers have evolved specialised anatomical structures for sequestration and produced hydrophobic oleoresins (resin) in massive amounts (Öztürk et al. 2011). These elements contribute to the unique aromatic properties of many conifer resins that make them so pleasurable in perfumes and incenses. A wide range of biological activities, including antimicrobial, anti-inflammatory, antifungal, antiviral, antioxidant, and anti-cancer activities, have been attributed to essential oils of these taxa (Bakkali et al. 2008, Adorjan and Buchbauer 2010, Bayala et al. 2014).

Essential oils obtained from the *Juniperus* L. berries and wood tar are used for the treatment of a variety of disorders, such as asthma, respiratory disorders, cough, colds, fevers, gonorrhoea, arthritis, hyperglycemia, bladder affections, chronic pyelonephritis, abdominal disorders, chronic eczema, and skin diseases (Fujita et al. 1995, Orhan et al. 2012, Bais et al. 2014, Darwish et al. 2020). *J. oxycedrus* L. is the most frequently used species of the family, and its essential oils are used for the treatment of respiratory diseases, including bronchitis, cough, cold and flu, shortness of breath, asthma tuberculosis, and pneumonia. *Juniperus* species are one of the important sources of cedar-wood oil, which is used in folk medicine as an antiseptic for wound healing and against anxiety, nervous tension, and stress-related conditions. Phytochemical studies indicated that α-pinene, besides cedrol and verbenol, is among the most abundant components of the essential oil of *J. excelsa* berries (Asili et al. 2008, Ehsani et al. 2012, Atas et al. 2012).

The chemical composition of the essential oils among *J. oxycedrus* subspecies and populations showed differences in the volatile profiles that come from their environmental and genetical inter-specific diversity, and this diversity can significantly affect the variability in biological activities and efficacy (Orhan et al. 2012, Ehsani et al. 2012, Darwish et al. 2020).

Pinaceae is the most abundant and widespread coniferous family, particularly in the Northern Hemisphere predominated by *Abies* Mill. (firs), *Picea* Mill. (spruces), and *Pinus* L. (pines). The genus *Pinus* is the largest genus of conifers, which is used by local people for the treatment of respiratory diseases. Herbal drugs from different parts of *Pinus* are used in the form of decoctions, ointments, bathing oils, or drugs for inhalation to treat various diseases, such as cough, bronchitis, shortness of breath, asthma and pulmonary ailments, hypertension, cardiac disease, wound healing, and muscle disorders of infectious in most countries (Mustafa et al. 2012, Grassmann et al. 2003, Mamedov and Craker 2001, Kızılarslan and Sevg 2013, Shuaib et al. 2013).

The main constituents of Pine essential oil are α-pinene, camphene, β-pinene, δ3-carene, β-myrcene, limonene, β-phellandrene, α-terpinolene, β-caryophyllene, germacrene D, and δ-cadinene (Kurti et al. 2019). Additionally, the major compounds of the essential oils in the needles were pinene, carene, and limonene, while the major components of essential oils in the cone were caryophyllene and carene (Basholli-Salihu et al. 2017). Genetic variation among taxa and environmental factors, such as geographic location, climate, season, stress during growth periods, maturity time, drying, and storage conditions cusses to variation in the chemical composition of the essential oils and subsequently influences their biological properties (Raut and Karuppayil, 2014).

The results of Basholli-Salihu et al. (2017) revealed that the essential oil of Pine twigs showed a significant anti-inflammatory activity than those of their needles or cones, and he suggested *α*-pinene as the major compound of essential oils in these taxa can be responsible for the secretion profile of cytokines and thus play an important role in the anti-inflammatory activity. Also, Yoon et al. (2009) research indicated that the essential oils of *Abies koreana*, rich in bornyl acetate, limonene and *α*-pinene, inhibited the IL-1b, IL-6, and TNF-*α* production in LPS-stimulated macrophages.

Angiosperm

Cymbopogon citratus (DC.) Stapf from the Poaceae family is an economically important plant, which is widely cultivated for its high-quality essential oil. The essential oil of *Cymbopogon* Spreng. has a characteristic strong lemony odour due to the presence of high content of aldehyde citral, geraniol, myrcene, geranyl acetate, caryophyllene, and monoterpene olefins, such as limonene (Shahi et al. 2005, Weiss 1997). Due to the presence of a large number of essential oils, this plant has been widely used in traditional medicines for the treatment of nervous, gastrointestinal disturbances, pneumonia, coughs, fevers, reducing cholesterol, uric acid, pancreas, kidney, bladder, and the digestive tract (Ojo 2006, Santin et al. 2009).

The Apiaceae plants are rich in phytochemical compounds, especially essential oils, and they are largely used as traditional remedies to treat gastrointestinal, reproductive, and respiratory diseases (Acimovic et al. 2015, Sayed-Ahmad et al. 2017). The family members are known for their characteristic odour due to the presence of schizogenous ducts containing oil, mucilage, or resin in all the parts of these plants especially seeds (Berenbaum 1990, Anastasopoulou et al. 2020, Barros et al. 2010). The chemical profile of the essential oils is diversified among the family, including different compounds such as monoterpene hydrocarbons, oxygenated, and aromatic monoterpenes, sesquiterpene hydrocarbons, oxygenated sesquiterpenes, phenylpropanoids, and aliphatic compounds (Kamte et al. 2018), which have many biological activities, such as antioxidant, hepatoprotective, antibacterial, vasorelaxant, apoptosis-inducing, anti-tumour, and cyclooxygenase inhibitory. *Coriandrum sativum* L. and *Foeniculum vulgare* Mill. is extremely aromatic and is subsequently used in perfumes, condiments, and liqueurs (Bahmani et al. 2015, Barros et al. 2010). Essential oil isolated from *Smyrnium olusatrum* L. and *Pimpinella anisum* L. exhibited antifungal and antimicrobial activity (Samojlik et al. 2010, Marongiu et al. 2012, Kosalec et al. 2005) but also cytotoxic effect (Quassinti et al. 2013). In addition, essential oils obtained from the Apiaceae family represent novel candidates as eco-friendly and natural insecticides (Spinozzi et al. 2021).

Some genera of the Asteraceae family such as *Artemisia* L., *Matricaria* L., *Helichrysum* Mill. and *Tussilago* L. are used for economical and pharmaceutical essential oil isolation. The flower buds and leaves of *Tusilago farfara* L. (coltsfoot) from Asteraceae have been recommended for cough, shortness of breath, phlegm, asthma, and other pulmonary disorders in Anatolia, Persian and Chinese Traditional Medicine (Baytop 1999, Avicenna 1988, Qu et al. 2018). The fatty alkanes include *n*-undecane, *n*-tetracosane, *n*-hexacosane, *n*-tetradecanol, *n*-nonadecane, plus phytol, and caryophyllene oxide are the most important phytochemicals identified from the essential oil of the flowers (Norani et al. 2019).

Artemisia annua L. and *A. absinthium* L. have a long history of human use in traditional medicine throughout the world. Artemisinin, as the most well-known metabolite, is a sesquiterpene lactone with antimalarial properties, which is isolated from this plant. The essential oil constituents of these plants generally contain esters of thujyl alcohol, thujone, thujane, pinene, cymene, camphene, cadinene, guaiazulene, cymene, 1,8-cineol, methylheptenone, *β*-phellandrene, caryophyllene oxide, *α*-terpineol, geraniol, caryophyllene, α-himachalene, and elemol (Singh et al. 1959, Kaul et al. 1979).

According to Kaul et al. (1976), the essential oil of *A. absinthium* even in low concentration is active against both sensitive and resistant strains of *some* bacteria. The genus *Matricaria* L. is one of the most popular ingredients herbal teas. Essential oils of this plant contain exhibited antibacterial activity against 25 different Gram-positive and Gram-negative bacteria (McKay 2006, Sokovic 2010).

EO extracted from *M. recutita* L. flowers contain terpenoid α-bisabolol and its oxides and azulenes, including chamazulene (Kamatou et al. 2010). It initiates disruption of the bacterial cell membrane, allowing permeation into the cell of exogenous solutes.

Radusiene and Judzentiene (2008) reported *trans*-caryophyllene, δ-cadinene, 1,8-cineole, and tetradecanoic acid were the major constituents in essential oils from inflorescences of the *H. arenarium* (L.) Moench grown in Lithuania. Whereas, Sesquiterpenes β-spatulenol, ledol, bicyclogermacrene, aromadendrene, and α-eudesmol are found as the main constituents of H. arenarium essential oil from China (Liu et al. 2019), and linalool, anethole, carvacrol, and α-muurolene were reported as the chemical composition of this species from Hungary (Czinner et al. 2000). The analysis of the data indicated that the chemical profile of the essential oil varies depending on both the method extraction and plant growing condition containing the soil and climatic conditions (Stankov et al. 2020).

Lamiaceae family is considered an economically and medicinally important group of aromatic plants with rich sources of essential oils. Essential oils are known as the origin of the complex components of secondary metabolites, especially terpenoids (Abd El-Gawad et al. 2019, Assaeed et al. 2020). The majority of species in this family such as *Lavanduala* L.*, Origanum* L.*, Ocimum* L.*, Rosmarinus* L.*, Thymus* L., and *Salvia* L. are reported to be used in the treatment of exhaustion, weakness, depression, strengthening of fragile blood vessels, skin allergies, and asthma (Naghibi et al. 2010, Raja, 2012). Thymol and carvacrol are the main phytochemical constituents of these family members. A wide range of biological activities of essential oils have been proved, such as phytotoxic (Abd-El-Gawad et al. 2019), antimicrobial (Deng et al. 2020), anti-inflammatory, antipyretic (Elshamy, et al. 2020), and hepatoprotective activities (Damtie et al. 2019). These metabolites (Kotan et al. 2014) show significant antibacterial activities and inhibit the growth of 14 different phytopathogenic bacterial strains. Also, carvacrol has an inhibitory effect on prostaglandin (Wagner et al. 1986a) and subsequently causes analgesia (Ferreira et al. 1978). Generally, essential oils revealed chemical variations depending on genetically determined properties, localities and environmental factors and show notable antimicrobial activity against respiratory pathogens (Vladimir-Knežević et al. 2012, Ozcan and Chalchat 2004).

The Lauraceae family have been widely used in food, timber, pharmaceutical, and perfumery industries besides their traditional medical benefits. These family members traditionally have been applied to different purposes, such as infectious, malaria, gastrointestinal infections, female genital infections, and rheumatism (Wmnhw et al. 2015). Generally, Sesquiterpene hydrocarbons among alcohols and aldehydes were also found in their essential oils. In the *Cinnamomum* species, phenylpropanoids mainly safrole, eugenol, linalool, camphor, benzyl benzoate, or cinnamaldehyde was reported as major components. These species also exhibited strong to moderate levels of antifungal, antimicrobial, antioxidant, and anti-inflammatory activity due to the presence of high phenolic content (Nam and Jantan 1993, Wmnhw et al. 2015, Wan Salleh et al. 2016).

Myrtaceae members contain *Eucalyptus* L'Hér., *Eugenia* P.Micheli ex L., *Syzygium P. Browne ex Gaertn.*, and *Melaleuca* L. are among the significant genera known for their essential oils used medicinally. The essential oils obtained from the leaves of *Eucalyptus* exhibit antimicrobial, antifungal, anti-inflammatory and expectorant properties (Batish et al. 2008, Ghaffaret al. 2015). Also, inhalation of *Eucalyptus* derivatives has been used to treat pharyngitis, bronchitis, and sinusitis. According to phytochemical analyses, the main constituents of *Eucalyptus* oils are *p*-cymene, α-pinene, 1,8-cineole, spathulenol, cryptone, and α-terpineol (Cermelli et al. 2008, Elaissi et al. 2012). The essential oil of *Syzygium samarangensehas* (Blume) Merr. and L.M. Perry contains α-pinene, γ-terpinene, β-caryophyllene, and caryophyllene oxide as major components (Gao et al. 2012, Lee et al. 2016).

In addition, *R. damascene* L. (Damask rose) from the Rosaceae family has been commonly used for the production of rose essential oil (Rusanov et al. 2005). The most therapeutic effects of *R. damascena* in ancient medicine are including treatment of abdominal and chest pain, strengthening the heart (Wood et al. 1839), treatment of menstrual bleeding and digestive problems (Ave-Sina, 1990), and reduction of inflammation, especially of the neck (Buckle et al. 1993).

The *Citrus* L. species of the family *Rutaceae* provide several popular edible fruits in the world. The *Citrus* species of the *Rutaceae* family are well known due to several common edible fruits of the genus. *Citrus* essential oils, widely used in perfumery and food industry, pharmaceutical and cosmetic products (Tranchida et al. 2012, Palazzolo et al. 2013, Othman et al. 2017), are obtained mainly from the fruit rind (flavedo). Volatile and semi-volatile compounds represent 85–99% of the entire oil in *Citrus* (Dugo and Mondello 2011, Tranchida et al. 2012, Sarrou et al. 2013) and contain limonene, a hydrocarbon monoterpene, as the most abundant compound (Jing et al. 2014). According to the number of identified compounds, sesquiterpene hydrocarbons are the most diverse group in this genus (Carmen Gonzalez-Mas et al. 2019).

References

Abd-El-Gawad, A.M., A.I. Elshamy, S. Al-Rowaily and Y.A. El-Amier. 2019. Habitat affects the chemical profile, allelopathy, and antioxidant properties of essential oils and phenolic enriched extracts of the invasive plant *Heliotropium Curassavicum*. Plants (Basel) 8(11): 482.

Acimovic, M., L. Kostadinovic, S. Popovic and N.S. Dojcinovic. 2015. Apiaceae seeds as functional food. Journal of Agricultural Sciences 60(3): 237–246.

Adorjan, B. and G. Buchbauer. 2010. Biological properties of essential oils: An updated review. Flavour Frag. J. 25: 407–426.

Ali, N.A.M. and I. Jantan. 1993. The essential oils of *Lindera pipericarpa*. J. Trop. For. Sci. 6(2): 124–30.

Anastasopoulou, E., K. Graikou, C. Ganos, G. Calapai and I. Chinou. 2020. Pimpinella anisum seeds essential oil from Lesvos island: Effect of hydrodistillation time, comparison of its aromatic profile with other samples of the Greek market. Safe use. Food Chem. Toxicol. 135: 110875.

Arabhosseini, A., W. Huisman, A. Boxtel and J. Müller. 2007. Long-term effects of drying conditions on the essential oil and colour of tarragon leaves during storage. J. Food Engineering 79(2): 561–566.

Asili, J., S.A. Emami, M. Rahimizadeh, B.S. Fazly-Bazzaz and M.K. Hassanzadeh. 2008. Chemical and antimicrobial studies of *Juniperus excelsa* subsp. *excelsa* and *Juniperus excelsa* subsp. *polycarpos* essential oils. J. Essent. Oil-Bearing Plants 11: 292–302.

Assaeed, A., A. Elshamy, A.E.-N. El Gendy, B. Dar, S. Al-Rowaily and A. Abd-ElGawad. 2020. Sesquiterpenes-rich essential oil from above ground parts of *Pulicaria somalensis* exhibited antioxidant activity and allelopathic effect on weeds. Agronomy 10(3): 399.

Atas, A.D., I. Goze, A. Alim, S.A. Cetinus, N. Durmus, N. Vural et al. 2012. Chemical composition, antioxidant, antimicrobial and antispasmodic activities of the essential oil of *Juniperus excelsa* subsp. *excelsa*. J. Essent. Oil-Bearing Plants 15: 476–483.

Ave-Sina, 1990. Law in Medicine. Interpreter; Sharaf khandy A, Teheran: Ministry of Guidance publication, pp. 129–131.

Avicenna, 1988. Qanoon Dar Teb, Translated to Persian by Abdorahman Sharafkandi. Soroosh Press, Tehran.

Babar, A., A.W. Naser, S. Saiba, A. Aftad, S.A. Khan, A. Firoz et al. 2015. Essential oils used in aromatherapy: A systemic review. Asian Pasific Journal of Tropical Biomedicine 2015: 1–9.

Bahmani, K., A.I. Darbandi, H.A. Ramshini, N. Moradi and A. Akbari. 2015. Agro- morphological and phytochemical diversity of various Iranian fennel landraces. Ind. Crop Prod. 77: 282–294.

Bais, S., N.S. Gill, N. Rana and S. Shandil. 2014. A phytopharmacological review on a medicinal plant: *Juniperus communis*. Int. Sch. Res. Not. 2014: 634723.

Bakkali, F., S. Averbeck, D. Averbeck and M. Idaomar. 2008. Biological effects of essential oils—A review. Food Chem. Toxicol. 46: 446–475.

Barradas, T.N. and K.G.D.H. Silva. 2020. Nanoemulsions of essential oils to improve solubility, stability and permeability: A review. Environ. Chem. Lett. 19: 1153–1171.

Barros, L., A.M. Carvalho and I.C.F.R. Ferreira. 2010. The nutritional composition of fennel (Foeniculum vulgare): Shoots, leaves, stems and inflorescences. Lebensmittel-Wissenschaft und- Technologie. 43(5): 814–818.

Basholli-Salihu, M., R. Schuster, A. Hajdari, D. Mulla, H. Viernstein, B. Mustafa et al. 2017. Phytochemical composition, anti-inflammatory activity and cytotoxic effects of essential oils from three *Pinus* spp. Pharmaceutical Biology 55(1): 1553–1560.

Batish, D., H. Singh, R. Kohli and S. Kaur. 2008. Eucalyptus essential oil as a natural pesticide. For. Ecol. Manage. 256: 2166–2174.

Bayala, B., I.H.N. Bassole, R. Scifo, C. Gnoula, L. Morel, J.M. Lobaccaro et al. 2014. Anticancer activity of essential oils and their chemical components—A review. Am. J. Cancer Res. 4: 591–607.

Baytop, T. 1999. Therapy with Medicinal Plants in Turkey (Past and Present), 2nd edition. Nobel Medical Bookhouse, Istanbul.

Berenbaum, M.R. 1990. Evolution of specialization in insect-umbellifer associations. Annu. Rev. Entomol. 35: 319–343.

Bilgiç, Ş. 2017. Hemşirelikte Holistik Bir Uygulama; Aromaterapi, Namık Kemal Tıp Dergisi, Tekirdağ 2017: 135–137.

Bilsland, D. and A. Strong. 1990. Allergic contact dermatitis from the essential oil of French marigold (*Tagetes patula*) in an aromatherapist. Contact. Dermat. 23: 55–56.

Böhme, K., J. Barros-Velázquez, P. Calo-Mata and S.P. Aubourg. 2014. Antibacterial, antiviral and antifungal activity of essential oils: Mechanisms and applications. pp. 51–81. *In*: T.G. Villa and P. Veiga-Crespo (eds.). Antimicrobial Compounds. Berlin, Germany: Springer

Buckle, D.R., J.R.S. Arch, N.E. Boering, K.A. Foster, J.F. Taylor, S.G. Taylor et al. 1993. Relaxation effect of potassium channel activators BRL 38227 and Pinacidil on guinea-pig and human airway smooth muscle, and blockade of their effects by Glibenclamide and BRL 31660. Pulm Pharmacol. 6: 77–86.

Carmen Gonzalez-Mas, M., J.L., Rambla, M.P. Lopez-Gresa et al. 2019. Volatile compounds in Citrus essential oils: A comprehensive review. Frontiers in Plant Science 10, 12: 1–18.

Cermelli, C., A. Fabio, G. Fabio and P. Quaglio. 2008. Effect of Eucalyptus essential oil on respiratory bacteria and viruses. Curr. Microbiol. 56: 89–92.

Czinner, E., E. Lemberkovics, E. Bihatsi-Karsai, G. Vitanyi and L. Lelik. 2000. Composition of the essential oil from the inflorescence of *Helichrysum arenarium* (L.) Moench. J. Essent. Oil Res. 12(6): 728–730.

Damtie, D., C. Braunberger, J. Conrad, Y. Mekonnen and U. Beifuss. 2019. Composition and hepatoprotective activity of essential oils from Ethiopian thyme species (*Thymus serrulatus* and *Thymus schimperi*). J. Essent. Oil Res. 31: 120–128.

Darwish, R.S., H.M. Hammoda, D.A. Ghareeb, A.S.A. Abdelhamid, E.M.B. El Naggar, F.M. Harraz et al. 2020. Efficacy-directed discrimination of the essential oils of three Juniperus species based on their *in-vitro* antimicrobial and anti-inflammatory activities. J. Ethnopharmacol. 259: 112971.

Degenhardt, J., T.G. Köllner and J. Gershenzon. 2009. Monoterpene and sesquiterpene synthases and the origin of terpene skeletal diversity in plants. Phytochemistry 70(15-16): 1621–1637.

Deng, W., K. Liu, S. Cao, J. Sun, B. Zhong and J. Chun. 2020. Chemical composition, antimicrobial, antioxidant, and antiproliferative properties of grapefruit essential oil prepared by molecular distillation. Molecules 25(1): 217.

Dhifi, W., S. Bellili, S. Jazi, N. Bahloul and W. Mnif. 2016. Essential oils' chemical characterization and investigation of some biological activities: A critical review. Medicines 3(4): 25.

Dugo, P. and L. Mondello. 2011. Citrus Oils: Composition, Advanced Analytical Techniques, Contaminants, and Biological Activity, Vol. 49. Boca Raton, FL: CRC Press.

Ehsani, E., A.K. Noghabi, M. Teimouri and A. Khadem. 2012. Chemical composition and antibacterial activity of two *Juniperus* species essential oils. Afr. J. Microbiol. Res. 6: 6704–6710.

Elaissi, A., Z. Rouis, S. Mabrouk, K. Salah, M. Aouni, M. Khouja et al. 2012. Correlation between chemical composition and anti-bacterial activity of essential oils from fifteen *Eucalyptus* species growing in the Korbous and Jbel Abderrahman Arboreta (North East Tunisia). Molecules 17: 3044–3057.

Elshamy, A.I., N.M. Ammar, H.A. Hassan, S.L. Al-Rowaily, T.R. Raga, A.E.-N. El Gendy et al. 2020. Essential oil and its nanoemulsion of *Araucaria heterophylla* resin: Chemical characterization, anti-inflammatory, and antipyretic activities. Ind. Crop. Prod. 148: 112272.

Ferreira, S.H., S. Moncada and J.R. Vane. 1978. Prostaglandins and the mechanism of analgesis produced by aspirin-like drugs. Br. J. Pharmacol. 49(1): 86–97.

Fujita, T., E. Sezik, M. Tabata, E. Yesilada, G. Honda, Y. Takeda et al. 1995. Traditional medicine in Turkey VII. Folk medicine in middle and west Black Sea regions. Econ. Bot. 49: 406–422.

Gao, Y., Q. Hu and X. Li. 2012. Chemical composition and antioxidant activity of essential oil from *Syzygium samarangense* (BL.) Merr. et Perry flower-bud. Spatula DD 2: 23–33.

Ghaffar, A., M. Yameen, S. Kiran, S. Kamal, F. Jalal, B. Munir et al. 2015. Chemical composition and in-vitro evaluation of the antimicrobial and antioxidant activities of essential oils extracted from seven *Eucalyptus* species. Molecules 20: 20487–20498.

González-Mas, M.G., J.L. Rambla, M.P. López-Gresa, M.A. Blázque and A. Granell. 2019. Volatile compounds in citrus essential oils: A comprehensive review. Front. Plant Sci. 10: 12.

Grassmann, J., S. Hippeli, R. Vollmann and E.F. Elstner. 2003. Antioxidative properties of the essential oil from *Pinus mugo*. J. Agric. Food Chem. 51: 7576–7582.

Henley, D.V., N. Lipson, K.S. Korach and C.A. Bloch. 2007. Prepubertal gynecomastia linked to lavender and tea tree oils. N. Engl. J. Med. 356: 479–485.

Kamatou, G.P.P. and A.M. Viljoen. 2010. A review of the application and pharmacological properties of α-bisabolol and α-bisabolol-rich oils. J. Am. Oil Chem. Soc. 87: 1e7.

Kamte, S.L.N., F. Ranjbarian, K. Cianfaglione, S. Sut, S. Dall'Acqua, M. Bruno et al. 2018. Identification of highly effective antitrypanosomal compounds in essential oils from the Apiaceae family. Ecotox. Environ. Safe. 156: 154–165.

Kaul, V.K., S.S. Nigam and A.K. Banerjee. 1979. Thin layer and gas chromatography studies of essential oil of *Artemisia absinthium* Linn. Indian Perfume 23: 1–7.

Kaya, D. and P. Ergönül. 2015. Uçucu Yağları Elde Etme Yöntemleri. Gıda 40(5): 303–310.

Kızılarslan, Ç. and E. Sevgi. 2013. Ethnobotanical uses of genus *Pinus* L. (Pinaceae) in Turkey. Indian Journal of Traditional Knowledge 12(2): 209–220.

Kosalec, I., S. Pepeljnjak and D. Kůstrak. 2005. Antifungal activity of fluid extract and essential oil from anise fruits (*Pimpinella anisum* L., Apiaceae). Acta Pharm. 55: 377–385.

Kotan, R., A. Cakir, H. Ozer, S. Kordali, R. Cakmakci, F. Dadasoglu et al. 2014. Antibacterial effects of Origanum onites against phytopathogenic bacteria: Possible use of the extracts from protection of disease cause by some phytopathogenic bacteria. Scientia Horticulturae 172: 210–220.

Kumar, R., S. Sharma, S. Sood, K.V. Agnihotri and B. Singh. 2013. Effect of diurnal variability and storage conditions on essential oil content and quality of damask rose (*Rosa damascena* Mill.) flowers in north western Himalayas. Scientia Horticulturae 154: 102–108.

Kurti, F., A. Giorgi, G. Beretta, B. Mustafa, F. Gelmini, C. Testa et al. 2019. Chemical composition, antioxidant and antimicrobial activities of essential oils of different *Pinus* species from Kosovo. Journal of Essential Oil Research 31(4): 263–275.

Lang, G. and G. Buchbauer. 2012. A review on recent research results (2008–2010) on essential oils as antimicrobials and antifungals. A review. Flavour and Fragrance Journal 27(1): 13–39.

Lee, P., H. Guo, C. Huang and C. Chan. 2016. Chemical composition of leaf essential oils of *Syzygium samarangense* (BL.) Merr. et Perry cv. Pink at three maturity stages. Int. J. App. Res. Nat. Prod. 9: 9–13.

Liu, X., X. King and G. Li. 2019. A process to acquire essential oil by distillation concatenated liquid-liquid extraction and flavonoids by solid-liquid extraction simultaneously from *Helichrysum arenarium* (L.) Moench inflorescences under ionic liquid- microwave mediated. Sep. Purif. Technol. 209: 164–174.

Mamedov, N. and L.E. Craker. 2001. Medicinal plants used for the treatment of bronchial asthma in Russia and Central Asia. Journal of Herbs, Spices and Medicinal Plants 8(2-3): 91–117.

Marongiu, B., A. Piras, S. Porcedda, D. Falconieri, M.A. Frau, A. Maxia et al. 2012. Antifungal activity and chemical composition of essential oils from *Smyrnium olusatrum* L. (Apiaceae) from Italy and Portugal. Nat. Prod. Res. 26: 993–1003.

McKay, D.L. and J.B. Blumberg. 2006. A review of the bioactivity and potential health benefits of chamomile tea (*Matricaria recutita* L.). Phytother. Res. 20(7): 519–530.

Mustafa, B., A. Hajdari, F. Krasniqi, E. Hoxha, H. Ademi, L.C. Quave et al. 2012. Ethnobotany of the Albanian alps in Kosovo. J. Ethnobiology Ethnomedicine 8: 6.

Naghibi, F., M. Mosaddegh, M. Mohammadi Motamed and A. Ghorbani. 2010. Labiatae family in folk medicine in Iran: From ethnobotany to pharmacology. Iran. J. Pharm. Res. 4: 63–79.

Nazzaro, F., F. Fratianni, L. De Martino, R. Coppola and V. De Feo. 2013. Effect of essential oils on pathogenic bacteria. Pharmaceuticals 6(12): 1451–1474.

Norani, M., M.T. Ebadi and M. Ayyari. 2019. Volatile constituents and antioxidant capacity of seven *Tussilago farfara* L. populations in Iran. Scientia Horticulturae 257: 108635.

Ojo, O.O., F.R. Kabutu, M. Bello and U. Babayo. 2006. Inhibition of paracetamol-induced oxidative stress in rats by extracts of lemongrass (*Cymbropogon citratus*) and green tea (*Camellia sinensis*) in rats. African Journal of Biotechnology 5(12): 1227–1232.

Orhan, N., E. Akkol and F. Ergun. 2012. Evaluation of antiinflammatory and antinociceptive effects of some *Juniperus* species growing in Turkey. Turk. J. Biol. 36: 719–726.

Othman, S.N.A.M., M.A. Hassan, L. Nahar, N. Basar, S. Jamil and S.D. Sarker. 2017. Essential oils from the Malaysian *Citrus* (Rutaceae) medicinal plants. Medicines 3: 13.

Ozcan, M. and J.C. Chalchat. 2004. Aroma profile of *Thymus vulgaris* L. growing wild in Turkey. Bulg. J. Plant Physiol. 30(3-4): 68–73.

Öztürk, M., İ. Tümen, A. Uğur, F. Aydoğmuş-Öztürk and G. Topçu. 2011. Evaluation of fruit extracts of six Turkish *Juniperus* species for their antioxidant, anticholinesterase and antimicrobial activities. J. Sci. Food Agric. 91: 867–876.

Palazzolo, E., V.A. Laudicina and M.A. Germanà. 2013. Current and potential use of *Citrus* essential oils. Curr. Org. Chem. 17: 3042–3049.

Qu, H., W. Yang and J. Li. 2018. Structural characterization of a polysaccharide from the flower buds of *Tussilago farfara*, and its effect on proliferation and apoptosis of A549 human non-small lung cancer cell line. Int. J. Biol. Macromol. 113: 849–858.

Quassinti, L., M. Bramucci, G. Lupidi, L. Barboni, M. Ricciutelli, G. Sagratini et al. 2013. *In vitro* biological activity of essential oils and isolated furanosesquiterpenes from the neglected vegetable *Smyrnium olusatrum* L. (Apiaceae). Food Chem. 138: 808–813.

Radusiene, J. and A. Judzentiene. 2008. Volatile composition of *Helichrysum arenarium* field accessions with differently coloured inflorescens. Biologia 54(2): 116–120.

Raja, R.R. 2012. Medicinally potential plants of Labiatae (Lamiaceae) family: An overview. Research Journal of Medicinal Plants 6(3): 203–213.

Raut, J.S. and S.M. Karuppayil. 2014. A status review on the medicinal properties of essential oils. Ind. Crops Prod. 62: 250–264.

Rusanov, K., N. Kovacheva, B.J. Vosman et al. 2005. Microsatellite analysis of Rosa damascena Mill. accessions reveals genetic similarity between genotypes used for rose oil production and old Damask rose varieties. Theoretical and Applied Genetics 111(4): 804–809.

Salleh, W.M.N.H.W., F. Ahmad, K.H. Yen and R.M. Zulkifli. 2015. A review on chemical constituents and biological activities of the genus *Beilschmiedia* (Lauraceae). Trop. J. Pharm. Res. 14: 2139–2150.

Salleh, W.M.N.H.W., F. Ahmad, K.H. Yen and R.M. Zulkifli. 2015. Chemical compositions and biological activities of essential oils of *Beilschmiedia glabra*. Nat. Prod. Commun. 10(7): 1297–300.

Samojlik, I., N. Lakić, N. Mimica-Dukić, K. Đaković-Svajcer and B. Bozin. 2010. Antioxidant and hepatoprotective potential of essential oils of coriander (*Coriandrum sativum* L.) and caraway (*Carum carvi* L.) (Apiaceae). J. Agric. Food Chem. 58: 8848–8853.

Santin, M.R., A.O. Dos Santos, C.V. Nakamura, B.P.D. Filho, I.C.P. Ferreira and T. Ueda-Nakamura. 2009. *In vitro* activity of the essential oil of *Cymbopogon citratus* and its major component (citral) on *Leishmania amazonensis*. Parasitol. Res. J. 105: 1489–1496.

Sarrou, E., P. Chatzopoulou, K. Dimassi-Theriou and I. Therios. 2013. Volatile constituents and antioxidant activity of peel, flowers and leaf oils of *Citrus aurantium* L. growing in Greece. Molecules 18: 10639–10647.

Sayed-Ahmad, B., T. Talou, Z. Saad, A. Hijazi and O. Merah. 2017. The Apiaceae: Ethnomedicinal family as source for industrial uses. Ind. Crop Prod. 109: 661–671.

Sell, C.S. 2006. The Chemistry of Fragrance. From Perfumer to Consumer. 2nd ed. The Royal Society of Chemistry; Cambridge, UK, pp. 329.

Shahi, A., M. Kaul, R. Gupta, P. Dutt, S. Chandra and G. Qazi. 2005. Determination of essential oil quality index by using energy summation indices in an elite strain of *Cymbopogon citratus* (DC) Stapf. Flavour Frag. J. 20: 118–121.

Shuaib, M., M. Ali, J. Ahamad, K.J. Naquvi and M.I. Ahmad. 2013. Pharmacognosy of *Pinus roxburghii*: A review. Journal of Pharmacognosy and Phytochemistry 2(1): 262–268.

Singh, G., T. Singh and K.L. Handa. 1959. Composition of *Artemisia absinthium* oil from plants growing in J and K. Indian Soap. J. 24: 305–307.

Soković, M., J. Glamočlija, P.D. Marin, D. Brkic and V. Griensven. 2010. Antibacterial effects of the essential oils of commonly consumed medicinal herbs using an *in vitro* model. Molecules 15: 7532e46.

Spinozzi, E., F. Maggi, G. Bonacucina, R. Pavela, M.C. Boukouvala, N.G. Kavallieratos et al. 2021. Apiaceae essential oils and their constituents as insecticides against mosquitoes—A review. Industrial Crops and Products 171: 113892.

Stankov, S., H. Fidan, N. Petkova, A. Stoyanova, I. Dincheva, H. Dogan et al. 2020. Phytochemical composition of Helichrysum arenarium (L.) Moench essential oil (aerial parts) from Turkey. Ukrainian Food Journal 9(3): 503–312.

Swamy, M.K., M.S. Akhtar and U.R. Sinniah. 2016. Antimicrobial properties of plant essential oils against human pathogens and their mode of action: An updated review. Evid. Based Complement. Alternat. Med. 2016: 3012462.

Tajidin, E.N., S.H. Ahmad, A.B. Rosenani, H. Azimah and M. Munirah. 2012. Chemical composition and citral content in lemongrass (*Cymbopogon citratus*) essential oil at three maturity stages. African Journal of Biotechnology 11: 2685–2693.

Tatlı, I.I. 2012. Doğal Aromaterapötik Yağlar ile Cilt Terapisi. Turkiye Klinikleri J. Cosm. Dermatol-Special Topics 5(4): 46–53.

Tayfun, K. 2019. Aromaterapi. Journal of Biotechnol. An Strategic Health Res. 3: 67–73.

Tranchida, P.Q., I. Bonaccorsi, P. Dugo, L. Mondello and G. Dugo. 2012. Analysis of *Citrus* essential oils: State of the art and future perspectives. A review. Flavour Frag. J. 27: 98–123.

Trease, G. and W. Evans. 2002. Phytochemicals. *In*: Pharmacognosy, 15th Edition, Saunders Publishers, London, 42–393.

Vladimir-Knezevic, S., I. Kosalec, M. Babac, M. Petrovic, J. Ralic, B. Matica et al. 2012. Antimicrobial activity of *Thymus longicaulis* C. Presl essential oil against respiratory pathogens. Cent. Eur. J. Biol. 7(6): 1109–1115.

Wagner, H., M. Wierer and R. Bauer. 1986. *In vitro* inhibition of prostaglandin biosynthesis by essential oils and phenolic compounds. Planta Med. Jun. (3): 184–187.

Wan Mohd Nuzul, H.S., A. Farediah, H.Y. Khong and M.Z. Razauden. 2016. Essential oil compositions of Malaysian Lauraceae: A mini review. Pharmaceutical Sciences 22(1): 60–67.

Weiss, E.A. 1997. Essential Oil Crops. CAB International, Wallingford, UK, pp. 86–103.

Wood, G. and F. Bache. 1839. The Dispensatory of the United States of America, 4th ed. Philadelphia: Griggand Elliot.

Yoon, W.J., S.S. Kim, T.H. Oh, N.H. Lee and C.G. Hyun. 2009. *Abies koreana* essential oil inhibits drug-resistant skin pathogen growth and LPS-induced inflammatory effects of murine macrophage. Lipids 44: 471–476.

7

Antimicrobial Activity of Essential Oils from Species Collected in Venezuelan Andean

*Janne Rojas** and *Alexis Buitrago*

Introduction

Essential oils are defined as volatile and fragrant metabolic secretions of plants and have been classified as secondary metabolites produced mainly by two out of three main pathways; mono-, sesqui-, and diterpenes are made through the mevalonate pathway, while phenylpropenes are produced via the shikimic acid pathway (Dilworth et al. 2017).

The ability of plants to accumulate essential oils is quite high in both Gymnosperms and Angiosperms, although the most commercially important essential oil plant sources are related to the Angiosperms. However, a limited number of families are considered big producers of essential oils and these include Myrtaceae, Lauraceae, Rutaceae, Lamiaceae, Asteraceae, Apiaceae, Cupressaceae, Poaceae, Zingiberaceae, and Piperaceae (Rojas and Buitrago 2015, 2013, Franz and Novak 2015).

Many years of investigation have revealed that essential oils play an important role in plants. The attraction of different insects in order to promote the dispersion of pollen and seeds, allopathic communication between plants, feeding deterrent, and repelling effects against herbivores are among the mechanisms plant use for survival. Thus, the detection of some biological properties needed for the survival of plants has also been the base for searching for similar activities to fight several microorganisms responsible for a number of infectious diseases in humans and animals (Franz and Novak 2015). In this regard, a lot of investigations have revealed that essential oils may have the ability to exhibit a wide range of biological properties, wherein antimicrobial activity is most frequently reported (Mangalagiri et al. 2021, Rojas-Vera 2020a, Aparicio et al. 2018).

Faculty of Pharmacy and Bioanalysis, University of Los Andes, Mérida, 5101, Venezuela.
* Correseponding author: janne.rojas24@gmail.com

It is well known that bacteria have the genetic ability to transmit and acquire resistance to drugs, which are utilized as therapeutic agents (Mangalagiri et al. 2021). Such a fact is cause for concern because of the number of patients in hospitals who have suppressed immunity and due to new bacterial strains, which are multi-resistant. Consequently, new infections can occur in hospitals resulting in high mortality. Even though pharmacological industries have produced a number of new antibiotics in the last three decades, resistance to these drugs by microorganisms has increased (Butnariu and Sarac 2018).

The microbial resistance is a matter of concern since is growing every day and the outlook for the use of antimicrobial drugs in the future is still uncertain. Therefore, actions must be taken to reduce this problem; controlling the use of antibiotic, ongoing research to better understand the genetic mechanisms of resistance, and to develop new drugs, either synthetic or natural, are some examples. For a long period of time, plants have been a valuable source of natural products for maintaining human health, especially in the last decade, with more intensive studies for natural therapies. The ultimate goal is to offer appropriate and efficient antimicrobial drugs to the patient (Butnariu and Sarac 2018, Dhifi et al. 2016).

According to World Health Organization medicinal plants would be the best source to obtain a variety of drugs. About 80% of individuals from developed countries use traditional medicine, which compounds are derived from medicinal plants. Therefore, such plants should be investigated to better understand their properties, safety, and efficiency (Franz and Novak 2015). The use of plant extracts and phytochemicals, both with known antimicrobial properties, might be of great significance in therapeutic treatments. In the last few years, a number of studies have been conducted in different countries to prove such efficiency (Rojas-Vera 2020a, Ouis and Hariri 2018, Pérez-Colmenares et al. 2017, Semnani et al. 2017, Zarkani 2016). Furthermore, many plants have been traditionally used because of their antimicrobial traits, which are due to compounds synthesized in the secondary metabolism of the plant. These products are known by their active substances, for example, phenolic compounds which are part of the essential oils as well as tannins (Butnariu and Sarac 2018).

The present chapter aims to summarize the potential antimicrobial agents of essential oils obtained by hydrodistillation from different plant species collected from the Venezuelan Andean. The chemical composition and antimicrobial activity of these essential oils are detailed.

Essential Oils

Essential oils are complex mixtures of volatile compounds present mostly in aromatic plants, mainly composed of terpenes biogenerated through the mevalonate pathway (Figure 1), such as monoterpenes and sesquiterpenes (both hydrocarbon and oxygenated) (Dhifi et al. 2016). Thousands of compounds belonging to the terpenes group have so far been identified in essential oils (Mokhtar et al. 2020); derivatives of alcohols (geraniol and linalool), ketones (menthone and *p*-vetivone), aldehydes (citronellal and sinensal), esters (*γ*-tepinyl acetate and cedryl acetate), and phenols (thymol and carvacrol) are some examples. However, essential oils also contain non-

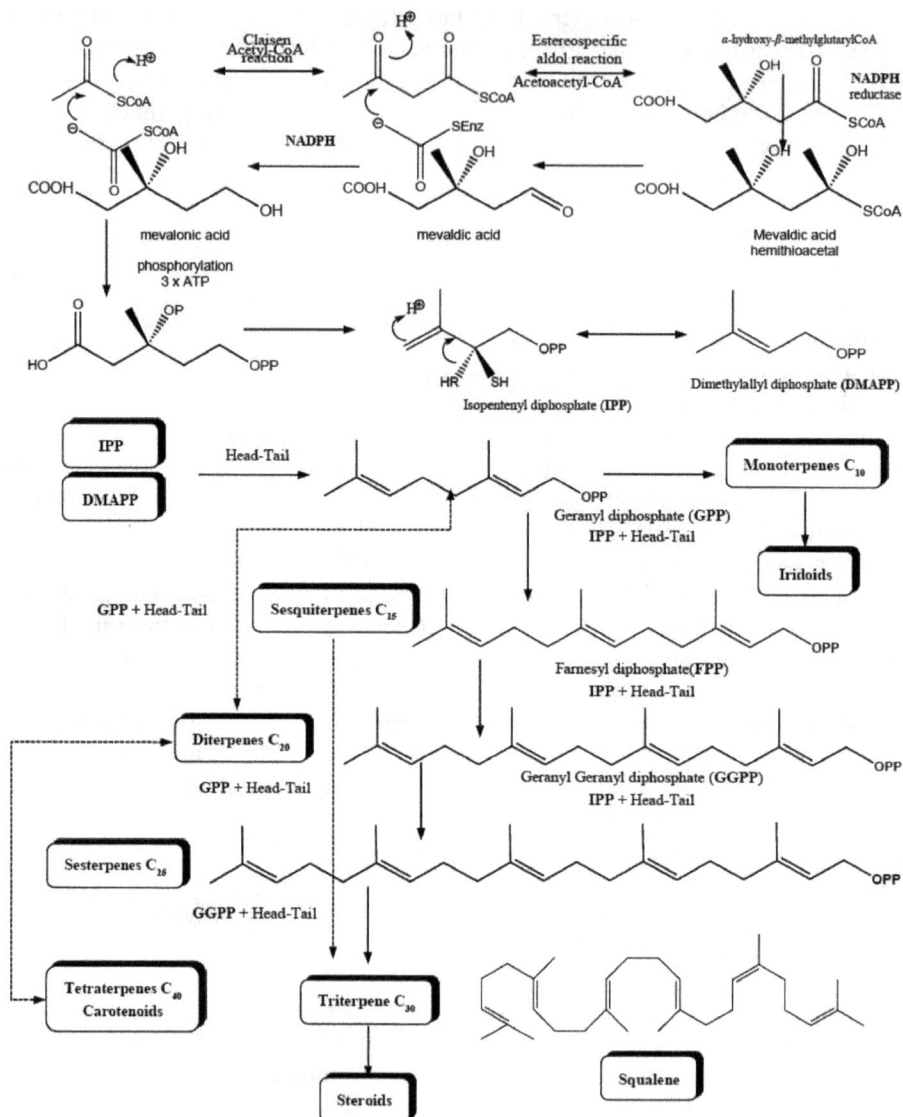

Figure 1: Biosynthesis of terpenes through mevalonate pathway (Marcano and Hasegawa 2002, Dewick 2002).

terpenic compounds biosynthesized through the shikimate pathway, such as eugenol, cinnamaldehyde, and safrole among others.

In addition, essential oils have a high variability of their composition, both in qualitative and quantitative terms. A number of factors may determine the essential oils' yield and composition, seasonal variations, plant organ, maturity stage of the plant, geographic origin and genetics are among these, although extrinsic factors that are related to the extraction method may also be affected (Rojas and Buitrago 2015, 2013).

Furthermore, volatile oils are responsible for different scents that plants emit, thus they are widely used in the cosmetics industry, perfumery, and aromatherapy. Therefore, essential oils are considered of great interest in the food and cosmetic industries, as well as in the human health field. However, despite their rich and complex composition, the use of essential oils remains limited to the cosmetics and perfumery domains. Thus, it is worthy to develop a better understanding of their chemistry and biological properties in order to learn new and valuable applications in human health, agriculture, and the environment as well since essential oils can be exploited as an effective alternative or can complement synthetic compounds for the chemical industry (Dhifi et al. 2016).

Essential Oils as Antimicrobial Agents

A wide range of investigations have been achieved concerning essential oils and a variety of biological properties, such as antibacterial, antifungal, antiviral, insecticidal, and antioxidant have also been described. Despite these biological potential, essential oils have been primarily used in food preservation, aromatherapy, and fragrance industries as described before. However, the increasing tolerance of several microorganisms against commonly used antibiotic drugs has encouraged researchers to find alternative ways for the treatment of such infections and aromatic plants, particularly essential oils seem promising (Mangalagiri et al. 2021, Dhifi et al. 2016).

A significant problem that researchers have to deal with in their investigations is the discordance between data obtained from the same species studied. To understand this variability, a number of factors have to be considered, which are climatic, seasonal and geographic conditions, harvest period, and distillation technique are among these. Besides, the plant maturity at the time of oil production and the existence of chemotypic differences can also drastically affect the oil composition. Fortunately, capillary-GC experiments can provide the exact composition of essential oils. Nevertheless, it is important to consider that essential oils are composed of a heterogeneous mixture of substances, thus biological activities are possible due to the synergism of these components although antagonism might also occur (Górniak et al. 2019, Szweda and Kot 2018, Dhifi et al. 2016).

Knowledge of methods for testing essential oils is, therefore, necessary to discover the spectrum of action of these natural products, their modes of action, and their therapeutic applications. It is also important to emphasize that over the last 50 years, the demand for essential oil products has gradually increased perhaps due to the public consciousness for health plus the fact that natural products are assumed to have superior quality; thus, there has been a progressive urge for incorporating natural materials into a wide range of foods, medicines, and lifestyle products, such as fragrances, culinary herbs, and aromatic teas among others. Pharmaceutical companies have also been motivated to develop new antimicrobial drugs in recent years, especially due to the constant emergence of microorganisms resistant to conventional antimicrobials, therefore they are sponsoring research to find novel compounds in order to synthesize new lead medicines (Mangalagiri et al. 2021, Górniak et al. 2019, Szweda and Kot 2018, Swamy et al. 2016, Dhifi et al. 2016).

Moreover, it has been stated that several components of the essential oils are important elements in defining fragrance, density, texture, colour, cell penetration, lipophilic or hydrophilic attraction, fixation on cell walls and membranes, and cellular distribution. This last characteristic is very important because the distribution of the oil in the cell determines the different types of radical reactions produced. In that sense, for biological purposes, researchers believe that it is more informative to study the entire oil rather than some of its components (Dhifi et al. 2016).

The antimicrobial properties of essential oils and their constituents have been considered and the mechanism of action has been studied in detail. An important feature of essential oils is their hydrophobicity, which allows them to partition into lipids of the bacteria cell membrane, disrupting the structure, and making it more permeable. This can then cause leakage of ions and other cellular molecules. Although a certain amount of leakage of bacterial cells can be tolerated without loss of viability, greater loss of cell contents or critical output of molecules and ions can lead to cell death (Mangalagiri et al. 2021, Górniak et al. 2019, Szweda and Kot 2018).

It has also been documented that the chemical structure of essential oils affects their mode of action concerning their antibacterial activity. The importance of the presence of hydroxyl group in the phenolic compounds, such as carvacrol and thymol, has been confirmed. However, the relative position of the phenolic hydroxyl group on the ring does not appear to influence the intensity of the antibacterial activity (Mangalagiri et al. 2021, Dhifi et al. 2016).

It is important to mention that essential oils are more active against Gram-positive than Gram-negative bacteria. The latter are less susceptible to the action of essential oils with the outer membrane surrounding the cell wall that restricts the diffusion of hydrophobic compounds through its lipopolysaccharide film. Furthermore, the antibacterial activity of essential oils is related to their chemical composition, the proportions of volatile molecules, and their interactions (Mangalagiri et al. 2021, Dhifi et al. 2016).

It is important to remember that an additive effect is observed when the combination is equal to the sum of the individual effects. Antagonism is observed when the effect of one or both compounds is less important when they are tested together than when used individually while a synergistic effect is observed when the combination of substances is greater than the sum of the individual effects (Górniak et al. 2019). Some studies have shown that the use of the whole essential oil provides an effect which is greater than that of the major components used together. This suggests that minor components are essential for activity and may have a synergistic effect (Górniak et al. 2019, Dhifi et al. 2016).

In this regard, a number of investigations with essential oils of aromatic species collected from different locations in Venezuelan Andean have been published and a variety of assays have also been tested in order to determine their biological activity of these. Chemical composition has been analyzed by GC and GC-MS techniques to establish the correct profile of each species evaluated (Obregón-Díaz et al. 2019, Araque et al. 2018, Pérez-Colmenares et al. 2017, Marcano et al. 2016, Buitrago et al. 2017, 2010, Mora et al. 2016, 2009a, Rojas et al. 2011a, 2011b, 2010, Hernández et al. 2015, 2013, Meccia et al. 2015). Studies carried out in different species collected from diverse locations in Mérida and Táchira States are described in Table 1; major components, as well as results of microbiological activity assays, are also presented. Examples of terpenes frequently characterized in these essential oils are detailed in Figure 2.

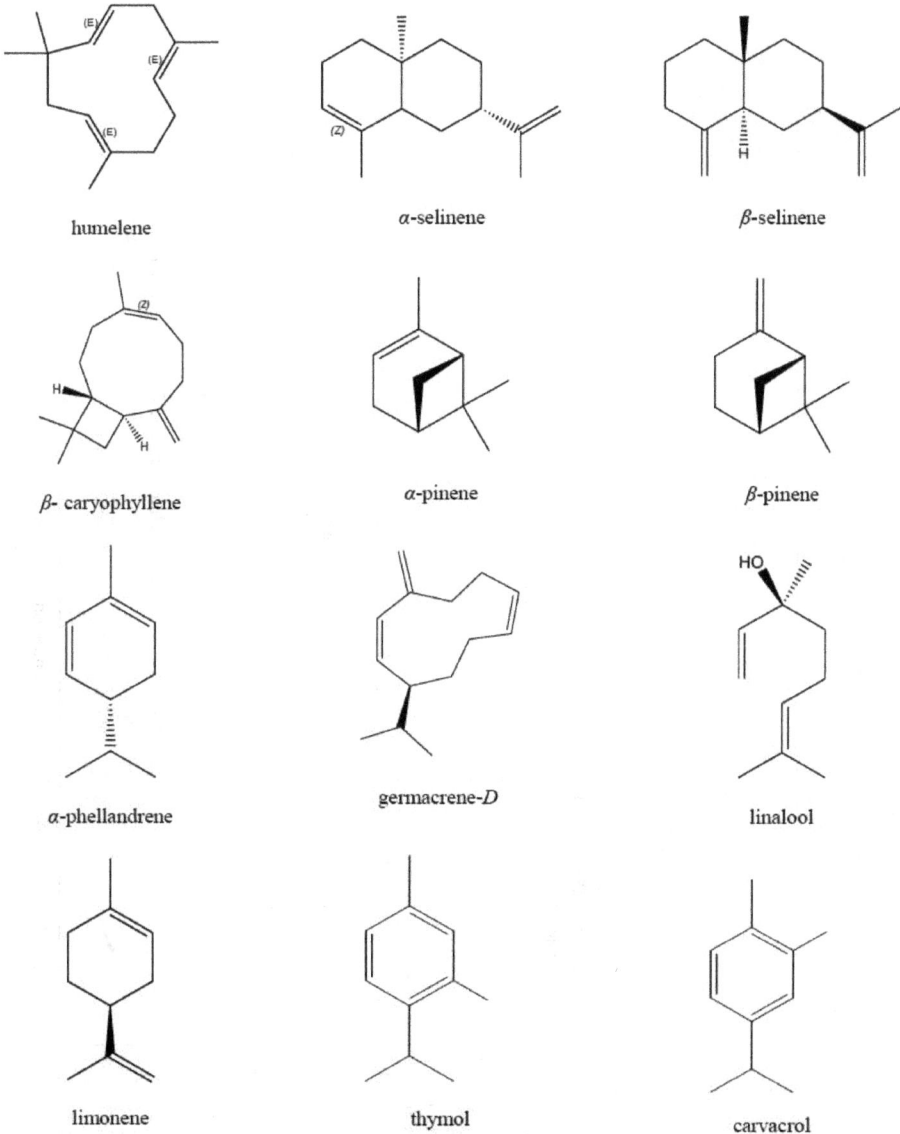

humelene

α-selinene

β-selinene

β- caryophyllene

α-pinene

β-pinene

α-phellandrene

germacrene-*D*

linalool

limonene

thymol

carvacrol

Figure 2: Monoterpenes and sesquiterpenes commonly isolated from essential oils collected in Venezuelan Andean (Marcano and Hasegawa 2002, Dewick 2002).

Table 1: Studies on the antimicrobial activity of essential oils obtained from different species collected from the Venezuelan Andean.

Species (Family)	Location	Chemical Composition	Antimicrobial Activity (MIC)	References
Ageratina jahnii (B.L. Rob.) R.M. King and H. Rob. *Ageratina pichinchensis* (Kunth) R.M. King and H. Rob (Asteraceae)	*A. jahnii:* Tostós at 2,547 m. a. s. l. *A. pichinchensis:* Las Lajas at 3621 m. a. s. l. Mérida state, Venezuela	*A. jahnii:* β-myrcene (31.6%), α-pinene (23.1%), limonene (7.8%), pentacosane (10.2%) *A. pichinchensis:* 8,9-epoxythymyl isobutyrate (21.2%), germacrene-*D* (20.8%), thymyl isobutyrate (15.8%), eupatoriochromene (5.5%), encecalol (4.9%)	*A. jahnii:* S. aureus (9.5 mg/mL) *A. pichinchensis:* E. faecalis (104 mg/mL)	Rojas-Vera 2020a
Austroeupatorium inulifolium (Kunth) R.M. King and H. Rob. (Asteraceae)	San Pedro at 3,100 m. a. s. l., Mérida state, Venezuela	α-pinene (6.94%), β-pinene (9.25%), β-caryophyllene (13.65%), germacrene-*D* (21.12%)	S. aureus 5000 (µg/mL) E. faecalis (10000 µg/mL) E. coli, K. pneumoniae, P. aeruginosa (78 µg/mL)	Lucena et al. 2019
Bursera. simaruba (L.) Sarg. *Bursera glabra* (Jacq.) and *Bursera inversa* Daly (Burseraceae)	*B. glabra:* El Morro, 1,900 m. a. s. l. *B. inversa:* El Vigía 36 m. a. s. l. *B. simaruba:* Ejido and Las Gonzales at 1,100 m. a. s. l. Mérida state, Venezuela	*B. simaruba:* α-pinene (52%), α-phellandrene (24%), germacrene D (11%) *B. glabra:* limonene (77.6%), cis-ocimene (7.93%) *B. inversa:* α-humulene (27.7%), β-caryophyllene (22.1%), germacrene B (16.3%)		Cáceres-Ferreira et al. 2019
Mangifera indica (L.) (Anacardiaceae)	Andes and Llanos	Mérida: β-selinene (22,56%), α-gurjunene (14,66%), β-caryophyllene (10 40%) Barinas: β-caryophyllene (36,32%), α-humulene (22,71%), α-gurjunene (21,43%) Portuguesa: β-caryophyllene (36,07%), α-gurjunene (22.55%), α-humulene (21.24%)	S. aureus (200 µL/mL) E. faecalis (300 µL/mL)	Aparicio-Zambrano et al. 2019a

m . a . s. l.: Meters above sea level. MIC: Minimal Inhibitory Concentration

Libanothamnus neriifolius (Bonpl. ex Humb) Ernst. (Asteraceae)	San José de Acequias at 2500 m. a. s. l., Mérida state, Venezuela	β-phellandrene (29.04%), α-phellandrene 19.86%), α-pinene (13.57%), α-tujene (12.35%)	*S. aureus* (50 μL/mL) *C. albicans* (700 μL/mL) *C. Krusei* (500 μL/mL)	Aparicio-Zambrano et al. 2019b
Railopezia marcescens (S.F. Blake) Cuatrec. (Asteraceae)	El Batallón at 3000 m. a. s. l., Táchira state, Venezuela	Germacrene-D (24.86%), α-pinene (22.38%), p-cimene (7.35%), α-phellandrene (6.34%)	*S. aureus* (0.25 mg/mL) *E. coli* (0.5 mg/mL) *K. pneumoniae*, *P. aeruginosa* (1 mg/mL)	Aparicio et al. 2018
Coespeletia moritziana (Sch. Bip. ex Wedd.) Cuatrec. *Espeletia schultzii* Wedd *Coespeletia timotensis* Cuatrec. (Asteraceae)	Pico el Águila at 4118 m. a. s .l., Mérida state, Venezuela	*C. moritziana:* β-f phellandrene (28.42%), α-pinene (25.16%), β-pinene (10.05%), kaurenal (4.63%), *E. schultzii:* α- phellandrene (15.34%), limonene (14.52%), mircene (12.1%), α-pinene (7.74%), *C. timotensis:* β- phellandrene (48.08%), α-pinene (21.68%), β-pinene (8.68%), γ-cadinene (3.98%)	*C. albicans* (410 μg/mL) *C. krusei* (510 μg/mL)	Cordero de Rojas et al. 2017
Zanthoxylum sp. nov (L.) (Rutaceae)	El Cobre at 2300 m. a. s. l., Táchira state, Venezuela	β-phellandrene (24.69%), *cis-p*-menth-2-en-1-ol (11.74%), linalool (10.34%), citronellal (9.87%), *trans-p*-2-menth-2-en-1-ol (8.68%)	*S. aureus, E. faecalis* (60 μg/mL) *E. coli* (50 μg/mL) *K. pneumoniae* (100 μg/mL)	González de Colmenares et al. 2017
Myrcianthes myrsinoides (Kunth) Grifo (Myrtaceae)	Los Frailes at 2173 m. a. s. l., Mérida state, Venezuela	*p*-terpinen-4-ol (32.22%), *o*-cymene (8.15%), spathulenol (7.59%), caryophyllene oxide (7.14%)	*B. cereus* (100 μg/mL), *B. subtilis* (200 μg/mL), *S. epidermidis* (400 μg/mL)	Araujo et al. 2017

m. a. s. l.: Meters above sea level. MIC: Minimal Inhibitory Concentration

Table 1 contd. ...

...*Table 1 contd.*

Species (Family)	Location	Chemical Composition	Antimicrobial Activity (MIC)	References
Railopezia lindenii (Schultz–Bip. Ex Wedd.) Cuatrec (Asteraceae)	San José (Pueblos del Sur): RL-A at 2870 m. a. s. l. and RL-B at 3048 m. a. s. l., Mérida state, Venezuela	RL-A: *trans*-caryophyllene (30.9%), caryophyllene oxide (12.3%), espathulenol (11.7 %), **α-pinene** (5.0%), ar-curcumene (4.3%) RL-B: *trans*-caryophyllene (18.7%), caryophyllene oxide (8.9%), espathulenol (13.5%), silfiperfol-6-ene (11.2%), α-pinene (4.5%), ar-curcumene (6,4%)		Pérez-Colmenares et al. 2017
Espeletia schultzii Wedd (Asteraceae)	Páramo de Ortiz at 2677 m. a. s. l. Trujillo state, Venezuela	*α*-pinene (50.11%), *β*-pinene (16.28%), *β*-myrcene (14.71)	*S. aureus* (280 µg/mL) *E. faecalis* (580 µg/mL)	Alarcón et al. 2016
Pimenta racemosa var. *racemosa* (Mill.) J.W. Moore (Myrtaceae)	"La Palmita" at 859 m. a. s. l., Táchira state, Venezuela	Light Oil (Lo): eugenol (60.4%), myrcene (11.7%), chavicol (6.0%), limonene (5.4%), linalool (4.4%) Heavy Oil (Ho): eugenol (82.9%), chavicol (9.3%), myrcene (1.5%), limonene (0.9%), linalool (0.7%)	*S. aureus*: (Lo: 60 µg/mL, Ho: 20 µg/mL), *E. faecalis*: (Lo: 100 µg/mL, Ho: 100 µg/mL), *E. coli* (Lo: 40 µg/mL, Ho: 20 µg/mL), *K. pneumoniae*: (Lo: 60 µg/mL, Ho: 20 µg/mL), *P. aeruginosa*: (Lo: 200 µg/mL, Ho: 400 µg/mL), MSRA: (Lo: 65 µg/mL, Ho: 45 µg/mL), *E. coli* ESBL and *E. cloacae* ESBL: (Lo: 50 µg/mL, Ho: 50 µg/mL), *C. albicans*: (Lo: 100 µg/mL, Ho: 100 µg/mL), *C. krusei*: (Lo: 200 µg/mL, Ho: 50 µg/mL)	Contreras-Moreno et al. 2016, 2014

m. a . s. l.: Meters Above Sea Level; MIC: Minimal Inhibitory Concentration; MRSA: Methicillin-Resistant *S. Aureus*; ESBL: Extended-Spectrum β-Lactamase

Species	Location	Composition	Activity	Reference
Vismia macrophylla Kunth (Hypericaceae)	Michelena, at 1200 m. a. s. l., Táchira state, Venezuela	Leaf oil (Lo): γ-bisabolene (44.4%), β-bisabolol (14.9%), Fruit oil (Fo): germacrene-*D* (12.1%), δ-cadinene (10.7%), γ-bisabolene (22.3%)	*S. aureus* (Lo: 100 µg/mL, Fo: 150 µg/mL), *E. faecalis* (Lo: 500 µg/mL, Fo: 250 µg/mL), *E. coli* (Fo: 740 µg/mL), *C. albicans, C. krusei* (Lo: 600 µg/mL)	Buitrago et al. 2015
Hyptis colombiana Epling (Lamiaceae)	San José Mucutuy at 3100 m. a. s. l., Mérida state, Venezuela	β-caryophyllene (29%), germacrene-*D* (31.5%)	*S. aureus* (20 mg/mL) *E. faecalis* (30 mg/mL).	Flores et al. 2015
Ruilopezia. bractreosa (Standl.) Cuatr. (Asteraceae)	Páramo de Ortiz at 3085 m. a. s. l., Trujillo state, Venezuela	β-myrcene (34.2%), α-pinene (24.3%), 7-epi-α-selinene (9.1%), β-pinene (8.5%), 6,9-guaiadiene (4.4%)	*S. aureus* (100 µg/mL) *E. faecalis* (600 µg/mL)	Alarcón et al. 2015
Vismia baccifera var. *dealbata* (Triana and Planch) (Hypericaceae)	Chiguará at 1,250 m. a. s. l., Mérida state, Venezuela	caryophyllene oxide (31.4%), β-caryophyllene (26.4%), α-zingiberene (12.6%)	*C. tropicalis, C. parapsilosis* and *C. neoformans* (1000 µg/mL), *C. krusei* (1.6 µg/mL), *C. glabrata* (200 µg/mL)	Vizcaya et al. 2014
Conyza bonariensis (L.) Cronquist (Asteraceae)	El Chama at 1,145 m. a. s. l., Mérida state, Venezuela	*trans*-β-farnesene (37.8%), *trans*-ocimene (20.7%), β-sesquiphellandrene (9.8%).	*B. cereus* (25 µg/mL), *S. epidermidis* (100 µg/mL) *C. albicans* (200 µg/mL)	Araujo et al. 2013
Hypericum laricifolium Juss (Hypericaceae)	Piñango road at 3900 m. a. s. l., Mérida state, Venezuela	α-pinene (20.2%), verticiol (13.4%), 3-methyl-nonane (12.3%), 2-methyl-octane (9.6%), nonane (7.6%)		Rojas et al. 2013a
Croton huberi Steyerm. (Euphorbiaceae)	Michelena at 1,200 m. a. s. l., Táchira state, Venezuela	β-caryophyllene (18.3%), germacrene-*D* (16.1%), alencene (8.3 %), caryophyllene oxide (7.3%), bicycle-germacrene (7.1%), *t*-muurolol (6.1%)		Rojas et al. 2013b

m. a. s. l.: Meters Above Sea Level; MIC: Minimal Inhibitory Concentration

Table 1 contd. ...

...*Table 1 contd.*

Species (Family)	Location	Chemical Composition	Antimicrobial Activity (MIC)	References
Piper hispidum Sw (Piperaceae)	Chiguará at 840 m. a. s. l., Mérida state, Venezuela	α-pinene (15.3%), β-pinene (14.8%), β-elemene (8.1%), caryophyllene oxide (7.8%), δ-3-carene (6.9%)	*S. aureus, S. epidermides, S. saprophyticus, B. cereus* and *B. subtilis* (6.25 µg/mL) and *E. faecalis* (15 µg/mL)	Morales et al. 2013
Monticalia imbricatifolia Schultz (Asteraceae)	Piñango Páramo at 2320 m. a. s. l, Mérida state, Venezuela	α-phellandrene (33.89%), β-phellandrene (19.28%), α-pinene (16.81%), β-pinene (10.97%)	*St. aureus* (20 µg/mL), *E. faecalis* (60 µg/mL), *E. coli* and *K. pneumonia* (50 µg/mL) *P. aeruginosa* (30 µg/mL)	Buitrago et al. 2012
Monticalia greenmaniana (Hieron) C. Jeffrey (Asteraceae)	Páramo San José at 3200 m. a. s. l., Mérida state, Venezuela	1-nonane: (flowers: 38.8%; stems 33.5%), α-pinene: (flowers: 29%; stems 14.8%), germacrene-*D*: (flowers: 15.6%; stems 18.6%; leaf: 50.7%)	*S. aureus* (75 µg/mL), *E. faecalis* (150 µg/mL), *E. coli* (1500 µg/mL), *K. pneumoniae* (3000 µg/mL) *P. aeruginosa* (6000 µg/mL)	Cárdenas et al. 2012

Lantana camara var. moritziana (Otto and Dietr.) López-Palacios (Verbenaceae)	Rubio at 101 m. a. s. l., Táchira state, Venezuela	germacrene-*D* (31.0%), β-caryophyllene (14.8%), α-phellandrene (6.7%), limonene (5.7%), 1,8-cineole (5.2%)	*E. faecalis* (350 mg/mL), *S. aureus* (400 mg/mL)	Rios et al. 2011
Ruta graveolens (L.) (Rutaceae)	Mérida at 1,500 m. a. s. l., Mérida state, Venezuela	2-undecanone (50.93%), 2-nonanone (16.85%), pregeijerene (8.72%)	*S. aureus, E. coli K. pneumonie* (200 µg/mL)	Rojas et al. 2011a
Vismia baccifera Triana and Planch. (Guttiferae)	La Hechicera at 1,680 m. a. s. l., Mérida state, Venezuela	*trans*-cadin-1,4-diene (36.6%), *cis*-cadin-1,4-diene (18.8%), β-caryophyllene (11.9%)	*S. aureus* (37 µg/mL), *E. faecalis* and *E. coli* (18 µg/mL) *K. pneumonia* and *P. aeruginosa* (9 µg/mL)	Rojas et al. 2011b
Allium schoenoprasum (L.) (Alliaceae)	Mucuchíes at 2893 m. a. s. l., Mérida state, Venezuela	*bis*-(2-sulfhydryethyl)-disulfide: (leaf: 72.06%; roots: 56.47%), 2,4,5-trithiahexane: (leaf 5.45%; roots: 15.90%), tris (methylthio)-methane: (leaf: 4.01%; roots: 12.81%)		Buitrago-Díaz et al. 2011

m. a . s. l.: Meters Above Sea Level; MIC: Minimal Inhibitory Concentration

Table 1 contd. ...

...*Table 1 contd.*

Species (Family)	Location	Chemical Composition	Antimicrobial Activity (MIC)	References
Myrcia fallax (Rich.) DC. (Myrtaceae)	Boca de Monte at 2292 m. a. s. l., Táchira state, Venezuela	Leaf: guaiol (31,0%) and carotol (9.9%) Flowers: guaiol (27.5%) and aristolone (24,5%)	*S. aureus* (50 mg/mL) and *E. faecalis* (400 μg/mL)	Alarcón et al. 2009
Minthostachys mollis (Kunth) Griseb Vaught var. *mollis* (Lamiaceae)	El Arenal-Pajarito at 3000 m. a. s. l., Trujillo state, Venezuela	pulegone (55.2%), *trans*-menthone (31.5%)	*S. aureus* and *E. coli* (16 μg/mL), *E. faecalis* and *P. aeruginosa* (64 μg/mL), *B. subtilis* (4 μg/mL), *K. pneumoniae* (8 μg/mL) *S. Typhi* (4 μg/mL)	Mora et al. 2009b
Euphorbia. caracasana Boiss *Euphorbia. cotinifolia* (L.) (Euphorbiaceae)	Miyoi Pueblo Llano at 2161 m. a. s. l., Mérida state, Venezuela	*E. caracasana* β-caryophyllene (33.7%), α-humulene (18.8%), aromadendrene (8.4%) *E. cotinifolia* β-caryophyllene (39.3%), germacrene-D (21.5%), α-copaene (9.3%)		Rojas et al. 2009
Monticalia andicola Turcz. (Asteraceae)	Collado del Condor" peak at 3.900 m. a. s. l., Mérida state, Venezuela	α-pinene (19.6%), β-pinene (10.5%), α-longipinene (6.5%), δ-3-carene (6.2%), cyperene (5.4%), β-phellandrene (5.2%)	*S. aureus* (37 μg/mL), *E. faecalis*: (150 μg/mL), *E. coli* (150 μg/mL), *K. pneumoniae* (10 μg/mL) *P. aeruginosa* (150 μg/mL)	Baldovino et al. 2009
Baccharis trinervis (Lam.) Pers. (Asteraceae)	La Hechicera at 1680 m. a. s. l., Mérida state, Venezuela	germacrene-D (20.4%), limonene (15.4%), δ-cadinene (5.2%), β-caryophyllene (4.8%), α-pinene (4.5%), bicyclogermacrene (4.0%)	*S. aureus* (80 μg/mL) *E. faecalis* (200 μg/mL)	Rojas et al. 2008

m. a . s. l.: Meters Above Sea Level; MIC: Minimal Inhibitory Concentration

Porophyllum ruderale (Jacq.) Cass. (Asteraceae)	San Juan de Lagunillas at 1099 m .a. s. l., Mérida state, Venezuela	mixture of limonene and β-phellandrene (50.3%), sabinene (20.2%), 1-undecene (4.7%), 4-terpineol (3.8%), α-pinene (2.9%)	*S. aureus* (20 µg/mL), *E. faecalis* (120 µg/mL), *E. coli* (100 µg/mL), *K. pneumonia* (130 µg/mL) *P. aeruginosa* (200 µg/mL)	Rondón et al. 2008
Lippia oreganoides (L.) (Verbenaceae)	La Huerta, San Juan de Lagunillas at 1059 m .a. s. l., Mérida state, Venezuela	thymol (61.9%), carvacrol (7.9%)	*S. aureus* MRSA (20 µg/mL), Extended spectrum β-lactamase producing *K.pneumoniae* (20 µg/mL) multiresistant *A. baumannii* (20 µg/mL)	Velasco et al. 2007
Baccharis latifolia Pers. and *B. prunifolia* H.B. and K. (Asteraceae)	La Culata at 2900 m .a. s. l., Mérida state, Venezuela	limonene (7.6%), germacrene-*D* (12.2%), 1,10-di-epi-cubenol (7.9%)	*B. latifolia* *S. aureus* (80 µg/mL), *E- faecalis* (90 µg/mL), *B. prunifolia* and *E- faecalis* (260 µg/mL)	Rojas et al. 2007
Tagetes patula (L.) (Asteraceae)	Albarregas at 1680 m .a. s. l., Mérida state, Venezuela	piperitone (33.77%), *trans-β*-ocymene (14.83%), terpinolene (13.87%), β-caryophyllene (9.56%)	*S. aureus* and *E. faecalis* (30 µg/mL), *E. coli* (60 µg/mL), *K. pneumoniae* (90 µg/mL), *P. aeruginosa* (130 µg/mL)	Rondón et al. 2006a

Table 1 contd. ...

...Table 1 contd.

Species (Family)	Location	Chemical Composition	Antimicrobial Activity (MIC)	References
Lasiocephalus longipenicillatus (Sch. Bip. Ex Sandwith) Cuatrec. (Asteraceae)	Piñango at 2320 m .a. s. l., Mérida state, Venezuela	α-pinene (48.3%), α-humulene (15.8%), germacrene-*D* (15.5%)	*S aureus* and *E faecalis* (2.5 µg/mL)	Rondón et al. 2006b
Salvia leucantha Cav (Lamiaceae.)	San Rafael de Mucuchíes at 3190 m .a. s. l., Mérida state, Venezuela	bornyl acetate (40.92%), azulene (12.94%), germacrene-*D* (8.26%), borneol (8.11%), bicyclogermacrene (5.34%)	*S. aureus* (100 µg/mL) *E. faecalis* (200 µg/mL)	Rondón et al. 2005
Verbessina turbacensis Kunth HBK (Asteraceae)	Pueblo Llano at 2161 m. a. s. l., Mérida state, Venezuela	germacrene-*D* (56.20%), biclogermacrene (8.16%) β-caryophyllene (8.14%), α-pinene (6.17%), α-eudesmol (6.13%)	*S aureus* and *E coli* (2 µg/mL), *K. pneumoniae* (5 µg/mL) *E. faecalis* 25 µg/mL)	Gualtieri et al. 2005

m. a . s. l.: Meters above sea level; MIC: Minimal Inhibitory Concentration; MRSA: Methicillin-Resistant *S. aureus*; ESBL: Extended-Spectrum β-Lactamase

It has been reported that essential oils containing mainly aldehydes or phenols, such as cinnamaldehyde, citral, carvacrol, eugenol, or thymol, are characterized by the highest antibacterial activity, followed by those containing terpene alcohols. However, the oils containing ketones or esters, such as β-myrcene, α-thujone, or geranyl acetate, had much weaker activity, while volatile oils containing terpene hydrocarbons are usually inactive (Mangalagiri et al. 2021, Górniak et al. 2019, Szweda and Kot 2018, Swamy et al. 2016, Dhifi et al. 2016).

It is also interesting to mention that several differences have been observed in the chemical profile of the essential oils collected from diverse locations; for example, the essential oil from leaves of *Vismia baccifera* var. *dealbata* collected from Chiguará (1250 m. a. s. l.) and La Hechicera (1800 m. a. s. l.), Mérida state, from Venezuela showed germacrene-*D* (15.8%), α-cadinol (14.5%), epi-α-cadinol (11.9%), β-caryophyllene (10.1%) and δ-cadinene (7.5%) as major constituents in the Chiguará sample, while β-caryophyllene (45.7%), valencene (12.3%), β-elemene (10.7%), α-humulene (8.9%) and germacrene-*D* (6.3%) were observed as major components in the La Hechicera sample. The difference observed might be due to the environmental conditions that could play an important role in the compounds produced by the plant species, although temperature and ground conditions might influence not only the yields but also the chemical composition of the oils (Buitrago et al. 2009).

Climate conditions might also affect the type and amount of components present in the plant analyzed. The essential oil of aerial parts of *Lasiocephalus longepenicillatus*, collected in two different climatic seasons, dry and rainy, showed germacrene-*D* (37.79%); α-pinene (26.36%) and α-humulene (12.29%) for the sample collected in the dry season, while the major components observed in the rainy season were α-humulene (33.54%), α-pinene (19.33%) and germacrene-*D* (17.82%) (Rondón et al. 2005b).

Another study indicated that the essential oil of *Valeriana parviflora* (Trevir) BM Vadillo, an endemic species of the Venezuelan Andes, was collected from "Paramo Piedra Blanca" at 4,000 m. a. s. l. in two different seasons (dry and rainy) of the year showed some differences as well. The essential oil obtained during the dry season was composed mainly of linalool (11.9%), eugenol (8.9%), *p*-menth-1-en-9-al (8.7%) and α-terpineol (7.7%), while the oil obtained from the rainy season showed *o*-xylol (16.2%), 3-methyl isovaleric acid (10.6%) and geranial (9.5%) as major compounds (Fernandez et al. 2015).

The leaves and inflorescences of five species of *Tagetes*, family Asteraceae, were collected from different locations in Mérida state, Venezuela, *Tagetes patula, T. caracasana, T. subulata* and *T erecta* at 1,600 m. a. s. l. above sea level, and *T. filifolia* collected in La Culata, Mérida state at 2,800 m. a. s. l. also showed differences in the composition of the oils. *T. caracasana* essential oil was mainly composed of *trans*-ocimenone (64.3%) and *cis*-tagetone (13.7%); *T. erecta* showed piperitone (35.9%) and terpinolene (22.2%), high amounts of *trans*-anethole (87.5%) and estragole (10.7%) were observed in *T. filifolia*, while *T. subulata* essential oil contained terpinolene (26.0%), piperitenone (13.1%) and limonene (10.8%). For *T. patula*, two different oil samples were analyzed in leaves and inflorescences.

The oil from the leaves showed terpinolene (20.9%) and piperitenone (14.0%) as the main components, while the oil from de inflorescence was composed mainly of β-caryophyllene (23.7%), terpinolene (15.6%) and *cis*-β-ocimene (15.5%) (Armas et al. 2012).

Essential oils of *Piper dilatatum* L.C. Rich. and *Piper tuberculatum* Jacq. collected from two different locations in Mérida (Venezuela), *Piper dilatatum* (Las Gonzáles, Campo Elías Municipality, 18°35'27,66" north latitude; 71°14'56,16" west longitude at 1.743 m. a. s. l.) and *P. tuberculatum* (Chiguará, Sucre Municipality, 18°29'23,81" north latitude, 71°32'18,73" west longitude at 904 m. a. s. l.) showed β-pinene (17.2%), α-pineno (10.8%) and *cis*-ocimene (10.3%) as major components for *P. dilatatum* species, while *P. tuberculatum* was composed mainly of dilapiol (72.4%). *P. tuberculatum* is considered a chemotaxonomy variety different from the species previously reported in the literature (Mora et al. 2008).

Some variances might also be detected from different parts of the plant species analyzed. *Physalis peruviana* L. (Solanaceae) a species native to the Andes region of South America was studied and the essential oil of leaves and stems of this species was isolated by hydrodistillation. The chemical composition was determined by GC-MS and Kovats indices measurement. The oil from leaves showed hexadecanoic acid (42.8%), (–)-hexadecene epoxide (28.8%), phytol (4.7%) and hexadecanoic acid methyl ester (2.1%) as major components, while in the stems oil were identified hexadecanoic acid (80.5%), pentadecanoic acid (3.3%) and tetradecanoic acid (2.6%) (Morillo et al. 2017).

The essential oils isolated by hydrodistillation from the leaves and flowers of *Magnolia grandiflora* L., analyzed by GC/FID and GC/MS showed β-pinene (17.73%) and germacrene-*D* (11.09%) as the main components for the leaves oil, whereas germacrene-*D* (28.56%), β-elemene (9.92%) and γ-acoradiene (9.66%) were major constituents in the flowers (Jiménez-Medina et al. 2007). *Tagetes lucida* collected from Cordero, Táchira State, Venezuela at 1,100 masl showed methyl chavicol (95.5%) and *trasn*-β-farnesene (2.5%) for the oil obtained from the flowers but the oil obtained from the leaves was composed almost completely by methyl chavicol (99.5%) (Visbal et al. 2010).

Different Biological Activities Assessed in Essential Oils of Species Collected from Venezuelan Andean

Essential oils are mainly used in aromatherapy for helping to reduce stress and anxiety, relieve headaches and improve the quality of sleep. It might also be useful as a base to elaborate fragrances, soaps, detergents, and some oils that have a gourmet use in the food industry as well as in the elaboration of Vermouth. However, antiviral and bactericide effects to treat skin diseases and as a rubefacient are other uses attributed to essential oils (Franz and Novak 2015).

In this sense, several investigations have been conducted on essential oils obtained from plants collected from different locations in Venezuelan Andean and diverse assays have been tested in order to determine the biological activities of these oils.

The essential oil obtain from *Origanum Majorana* L. is a non-toxic oil mainly composed of variable quantities of terpenes, specially terpinene, origanol, sabinene, and lesser quantities of sesquiterpenes and is mainly used in gastronomy and natural medicine. An investigation carried out by Meza et al. (2007) showed that oil obtained from the species collected in Táchira (Venezuela) contains lots of esters and has potential use in cosmetics. Besides, this type of oil has also bactericidal activity since it inhibits the growth of microorganisms such as *Staphylococcus aureus*, *Escherichia coli*, and *Pseudomonas* sp.

Vismia genus belongs to the Hypericaceae family and comprises 57 species of which 17 have been located in Venezuela. Previous investigations have been carried out in extracts as well as pure isolated compounds, revealing antimicrobial, antioxidant, and anti-HIV among other biological activities. Cytotoxic activity of essential oils from leaves of *Vismia baccifera* Triana and Planch and *Vismia macrophylla* Kunth collected in three different locations of the Venezuelan Andean region was analyzed following the MTT (3-[4,5-dimethylthiazol-2-yl]-2,5-diphenyltetrazolium bromide) assay. *β*-caryophyllene and *trans*-caryophyllene were present as major components in *Vismia baccifera*, while *γ*-bisabolene was the main component in the *Vismia macrophylla* sample. Anticancer activity was observed in *V. baccifera* essential oil against SKBr3 (breast cancer cells) 4.99 µg/mL, MCF-7 (breast adenocarcinoma) 20.49 µg/mL and PANC-1 (pancreatic carcinoma) 12.26 µg/mL. High selectivity against the SKBr3 cell line was observed in the assay with no activity against non-tumor cells (Rojas-Vera et al. 2020b).

Coniza bonariensis essential oil proved to be active against HeLa (IC_{50}: 1.41 µg/mL) cells after 48 hours of exposure. The cytotoxic activity always decreased when the oil was incorporated in the growth log phase, indicating that the mechanism of action must be through the biosynthesis of Ecadherins. It is important to emphasize that *C. bonariensis* essential oil showed a rather selective cytotoxic activity against HeLa and MCF-7 cell lines, showing indexes of 20.6 and 1.4, respectively, as demonstrated by their higher IC_{50} values against the non-tumor mammalian Vero cells (> 20 µg/mL) (Araujo et al. 2013).

Piper hispidum essential oil turned out to be active against the tested tumor cell lines, especially on HeLa cells (IC_{50}: 18.6 µg/mL) after 48 hours of exposure. Cytotoxic activity was also determined against HeLa (cervix carcinoma), A-459 (lung carcinoma), MCF-7 (breast adenocarcinoma) human cancer cell lines, and against normal Vero cells (African green monkey kidney), exhibiting potent antiproliferative effects with IC_{50} values ranging from 18.6 to 37.7 µg/mL with no activity observed against Vero cells (Morales et al. 2013).

Conclusion

Essential oils have been used for thousands of years as alternative medicine, pharmaceutical products, herbal therapies, and food preservation. The rich source of biologically active compounds as well as the variety of activities makes essential oils of great interest. In this sense, aromatic plants growing in Venezuelan Andean have revealed a selection of activities including antimicrobial, antifungal, and cytotoxic properties, thus essential oils obtained from these species might be considered a natural source of active compounds. However, more studies are required in order to obtain more profound and accurate information.

References

Alarcón, L., A. Peña, J. Velasco, J.G. Baptista, L. Rojas, R. Aparicio et al. 2015. Chemical composition and antibacterial activity of the essential oil of *Ruilopezia bracteosa*. Nat. Prod. Commun. 10(4): 655–656.

Alarcón, L., A. Peña, J. Velasco, A. Usubillaga, B.Z. Contreras-Moreno, L. Rojas et al. 2016. Composición química y evaluación de la actividad antimicrobiana del aceite esencial de *Espeletia Schultzii* Wedd (Asteraceae) recolectada en el estado Trujillo-Venezuela. Revista ACADEMIA. 15(35). N° 35: 69–79.

Alarcón, L.D., A.E. Peña, N. Gonzales de C., A. Quintero, M. Meza, A. Usubillaga et al. 2009. Composición y actividad antibacteriana del aceite esencial de (Rich.) DC. de Venezuela *Myrcia fallax*. Rev. Soc. Quim. Peru. 75(2): 221–227.

Aparicio-Zambrano, R., J. Velasco-Carrillo, R. Paredes-Uzcategui and L. Rojas-Fermin. 2019a. Chemical characterization and antibacterial activity of the essential oil of *Mangifera indica* L. in three regions of Venezuela. Rev. Colomb. Quim. 48(3): 13–18.

Aparicio-Zambrano, R., L. Rojas-Fermín, J. Velasco, A. Usubillaga, M. Sosa and J. Rojas. 2019b. Caracterización química y actividad antimicrobiana del aceite esencial de las hojas de *Libanothamnus neriifolius* (Asteraceae). Rev. Peru. Biol. 26(1): 95–100.

Aparicio, R.L., L.B. Rojas, A. Usubillaga and M.E. Lucena. 2018. Caracterización química y actividad antibacteriana del aceite esencial de *Ruilopezia marcescens* (S.F. Blake) Cuatrec. Revista Facultad de Farmacia 81(1-2): 6–12.

Araque, E., D. Urbina, M. Morillo, L. Rojas-Fermín and J. Carmona. 2018. Estudio de la composición química de los aceites esenciales de las hojas y flores de *Leonotis nepetifolia* (L.) R. Br. (Lamiaceae). Rev. Fac. Farm. 60(2): 25–30.

Araujo, L., L.M. Moujir, J. Rojas, L. Rojas, J. Carmona and M. Rondón. 2013. Chemical composition and biological activity of *Conyza bonariensis* essential oil collected in Mérida, Venezuela. Nat. Prod. Commun. 8(8): 1175–1178.

Araujo, L., M. Rondón, A. Morillo, E. Páez and L. Rojas-Fermín. 2017. Antimicrobial activity of the essential oil of *Myrcianthes myrcinoides* (Kunth) Grifo (Myrtaceae) collected in the Venezuelan Andes. Pharmacologyonline 2: 200–204.

Armas, K., J. Rojas, L. Rojas and A. Morales. 2012. Comparative study of the chemical composition of essential oils of five *Tagetes* species collected in Venezuela. Nat. Prod. Commun. 7(9): 1225–1226.

Baldovino, S., J. Rojas, L.B. Rojas, M. Lucena, A. Buitrago and A. Morales. 2009. Chemical composition and antibacterial activity of the essential oil of *Monticalia andicola* (Asteraceae) collected in Venezuela. Nat. Prod. Commun. 4(11): 1601–1604.

Buitrago, D., A. Morales, L. Rojas-Fermin, R. Aparicio and P. Meléndez. 2017. Composición química del aceite esencial de *Achyrocline satureioides* (Lam.) DC de los Andes Venezolanos. Rev. Fac. Farm. 59(1): 22–25.

Buitrago, A., J. Rojas, L. Rojas, J. Velasco, A. Morales, Y. Peñaloza et al. 2015. Essential oil composition and antimicrobial activity of *Vismia macrophylla* leaves and fruits collected in Táchira-Venezuela. Nat. Prod. Commun. 10(2): 375–377.

Buitrago, A., L.B. Rojas, J. Rojas, D. Buitrago, A. Usubillaga and A. Morales. 2009. Comparative study of the chemical composition of the essential oil of *Vismia baccifera* var. *dealbata* (Guttiferae) collected in two different locations in Merida-Venezuela. J. Essent. Oil-Bear. Plants 12(6): 651–655.

Buitrago, D., L.B. Rojas, J. Rojas and A. Morales. 2010. Volatile compounds from *Tagetes pusilla* (Asteraceae) collected from the Venezuela Andes. Nat. Prod. Commun. 5(8): 1283–1284.

Buitrago, D., J. Velasco, T. Díaz and A. Morales. 2012. Chemical composition and antibacterial activity of the essential oil of *Monticalia imbricatifolia* Schultz (Asteraceae). Rev. Lat. Quím. 40(1): 13–18.

Buitrago-Díaz, A., J. Rojas-Vera, L. Rojas-Fermín, A. Morales-Méndez, R. Aparicio-Zambrano et al. 2011. Composition of the essential oil of leaves and roots of *Allium schoenoprasum* L. (Alliaceae). Bol. latinoam. Caribe Plantas Med. Aromát. 10(3): 218–221.

Butnariu, M. and I. Sarac. 2018. Essential oils from plants. J. Biotechnol. Biomed. Sci. 1(4): 35–43.

Cáceres Ferreira, W., M. Rengifo Carrillo, L. Rojas and C. Rosquete Porcar. 2019. Chemical composition of essential oils from *B. simaruba* (L.) Sarg. fruits and the resins from three *Bursera* species: *B. simaruba* (L.) Sarg, *B. glabra* Jack and *B. inversa* Daly. Avances en Química 14(1): 25–29.

Cárdenas, J., J. Rojas, L. Rojas-Fermin, M. Lucena and A. Buitrago. 2012. Essential oil composition and antibacterial activity of *Monticalia greenmaniana* (Asteraceae). Nat. Prod. Commun. 7(2): 243–244.

Contreras-Moreno, B., J. Rojas, M. Celis, L. Rojas, L. Méndez and L. Landrum. 2014. Componentes volátiles de las hojas de *Pimenta racemosa* var. *racemosa* (Mill.) J.W. Moore (Myrtaceae) de Táchira-Venezuela. Bol. latinoam. Caribe Plantas Med. Aromát. 13(3): 305–310.

Contreras-Moreno, B.Z., J.J. Velasco, J.C. Rojas, L.C. Méndez and M.T. Celis. 2016. Antimicrobial activity of essential oil of *Pimenta racemosa* var. *racemosa* (Myrtaceae) leaves. J. Pharm. Pharmacogn. Res. 4(6): 224–230.

Cordero de Rojas, Y., C. Díaz, J. Velasco, L. Rojas-Fermín, R. Aparicio, A. Usubillaga et al. 2017. Composición química y efecto antifúngico de los aceites esenciales de tres especies de frailejones de Los Andes venezolanos. Revista Facultad de Farmacia 80(1-2): 60–67.

Dewick, P. 2002. Medicinal Natural Products. John Wiley and Sons, Nottingham, UK.

Dhifi, W., S. Bellili, S. Jazi, N. Bahloul and W. Mnif. 2016. Essential oils' chemical characterization and investigation of some biological activities: A critical review. Medicines 3(25): 1–16.

Dilworth, L.L., C.K. Riley and D.K. Stennett. 2017. Plant constituents: Carbohydrates, oils, resins, balsams and plant hormones. pp. 61–80. *In*: S. Badal and R. Delgoda (eds.). Pharmacognosy Fundamentals, Applications and Strategies. Academic Press, London, UK.

Fernandez, S., M. Rondón, J. Rojas, A. Morales and L. Rojas-Fermin. 2015. Comparison of the chemical composition of *Valeriana parviflora* essential oils collected in the Venezuelan Andes in two different seasons. Nat. Prod. Commun. 10(4): 657–659.

Flores, M., L. Rojas, R. Aparicio, M.E. Lucena and A. Usubillaga. 2015. Essential oil composition and antibacterial activity of *Hyptis colombiana* from the Venezuelan Andes. Nat. Prod. Commun. 10(10): 1751–1752.

Franz, C. and J. Novak. 2015. Sources of essential oils from. pp. 43–86. *In*: K. Hüsnü, C.B. and G. Buchbauer (eds.). Handbook of Essential Oils, Science, Technology, and Applications CRC Press, London, UK.

González de Colmenares, N., R. Aparicio, C. Araque, J. Velasco and A. Usubillaga. 2017. Composición química y actividad antibacteriana del aceite esencial de los frutos de una nueva especie del género *Zanthoxylum* (Sección Tobinia-Rutaceae) de Venezuela. Rev. Fac. Agron. 34: 158–174.

Górniak, I., R. Bartoszewski and J. Kroliczewski. 2019. Comprehensive review of antimicrobial activities of plant flavonoids. Phytochemistry Rev. 18: 241–272.

Gualteri, M., M. Araque, A. Morales, M. Rondón, J. Rojas, L. Araujo et al. 2005. Chemical compostion and antibacterial activity of the essential oil of *Verbesina turbacensis* Kunth HBK. Rev. Lat. Quím. 3(33): 128–131.

Hernández, J., I. Bracho, L.B. Rojas-Fermin, A. Usubillaga and J. Carmona. 2013. Chemical composition of the essential oil of *Erechtites valerianaefolia* from Mérida, Venezuela. Nat. Prod. Commun. 8(10): 1477–1478.

Hernández, J., L.B. Rojas-Fermin, J. Amaro-Luis, L. Pouységu, S. Quideauc and A. Usubillaga. 2015. Chemical composition of the essential oil of *Gynoxys meridana* from Mérida, Venezuela. Nat. Prod. Commn. 10(4): 653–654.

Jiménez-Medina, D., A. Cordero-Gallardo, L.B. Rojas and M. Rodríguez-A. 2007. Estudio de los componentes volátiles de las hojas y flores de *Magnolia grandiflora* L., que crece en el estado Mérida, Venezuela. Rev. Fac. Farm. 49(1): 2–4.

Lucena, M.E., M. Escalante Contreras, V. González Moreno, L. Rojas-Fermín, Y. Cordero de Rojas, F.J. Ustáriz Fajardo et al. 2019. Composición y actividad antibacteriana del aceite esencial de

Austroeupatorium inulifolium (Kunth) King and Robinson (Asteraceae). Rev. Cuba. de Farm. 52(4): e369.

Mangalagiri, N.P., S.K. Panditi and N.L.L. Jeevigunta. 2021. Antimicrobial activity of essential plant oils and their major components. Heliyon 7(4): e06835. Doi: 10.1016/j.heliyon.2021.e06835.

Marcano, D. and M. Hasegawa. 2002. Fitoquímica Orgánica, Consejo de Desarrollo Científico y Humanístico de la Universidad Central de Venezuela, Caracas, Venezuela.

Marcano-Pacheco, E., A. Padilla-Baretic, L. Rojas-Fermin, F.D. Mora-Vivas and H. Ferrer-Pereira. 2016. Chemical composition of wood essential oil of *Aniba cinnamomiflora* C. K. Allen from Venezuelan Andes. Colom. For. 19(2): 233–238.

Meccia, G., P. Vit, L.B. Rojas, J. Carmona, B. Santiago and A. Usubillaga. 2015. Composición química del aceite esencial de hojas frescas de *Annona muricata* L., de Mérida, Venezuela. Rev. Fac. Farm. 57(2): 2–7.

Meza, M., N. González de C. and A. Usubillaga. 2007. Composición del aceite esencial de *Origanum majorana* L. extraído por diferentes técnicas y su actividad biológica. Rev. Fac. Agron. 24: 725–738.

Mokhtar, M.M., L. Jianfeng, Du Z. and C. Fangqin. 2020. Review of the chemical separation. International Journal of Advanced Engineering, Management and Science 6(9): 438–444.

Mora, F.D., B. Silva, V. Hernández, L.B. Rojas and J. Carmona. 2016. Chemical composition of the essential oil of *Morella parvifolia* (Benth.) Parra-O from the Venezuelan Andes. Emir. J. Food Agric. 28(4): 288–290.

Mora, F.D., J. Peña, L.B. Rojas, A. Usubillaga and P. Meléndez. 2008. Composición química de los aceites esenciales de *Piper dilatatum* L.C. Rich. y *Piper tuberculatum* Jacq. de Mérida, Venezuela. CIENCIA 16(3): 365–369.

Mora, F.D., M. Araque, L.B. Rojas, R. Ramírez, B. Silva and A. Usubillaga. 2009b. Chemical composition and *in vitro* antibacterial activity of the essential oil of *Minthostachys mollis* (Kunth) Griseb Vaught from the Venezuelan Andes. Nat. Prod. Commun. 4(7): 997–1000.

Mora, V.F.D., L.B. Rojas, A. Usubillaga, J. Carmona and B. Silva. 2009a. Composición química del aceite esencial de *Myrcianthes fragrans* (Sw.) Mc Vaught de los Andes venezolanos. Rev. Fac. Farm. 51(1): 20–23.

Morales, A., J. Rojas, L.M. Moujir, L. Araujo and M. Rondón. 2013. Chemical composition, antimicrobial and cytotoxic activities of *Piper hispidum* Sw., essential oil collected in Venezuela. J. App. Pharm. Sci. 3(06): 16–20.

Morillo, M., V. Marquina, L. Rojas-Fermín, R. Aparicio, J. Carmona and A. Usubillaga. 2017. Estudio de la composición química del aceite esencial de hojas y tallos de *Physalis peruviana* L. Revista ACADEMIA 16(38): 85–93.

Obregón-Díaz, Y., A. Pérez-Colmenares, K. Obregón-Alarcón, R. Aparicio-Zambrano, L. Rojas-Fermín, A. Usubillaga et al. 2019. Volatile constituents of the leaves of *Kalanchoe pinnata* from the Venezuelan Andes. Nat. Prod. Commun. 14(5): 1–3.

Ouis, N. and A. Hariri. 2018. Antioxidant and antibacterial activities of the essential oils of *Ceratonia silique*. Banats J. Biotechnol. 9(17): 13–23.

Pérez-Colmenares, A., L. Rojas-Fermín and R. Aparicio-Zambrano. 2017. Análisis comparativo de los componentes volátiles *de Ruilopezia lindenii* (Schultz-Bip. Ex wedd.) Cuatrec., recolectada a diferentes altitudes en Mérida-Venezuela. Revista ACADEMIA 16(38): 39–44.

Rios Tescha, N., F. Mora, L. Rojas, T. Díaz, J. Velasco, C. Yánez et al. 2011. Chemical composition and antibacterial activity of the essential oil of *Lantana camara* var. *moritziana*. Nat. Prod. Commun. 6(7): 1031–1034.

Rojas-Vera, J. 2020a. Essential oil composition and antibacterial activity of two *Ageratina* species collected in Mérida-Venezuela. Acta Chimica and Pharmaceutica Indica 10(2): Doi: 10.37532/2277-288X.2020.10(1).140.

Rojas-Vera, J., A. Buitrago-Díaz, F.A. Arvelo, F.J. Sojo, A.I. Suarez and L. Rojas. 2020b. Essential oil composition and cytotoxic activity in two species of the plant genus *Vismia* (Hypericaceae) from the Venezuelan Andes. Rev. Biol. Trop. 68(3): 884–891.

Rojas, J. and A. Buitrago. 2013. Essential oils and antimicrobial activity. pp. 71–129. *In*: J.N. Govil (ed.). Recent Progress in Medicinal Plants, Volume 37: Essential Oils II. Studium Press, Houston, TX, USA.

Rojas, J. and A. Buitrago. 2015. Essential oils and their products as antimicrobial agents: progress and prospects. pp. 253–278. *In*: M.C. Teixeira Duarte and R. Mahendra (eds.). Therapeutical Medicinal Plants: From Lab to The Market. CRC Press, Taylor and Francis Group, New York, NY, USA.

Rojas, J., A. Buitrago, L.B. Rojas and A. Morales. 2013a. Chemical composition of *Hypericum laricifolium* Juss. Essential oil collected from Mérida-Venezuela. Med. Aromat. Plants 2: 132. Doi: 10.4172/2167-0412.1000132.

Rojas, J., A. Buitrago, L.B. Rojas, J. Cárdenas and J. Carmona. 2013b. Chemical composition of the essential oil of *Croton huberi* Steyerm. (Euphorbiaceae) collected from Táchira-Venezuela. J. Essent. Oil-Bear. Plants 16(5): 646–650.

Rojas, J., A. Buitrago, L.B. Rojas, A. Morales and S. Baldovino. 2010. Chemical composition of the essential oil of leaves and roots of *Ottoa oenanthoides* (Apiaceae) from Mérida, Venezuela. Nat. Prod. Commun. 5(7): 115–116.

Rojas, J., A. Buitrago, L. Rojas, A. Morales, M. Lucena and S. Baldovinoa. 2011b. Essential oil composition and antibacteral activity of *Vismia baccifera* fruits collected from Mérida, Venezuela. Nat. Prod. Commun. 6(5): 699–700.

Rojas, J., J. Velasco, L.B. Rojas, T. Díaz, J. Carmona and A. Morales. 2007. Chemical composition and antibacterial activity of the essential oil of *Baccharis latifolia* Pers. and *B. prunifolia* H.B. and K. (Asteraceae). Nat. Prod. Commun. 2(12): 1245–1248.

Rojas, J., S. Baldovino, M. Vizcaya, L.B. Rojas and A. Morales. 2009. The chemical composition of the essential oils of *Euphorbia caracasana* and *E. cotinifolia* (Euphorbiaceae) from Venezuela. Nat. Prod. Commun. 4(4): 571–572.

Rojas, J., T. Mender, L. Rojas, E. Guillen, A. Buitrago, M. Lucena et al. 2011a. Estudio comparativo de la composición química y actividad antibacteriana del aceite esencial de *Ruta graveolens* L. recolectada en los estados Mérida y Miranda, Venezuela. Avances en Química 6(3): 89–93.

Rojas, J., J. Velasco, A. Morales, L. Rojas, T. Díaz, M. Rondón et al. 2008. Chemical composition and antibacterial activity of the essential oil of *Baccharis trinervis* (Lam.) Pers. (Asteraceae) collected in Venezuela. Nat. Prod. Commun. 3(3): 369–372.

Rondón, M., J. Velasco, J. Hernández, M. Pecheneda, J. Rojas, A. Morales et al. 2006a. Chemical composition and antibacterial activity of the essential oil of *Tagetes patula* L. (Asteraceae) collected from the Venezuela Andes. Rev. Lat. Quím. 34(1-3): 32–36.

Rondón, M., J. Velasco, A. Morales, J. Rojas, J. Carmona, M. Gualtieri et al. 2005a. Composition and antibacterial activity of the essential oil of *Salvia leucantha* Cav. cultivated in Venezuela Andes. Rev. Lat. Quím. 33(2): 55–59.

Rondón, M., M. Araque, A. Morales, M. Gualtieri, J. Rojas, K. Veres et al. 2006b. Chemical composition and antibacterial activity of the essential oil of *Lasiocephalus longipenicillatus* (*Senecio longipenicillatus*). Nat. Prod. Commun. 1(2): 113–115.

Rondón, M.E., A. Morales, D. Buitrago, J. Rojas and M. Gualtieri. 2005b. Comparative study of the chemical composition of the essential oil of the *Lasiocephalus longepenicillatus* (Schultz-Bip. ex Sandw.) Cuatrec. (*Senecio longepenicillatus*) in two seasons of the year. CIENCIA 13(4): 440–442.

Rondón, M.E., J. Delgado, J. Velasco, J. Rojas, L.B. Rojas, A. Morales et al. 2008. Chemical composition and antibacterial activity of the essential oil from aerial parts of *Porophyllum ruderale* (Jacq.) Cass. collected in Venezuela. CIENCIA 16(1): 5–9.

Semnani, S.N., N. Hajizadeh and H. Alizadeh. 2017. Antibacterial effects of aqueous and organic quince leaf extracts on gram–positive and gram–negative bacteria. Banats J. Biotechnol. 8(16): 54–61.

Swamy, M.K., M.S. Akhtar and U.R. Sinniah. 2016. Antimicrobial properties of plant essential oils against human pathogens and their mode of action: An updated review. Evidence-Based Compl. Alternat. 2016. Article ID 3012462.

Szweda, P. and B. Kot. 2018. Antibacterial activity of essential oils and plant extracts against *S. aureus*. pp. 203–224. *In*: S. Enany and L.E. Crotty Alexander (eds.). Frontiers in *Staphylococcus aureus*. InTechOpen, London, UK.

Velasco, J., J. Rojas, P. Salazar, M. Rodríguez, T. Díaz, A. Morales et al. 2007. Antibacterial activity of the essential oil of *Lippia oreganoides* against multiresistant bacterial strains of nosocomial origin. Nat. Prod. Commun. 2(1): 85–88.

Visbal, T., L. Rojas, Y. Cordero, J. Carmona, M. Morillo and A. Usubillaga. 2010. Componentes volátiles de *Tagetes lucida* Cav. (Asteraceae) (Cordero edo. Táchira Venezuela). Rev. Fac. Farm. 52(1): 2–4.

Vizcaya, M., C. Pérez, J. Rojas, L. Rojas-Fermín, C. Plaza, A. Morales et al. 2014. Composición química y evaluación de la actividad antifúngica del aceite esencial de corteza de *Vismia baccifera* var. *dealbata*. Rev. Soc. Ven. Microbiol. 34(2): 86–90.

Zarkani, A.A. 2016. Antimicrobial activity of *Hibiscus sabdariffa* and *Sesbania grandiflora* extracts against some G–ve and G+ve strains. Banats J. Biotechnol. 7(13): 17–23.

8

Chemical Composition and Antimicrobial Activity of Plant Essential Oils from the Phytolaccaceae and Petiveriaceae Families

Suelen Pereira Ruiz Herrig,[1,]* Evellyn Claudia
Wietzikoski Lovato,[2] Juliana Silveira do Valle,[1]
Zilda Cristiani Gazim,[1] Camila Frederico,[1]
Ana Daniela Lopes,[1] Giani Andrea Linde,[3]
Nelson Barros Colauto[4] and Maria Graciela Iecher Faria[1]

Introduction

Antimicrobial resistance is a public health challenge around the world, even in developing countries. The spread of multidrug-resistant bacteria is recurrent and even with advances in antibiotic therapy, infectious complications remain an important cause of mortality (El-Tarabily et al. 2021). According to US Centers for Disease Control and Prevention, approximately 2 million people are infected with bacterial multiple drug resistance (MDR) each year, with an average of 23,000 deaths (Cheng et al. 2019). There is, therefore, a need for the development of alternative sources that could treat these bacterial infections.

---

[1] Graduate Program in Biotechnology Applied to Agriculture, Universidade Paranaense, Umuarama, PR, Brazil.

[2] Graduate Program in Medicinal Plants and Phytotherapeutics in Basic Attention, Universidade Paranaense, Umuarama, PR, Brazil.

[3] Graduate Program in Food, Nutrition and Health, Federal University of Bahia, Salvador, BA, Brazil.

[4] Graduate Program in Food Science, Federal University of Bahia, Salvador, BA, Brazil.

Emails: gracielaiecher@prof.unipar.br; jsvalle@prof.unipar.br; camila.frederico@edu.unipar.br; anadanielalopes@prof.unipar.br; cristianigazim@prof.unipar.br; evellyn@prof.unipar.br; gianilindecolauto@gmail.com; nelsonbcolauto@gmail.com

* Corresponding author: suelenruiz@prof.unipar.br

Natural products, such as essential oils, have become a research target to identify new bioactive compounds with antibacterial activity as an alternative for the development of new drugs against bacterial resistance (Chouhan et al. 2017). Filamentous and yeast-like fungi have also been resistant to available antifungal agents, mainly azoles (Pérez-Cantero et al. 2020, Chen et al. 2021), and essential oils are also a promising alternative source of new antifungal agents.

Essential oils are compounds naturally present in plants, which are made up of a complex mixture of volatile polar and non-polar substances and exert functions related to their defense mechanisms (Dhifi et al. 2016). The chemical compounds of essential oils can vary according to the plant species and also to the part of the plant used for essential oil extraction, such as branches, leaves, flowers, seeds, rhizomes, or fruits (Ribeiro-Santos et al. 2018). These oils are mainly composed of monoterpenes, diterpenes, sesquiterpenes, and phenylpropanoids with therapeutic properties, such as antifungal, antibacterial, antioxidant, anti-inflammatory, antitumor, antilipemic, immunomodulatory, antinociceptive, cytotoxic, and antiviral (Costa et al. 2015, Seol and Shin 2018, Pavitra et al. 2019, Filho et al. 2020).

Essential oil chemical composition can show qualitative and quantitative variation, which may be related to genetic or intrinsic factors and extrinsic or abiotic factors related to plant interaction with the environment and maturation stage (Dhifi et al. 2016). Environmental stimuli in which the plant is located can redirect metabolic routes and promote the biosynthesis of different compounds. Among these stimuli, we can highlight the plant-microorganism, plant-insect and plant-plant interactions, age and stage of development, abiotic factors such as light, temperature, rainfall, nutrition, and collecting time as well as harvest and post-harvest techniques (Morais 2009), which will directly affect the biological activity of the essential oil.

The essential oil of some species of the Petiveriaceae and Phytolaccaceae families has shown antibacterial, antifungal, anti-inflammatory, and insecticidal activities (Raimundo et al. 2018, Matebie et al. 2019, Bortolucci et al. 2021, Raimundo et al. 2021). Although not fully elucidated, the antimicrobial activity of essential oil may be related to its ability to disrupt and penetrate the cell wall, causing a loss of function and becoming an alternative for the development of new antimicrobial drugs (Adelakun et al. 2016).

Petiveriaceae and Phytolaccaceae Families

After the recognition of the genus Microtea Sw. and the subfamily Rivinoideae as distinct families, Microteaceae Schäferh and Borsch (Stevens 2001, Schäferhoff et al. 2009) and Petiveriaceae C. Agardh (Stevens onwards, APG IV 2016) respectively, the Phytolaccaceae family is now represented by five genera; *Agdestis* Moc. and Sessé, *Anisomeria* D. Don, *Ercilla* A. Juss., *Nowickea* J. Martínez and JA McDonald, and *Phytolacca* L. and about 33 species are distributed mainly in South Africa and tropical and subtropical America (Stevens 2001, Xu and Deng 2017). In Brazil, this family is represented only by the genus *Phytolacca* with three native species and some are known for their medicinal properties (Steinmann 2010).

The Petiveriaceae family has nine genera and approximately 22 species (Christenhusz and Byng 2016, Powo 2021). It is formed by herbaceous, shrub, or arboreal species, which are distributed from tropical to subtropical America (Savolainen et al. 2000). However, information on the Petiveraceae family is scarce, but some species are reported to have a medicinal activity, such as *Gallesia integrifolia* (Spreng.) Harms and *Petiveria alliacea* L.

Gallesia integrifolia is a native and endemic plant in Brazil and is found in several Brazilian regions, such as the Northeast (Bahia, Ceará, Paraíba, and Pernambuco), Midwest (Mato Grosso), North (Acre and Amazon), Southeast (Minas Gerais, Rio de Janeiro, and São Paulo), and South (Paraná) (Forzza et al. 2012). The synonyms are *Crateva gorarema* Vell., *Gallesia gorarema* (Vell.) Moq., *Gallesia integrifolia* var. ovata (O.C. Schmidt) Nowicki, *Gallesia ovata* O.C. Schmidt, *Thouinia integrifolia* Spreng., *Gallesia acorododendron, Gallesia gorarema,* and *Crataeva gorarema* (Akisue et al. 1986, Guarim-Neto and de Morais 2003, Tropicos.org 2015). It is a tree species with 15 to 30 m in height, a broad and dense crown with a trunk diameter of 70–140 cm. All parts of the plant release a strong garlicky odor, hence the popular name *pau d'alho* in Portuguese from Brazil (Lima et al. 2010). It is used in folk medicine, mainly the bark to prepare teas for treatments of flu, cough, pneumonia, worms, gonorrhea, prostate tumors, and rheumatism (Lorenzi 2002). Tea from leaves and stem bark is used to treat ulcers (Grandtner and Chevrette 2013) and boiled leaves and stem bark are used to treat intestinal worms, respiratory, and lymphatic diseases. Freshly crushed leaves are used topically to treat abscesses, orchitis, and gonorrhea (Balbach 1993). The infusion of roots is used to treat rheumatism and ulcers in some Brazilian communities and the essential oil is used to treat gonorrhea (Barbosa et al. 1999, Muñoz et al. 2000, Biesk et al. 2012).

Petiveria alliacea L. is found in tropical regions of the Americas such as the Amazon rainforest, Central America, Caribbean islands, and Mexico, and some regions of Africa (Luz et al. 2016). It preferably develops in sub-humid and shady areas (Rocha et al. 2006) and is characterized as a perennial shrub, which can reach up to 5–150 cm in height. It has a rigid and straight stem, branched with long, erect, delicate, and ascending branches (Almanza 2012, Duarte and Lopes 2005, Rzedowski and de Rzedowski 2000). Its synonyms are *Petiveria foetida* Salisb., *Petiveria alliacea* var. *grandifolia* Moq., *P. alliacea* var. *octandra* (L.) Moq., *Petiveria hexandria* Sessé and Moc., *Petiveria ochroleuca* Moq., *Petiveria octandra* L., and *Petiveria paraguayensis* D. Parodi (Tropicos.org 2015). It is popularly known as *mucuracaá, guiné, pipi, tipi, pênis-de-coelho, apacin, anamú, zorrillo, amansa-senhor, erva-de-olho, caá, embayayendo,* and *ouoembo* (Andrade et al. 2012, Duarte and Lopes 2005, Lima et al. 1991) and produces a strong odor that is usually associated with garlic (Luz et al. 2016). It is used in folk medicine as antirheumatic, antispasmodic, antifungal, analgesic (Silva et al. 2018), diuretic, anthelmintic, antiemetic, abortive, antipyretic, antitumor, and hypoglycemic effect among others (Camargo 2007, Gomes et al. 2005, Gomes et al. 2008).

Phytolacca dodecandra (L'Herit) is a plant native to sub-Saharan Africa and Madagascar. It has a woody climbing plant habit with an average length of stems that reaches 5 to 8 m. It grows very quickly, especially during the rainy season, with erect, racemic, dioecious flowering stalks, and red berries (Adams et al. 1989, Lemma 1970). Its synonyms are *Phytolacca abyssinica* Hoffm., *Phytolacca dodecandra* var. *brevipedicellata* H. Walter, and *Pircunia abyssinica* Moq. (Tropicos.org 2015). It is known as *endod* in Ethiopia and *chihakahaka* in Tanzania (Legère 2009). The roots, root barks, leaves, and berries of this plant are used to treat liver problems, and it has been reported with pharmacological activities, such as antirabies, abortifacient, analgesic, anti-inflammatory, antibacterial, antioxidant, antimalarial, insecticide (Meharie and Tunta 2021), and molluscicide (McCullough et al. 1980, Webbe and Lambert 1983). Berry, seed, leaf, and root extracts have also been traditionally used as a purgative, anthelmintic, laxative, emetic, diuretic, and antidiarrheal for humans and a purgative for animals (Schmelzer and Gurib-Fakim 2008). The powdered dried fruit forms a foaming detergent when mixed in water (Schmelzer and Gurib-Fakim 2008), which is used as a detergent for cleaning clothes in Ethiopia, Somalia, and Uganda (Lemma 1965) and African soap maker (Legère 2009).

Essential Oils and Chemical Composition

Gallesia integrifolia

Raimundo et al. (2018) analyzed the chemical composition of *Gallesia integrifolia* essential oil obtained from fresh fruits in southern Brazil, using the hydrodistillation technique. Thirty-five compounds were found and 34 were identified of which 68% belonged to the organosulfur class. The major compounds were dimethyl trisulfide (15.5%), 2,8-dithianonane (52.6%), and lenthionine (14.7%).

Raimundo et al. (2021) investigated the essential oil composition of fruits, leaves, and flowers of *G. integrophilia* in Umuarama, Paraná, Brazil. Fifty-six essential oil compounds were obtained by gas chromatography coupled with mass spectrometry (CG-MS), 31 were identified from fruits, 47 from leaves, and 42 from flowers. The major compounds were from the sulfur compound class, 99.2%, 95.3%, and 95.9% from fruits, leaves, and flowers, respectively. The major essential oil compounds were 2,8-dithianonane (52.6%) in fruits, 3,5-dithiahexanol-5,5-dioxide (38.9%) in leaves, and methionine, ethyl ester (45.3%) in flowers. Other major compounds have been found, such as dimethyl trisulfide (15.3%) and lenthionine (14.7%) in fruits, 1,3,5-trithiane (13.7%) and N-ethyl-1,3-dithioisoindole (12.6%) in leaves, and methyl p-tolyl sulfide (17.1%) and N-ethyl-1,3-dithioisoindole (13.4%) in flowers.

Bortolucci et al. (2021) extracted the essential oil from *G. integrifolia* fresh fruits from a Southern region of Brazil by hydrodistillation for 3 hours, identified 29 compounds with predominance of organosulfur compounds (99.7%) and the major compounds were 2,8-dithianonane (52.9%), dimethyl trisulfide (15.2%), and lenthionine (14.8%).

The inner bark essential oil extracted from *G. integrifolia* from the central-west region of Brazil was by hydrodistillation technique and identified 20 compounds, the major compounds were α-santalene (18.9%), phytol (11.8%), bis disulfide bis

(9.9%), methyl disulfide methyl (6.9%), β-bisabolene (6.3%), β-sesquiphellandrene (4.5%), zingiberene (2.8%), and α-bergamotene (2.2%) (Arunachalam et al. 2017).

Venn diagrams were constructed in order to expand the analysis of the compounds identified in essential oils from different plant parts of *G. integrifolia*. Among the compounds identified in *G. integrifolia* essential oils, 21 (23.6%) were identified exclusively in the fruits and 16 (18%) in the stem bark (Figure 1). Only four (4.5%) occur only in leaves and three (3.4%) only in flowers. Flowers and leaves share 15 (16.9%) compounds, while fruits, flowers, and leaves share 18 compounds (20.2%), but do not occur in stem bark. The compounds dimethyl disulfide, dimethyl trisulfide, limonene, and phytol are present in fruits, flowers, leaves, and stem bark.

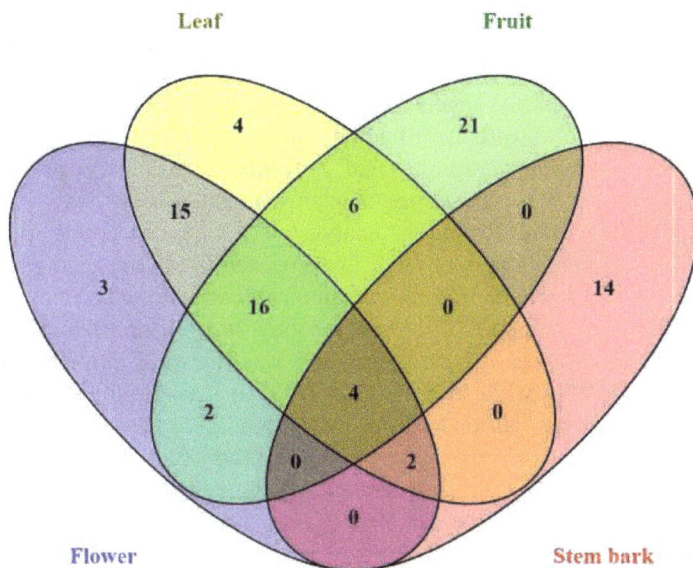

Figure 1: Venn diagram showing the number of essential oil compounds identified from different parts of *Gallesia integrifolia*. Data compiled from Arunachalam et al. (2017), Raimundo et al. (2018), Raimundo et al. (2021), and Bortolucci et al. (2021).

Petiveria alliacea

The essential oil of *P. alliacea* can be obtained from leaves, stems, roots, and inflorescences and has a yellow color with a strong odor due to its organosulfur content (Luz et al. 2016). Ayedoun et al. (1998) identified 13 compounds of the essential oil from *P. alliacea* roots collected in Benin, West Africa, in which the most abundant were benzaldehyde (48.3%), dibenzyl disulfide (23.3%), dibenzyl trisulfide (9.4%), cis-stilbene (1.3%), trans-stilbene (6.8%), benzyl alcohol (2.5%), 2,5-dimethoxy-p-cymene (2.0%), phytol (1.5%), and benzyl benzoate (0.6%).

Zoghbi et al. (2002) extracted the essential oil from *P. alliacea* inflorescences and identified the volatile compounds by CG-MS. The major compounds were benzaldehyde (54.8%), benzyl thiol (20.3%), and dibenzyl disulfide (18.0%). In addition to that cumin alcohol (2.5%), benzyl alcohol (0.6%), (Z)-hexen-3-yl benzoate (0.6%), phenylacetaldehyde (0.3%), benzyl benzoate (0.3%), dillapiole (0.1%), and palmitic acid (0.1%) were also identified

Neves et al. (2011) analyzed the essential oils obtained by the hydrodistillation technique from leaves, flowers, stems, and roots of *P. alliacea* collected in northeastern Brazil. Eighteen compounds were identified, which is the largest variety in the root. Carvacrol was identified in all plant parts, and it was the major compound in leaves (50.9%) and stem (48.3%). In flowers, the major compounds were (Z)-3-hexenyl benzoate (30.5%) and carvacrol (29.7%). In roots, the main component was benzyl alcohol (46.6%). High levels of dibenzyl disulfide were found in leaves (17.6%), stems (23.1%), flowers (15.7%), and roots (19.1%). Benzyl benzoate (0.6%) and palustrol (2.6%) were found exclusively in the stem while dillapiole (6.5%) and 1-phenylethyl anthranilate (6.7%) were found exclusively in flowers. The benzyl thiol was identified only in the stem (9.0%) and root (0.8%). Compounds such as benzyl formate (0.4%), ethyl benzoate (3.7%), 1-phenylethyl acetate (0.7%), (E)-cinnamaldehyde (2.8%), eugenol (2.3%), vanillin (3.1%), cis-stilbene (2.6%), and trans-stilbene (6.2%) were identified only in roots.

Roots, leaves, and micropropagated *P. alliacea* inflorescences were used to analyze the phytochemical profiles of plants obtained by *in vitro* and *ex vitro* cultivation (Castellar et al. 2014). The analysis of volatile compounds revealed the presence of 40 different compounds and 28 were identified. The roots produced via *ex vitro* showed 31 compounds, and benzaldehyde was the only compound in all structures as a major compound in *in vitro* roots (34.0%), *ex vitro* roots (55.1%), and *ex vitro* inflorescences (32.5%). Guaiacol was the most abundant compound *in vitro* leaves (13.6%) while heneicosane (32.4%) was the major compound in *ex vitro* leaves. The hydrocarbon undecane was found in *in vitro* leaves (13.8%) and inflorescences (14.7%), and pentadecane in *in vitro* roots (29.4%). The root essential oil had the highest number of compounds and the major ones were benzaldehyde (48.3%) and dibenzyl disulfide (23.3%) (Castellar et al. 2014). Kerdudo et al. (2015) analyzed the essential oil obtained by hydrodistillation from *P. alliacea* aerial parts from Martinique and the major compounds identified were toluenethiol (2.3–23.0%), phytol (6.4–40.0%), dibenzyl disulfide (13.2–35.3%), and benzaldehyde (0.8–31.3%).

Oluwa et al. (2017) investigated the essential oil from *P. alliacea* dried leaves, collected in southwestern Nigeria, obtained by hydrodistillation and reported as the major compounds phytol (56.1%), citronellol (16.0%), and (Z, Z)-α-farnesol (14.6%).

Extraction by steam distillation of essential oil from *P. alliacea* leaves was described by Olomieja et al. (2020) in which 12 compounds were identified by GC-MS and the major ones were benzyl alcohol (4.0%) and benzaldehyde (3.0%), 2-methyltetracosane (2.0%), (Z)-7-hexadecenal (1.6%), 2-hexyldecan-1-ol (1.2%), and 2-octyldecan-1-ol (0.9%). These authors described for the first time in *P. alliacea* sulfur heterocyclic compounds, such as 1,2,3-trithiolane (0.6%), 1,2,5-trithiepane (1.1%), 1,2,5,6-tetrathiocane (1.3%), and benzenecarbothioic acid (2.0%).

In *P. alliacea* (Figure 2), Venn diagrams showed that the essential oils from the root and leaves had the highest number of exclusively identified compounds with

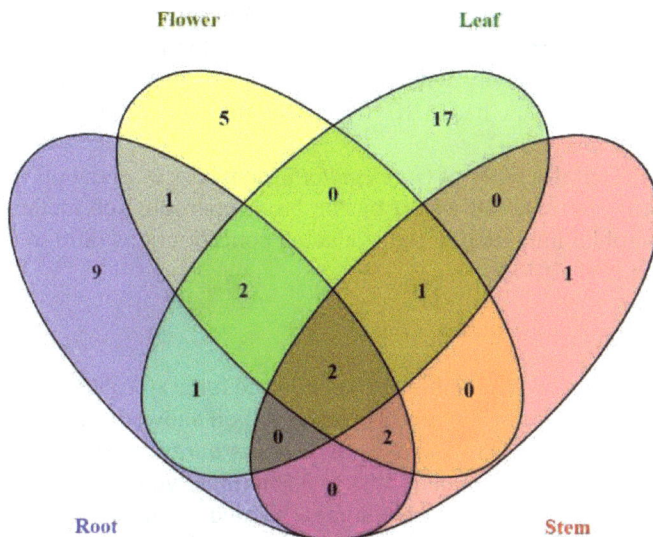

Figure 2: Venn diagram showing the number of essential oil compounds identified from different parts of *Petiveria alliacea*. Data compiled from Ayedoun et al. (1998), Zoghbi et al. (2002), Neves et al. (2011), Kerdudo et al. (2015), Oluwa et al. (2017), and Olomieja et al. (2020).

nine (22%) different compounds in the root and 17 (41.5%) different compounds in the leaves. The flowers (or inflorescences) have five (12.2%) exclusive compounds. The only compounds identified in essential oils from all plant parts were carvacrol and dibenzyl disulfide.

Phytolacca dodecandra

The chemical composition of leaf essential oil from four *P. dodecandra* samples collected in different locations in Ethiopia was evaluated by Matebie et al. (2019). These authors identified six compounds in the essential oils and three of them were the same in four samples such as phytone, pentacosane, and hexadecanoic acid but with different amounts of each compound, according to the collection location. The major compounds in sample-1 were phytol (21.6%) and hexadecanoic (23.9%), in sample-2 were phytone (21.2%) and hexadecanoic acid (20.1%), in sample-3 were phytone (17.4%) and phytol (26.3%), and in sample-4 were (+)-spathulenol (9.1%) and hexadecanoic acid (13.5%) (Matebie et al. 2019).

In vitro Antimicrobial Activity of the Essential Oils

Gallesia integrifolia

Antifungal activity of fresh fruit essential oil from *G. integrifolia* were evaluated by the broth microdilution method against fungi, such as *Aspergillus fumigatus, Aspergillus niger, Aspergillus ochraceus, Aspergillus versicolor, Penicillium funiculosum, Penicillium ochrochloron, Penicillium verrucosum* var. *cyclopium* and *Trichoderma viride*. The minimum inhibitory concentration (MIC) ranged from

10 to 90 µg/mL and the minimum fungicidal concentration (MFC) ranged from 20 to 180 µg/mL (Raimundo et al. 2018). According to the authors, the essential oil had an antimicrobial activity with lower values than positive controls bifonazole and ketoconazole, which had MIC from 100 to 2,500 µg/mL and MFC from 200 to 3,500 µg/mL.

The antibacterial activity of *G. integrifolia* stem bark essential oil was evaluated *in vitro* against *Helicobacter pylori* by the broth microdilution method, but it did not show microbial inhibition at the evaluated concentrations (800 to 0.39 µg/mL) (Arunachalam et al. 2017).

Petiveria Alliacea

The essential oil of *P. alliacea* inhibited bacteria and fungi at 0.5% (mass/volume) and 0.05% (mass/volume), respectively. The essential oil showed antimicrobial activity against *Candida albicans* (100%), *Staphylococcus aureus* (99%), *Escherichia coli* (66%), and *Salmonella arizonae* (81%) at 0.5% (mass/volume), after 24 h contact but without activity at 0.05% (mass/volume) (Kerdudo et al. 2015). Moreover, when evaluated against resistant strains of *E. coli* (J96) and *S. aureus* (1799) showed 61% and 51%, respectively, of antimicrobial activity at 1.0% (mass/volume) of essential oil (Kerdudo et al. 2015).

Phytolacca Dodecandra

The antimicrobial activity of leaf essential oil from four samples (identified as samples 1 to 4) of *P. dodecandra* from different locations in Ethiopia was evaluated by broth microdilution method against *Escherichia coli* (ATCC 25922), *Staphylococcus aureus* (ATCC 25923), *Bacillus subtilis* (ATCC 6633), and *Candida albicans* (ATCC 10231) with MIC from 64 to > 128 µg/mL (Matebie et al. 2019). These authors reported that *E. coli* and *S. aureus* had greater sensitivity with MIC of 64 µg/mL for both when *P. dodecandra* essential oil was used from sample-1.

Antimicrobial Activity of Major Compounds in the Essential Oil

The antimicrobial activity of essential oils is related to the chemical composition, functional groups of the active compounds, as well as potential synergistic effects (Dorman and Deans 2000). Most compounds of Petiveraceae and Phytolaccaceae families have essential oil with organosulfur compounds. Sulfur-containing compounds have a broad spectrum of antimicrobial activity (Silva et al. 2018), and this antimicrobial activity may be related to disulfide bonds (Avato et al. 2000). Lenthionine or 1, 2, 3, 5, 6-pentathiepane is a cyclic sulfur-containing compound also produced by *Lentinula edodes* (Chen and Ho 1986). The MIC value of lenthionine by broth microdilution method was 3.12 µg/mL against *Trichophyton rubrum* and *Trichophyton mentagrophytes*, 6.25 µg/mL against *Candida albicans*, 6.25 µg/ml against *Saccharomyces cerevisiae*, and *Cryptococcus neoformans* (Morita and Kobayashi 1967). Lenthionine also showed antimicrobial activity by broth microdilution method against *E. coli* (128 µg/mL), *B. cereus* (64 µg/mL), *Micrococcus luteus* (64 µg/mL), *Enterococcus faecalis* (128 µg/mL), *S. cerevisiae* (8 µg/ml), *C. albicans* (8 µg/mL), *A. niger* (4 µg/mL), and *Fusarium solani* (2 µg/mL) (Kupcová et al. 2018).

Dimethyl disulfide and dimethyl trisulfide are volatile organic compounds known for their odors (Urru et al. 2011). Dimethyl trisulfide was evaluated by broth microdilution method and showed activity against *Enterobacter aerogenes* (MIC of 310 µg/mL), *Escherichia coli* (MIC of 310 µg/mL), *Salmonella enterica* serovar Typhimurium (MIC of 310 µg/mL), *Shigella sonnei* (MIC of 20 µg/ml), and *Listeria monocytogenes* (MIC of 40 µg/ml) (Ross et al. 2001). Leontiev et al. (2018) reported the antimicrobial activity of dimethyl trisulfide by broth microdilution method against *Escherichia coli* (MIC and minimal bactericidal concentration (MBC) of 64 µg/mL), *Pseudomonas fluorescens* (MIC of 16 µg/mL and MBC of 32 µg/mL), *Pseudomonas syringae* pv. *phaseolicola* (MIC and MBC of 16 µg/mL), *Micrococcus luteus* (MIC and MBC of 64 µg/mL), and *S. cerevisiae* (MIC and MFC of 16 µg/mL).

The dimethyl disulfide compound had MIC > 500 µg/mL and dimethyl trisulfide MIC of 20 µg/mL and 40 µg/mL against *Candida utilis* (ATCC 42416) and *S. aureus* (B31), respectively (Kim et al. 2004). According to these authors, the increase in the number of sulfur atoms to disulfides, in dimethyl trisulfide, potentiated the antimicrobial activity against *C. utilis* with MIC of 15 µg/mL, but against *S. aureus* MIC was > 500 µg/mL (Kim et al. 2004).

Benzyl sulfide in *P. alliaceae* roots was reported to have antifungal activity by broth microdilution method with MIC of 256 µg/mL against *A. flavus*, 256 µg/mL against *Mucor racemosus*, 64 µg/mL for *Pseudallescheria boydii*, 128 µg/mL for *C. albicans*, 128 µg/mL for *Candida tropicalis*, and 128 µg/mL for *Issatchenkia orientalis* (Kim et al. 2006). The compounds dibenzyl disulfide and dibenzyl trisulfide had potential antifungal activity by TLC bioautography assay method at 0.1 µg and 1.0 µg, respectively, against *Cladosporium sphaerospermum* and 1.0 µg for both compounds against *Cladosporium cladosporioides* (Benevides et al. 2001). For the mutant strains of *S. cerevisiae*, it presented a MIC from 11 to 412 µg/mL by IC_{12} method, that is the concentration required to inhibit microbial growth in 12 mm diameter plates (Benevides et al. 2001).

Phytol is an acetyl monounsaturated alcohol (Rontani and Volkman 2003) used to produce chlorophyll metabolism in plants, which can increase the inactivation of proteins and enzymes in bacterial cells (Ghaneian et al. 2015). The antimicrobial activity of phytol by broth microdilution method showed antibacterial activity with MIC from 3 to 38 µg/mL and MBC from 13 to 52 µg/mL, and *Listeria monocytogenes* was the most sensitive one; antifungal activity with MIC values was from 8 to 16 µg/mL and MFC was from 90 to 520 µg/mL, and *T. viride* was the most sensitive one (Pejin et al. 2014). Ghaneian et al. (2015) reported that phytol showed antimicrobial activity against *E. coli*, *C. albicans*, and *A. niger* with MIC_{50} of 62.5 µg/mL and MIC_{50} greater than 1,000 µg/mL against *S. aureus*.

Phytol was evaluated by broth microdilution method against *Pseudomonas aeruginosa* and had MIC of 20 µg/mL, and the bactericidal activity assessed by the time-to-death kinetics occurred after 6 h of contact with the compound (Lee et al. 2016). Inoue et al. (2005) reported that the inhibitory action of phytol (0.15 to 160 µg/mL) by the kinetic curve of death against *S. aureus* (FDA2090P), which increased until it reached 10 µg/mL but without significant concentration increase in the inhibitory action (Inoue et al. 2005).

Citronellol is monoterpene alcohol found in plant essential oils that has several pharmacological properties (Santos et al. 2019). Antibacterial activity by broth microdilution method against broad-spectrum beta-lactamase-producing *E. coli* strains had MIC from 256 to 512 µg/mL (Lima et al. 2020). Cardoso et al. (2020) evaluated the antimicrobial activity of citronellol by broth microdilution method with MIC of 64, 256, and 256 µg/mL for *Trichophyton rubrum*, *Microsporum canis*, and *Microsporum gypseum*, respectively. On the other hand, citronellol evaluated by broth microdilution method against *Trichophyton rubrum* strains had a MIC of 8 to 1,024 µg/mL (Pereira et al. 2015). The β-citronellol isomer was evaluated against *Candida glabrata* and *C. tropicalis* and showed MIC of 400 µg/mL and 600 µg/mL and MFC of 100 µg/mL and 200 µg/mL, respectively (Sharma et al. 2020). Citronellol showed antifungal activity against *Cladosporium* spp. with MIC from 256 to 512 µg/mL and MFC from 256 to 2.048 µg/mL (Santos et al. 2017). Citronellol evaluated by broth microdilution method against *E. coli* (CECT 434) and *S. aureus* (CECT 976) had MIC of 5 and 375 µg/mL and MBC of 15 and 400 µg/mL, respectively (Lopez-Romero et al. 2015).

The antimicrobial activity of citronellol carried out by *in vitro* broth microdilution method against dermatophyte fungi, such as *Trichophyton rubrum*, *Microsporum canis*, and *Microsporum gypseum*, presented MIC of 64, 256, and 256 µg/mL, respectively (Cardoso et al. 2020). However, these authors also reported that dermatophytes develop resistance when subjected to successive sub-inhibitory concentrations of treatments with citronellol.

The antifungal activity of citronellol evaluated against *A. flavus* and *A. niger* showed MIC of 0.78 and 0.79 µg/mL, respectively (Shin 2003). Shin and Lim (2004) reported the antifungal activity of citronellol by disk diffusion method against *Trichophyton* spp. resulted in MIC from 500 to 2,000 µg/mL and MFC from 1,000 to 4,000 µg/mL. They also reported that citronellol had a synergistic effect when combined with ketoconazole with a fractional inhibitory index of 0.18 (Shin and Lim 2004). Pereira et al. (2015) demonstrated the antifungal activity of citronellol against *Trichophyton rubrum*, isolated from nails of patients with dermatophytosis, whose MIC ranged from 8 to 1,021 µg/mL. The compound reduced dry mycelium and germinated conidia, causing morphological changes, such as reduced conidiogenesis, abnormal hyphae, and conidia formation, and also increased intracellular material released, probably by inhibiting ergosterol biosynthesis (Pereira et al. 2015).

Kotan et al. (2007) reported that the antibacterial activity of β-citronellol isomer at 30,000 µg/mL by disk diffusion method showed an inhibition halo against *S. aureus* (7 mm), *Streptococcus pyogenes* (7 mm), *Klebsiella pneumoniae* (8 mm), *Pseudomonas aeruginosa* (7 mm), *Enterobacter cloacae* (7 mm), *Proteus vulgaris* (8 mm) and *Salmonella Typhimurium* (9 mm). Mulyaningsih et al. (2011) reported that the antibacterial activity of citronellol by broth microdilution method against

Gram-negative bacteria—including *E. coli* (ATCC 25922) and *P. aeruginosa* (ATCC 27853), *K. pneumoniae* (ATCC 700603), and *Acinetobacter baumannii* (ATCC BAA 747)—had MIC from 0.125 to > 8.0 µg/mL and MBC from 0.25 to > 8.0 µg/mL against Gram-positive bacteria, including *S. aureus* MRSA strain (NCTC 10442), clinical isolates, and VRE (vancomycin-resistant *Enterococcus*, ATCC 51299) that showed MIC and MBC from 0.125 to 8.0 µg/mL. It suggests that citronellol can control Gram-positive and Gram-negative bacteria.

β-bisabolene, bisabolene isomer, is a terpenoid hydrophobic compound (Rodrigues and Lindberg 2021) and it is evaluated by broth microdilution method in *S. aureus* (ATCC 25923) with MIC of 8 µg/mL (Nascimento et al. 2007). When evaluating the synergistic effect of bisabolene on the MIC value with ampicillin (MIC of 0.0125 µg/mL), it resulted in MIC of 0.0125 µg/mL and maintained the antibiotic activity (Nascimento et al. 2007).

Aldehydes are considered strong inhibitors of fungal spores, mycelia, and bacterial cell growth (Utama et al. 2002). Benzaldehyde is a natural aldehyde, a constituent of essential oils, and is synthesized by plants through the transcyclization of acetyl-CoA (Neto et al. 2021). Benzaldehyde and its derivatives are considered environmentally safe antimicrobial compounds (Alamri et al. 2012) and are used for applications in food, beverage, and industrial flavorings (Verma et al. 2017). Ullah et al. (2015) verified by broth microdilution method that benzaldehyde has antimicrobial activity against Gram-negative and Gram-positive bacteria—such as *Pantoea conspicua* (RSC-6, MIC of 10 mM or 1,060 µg/mL), *Enterobacter cowanii* (RSC-3, MIC of 8.0 mM, or 850 µg/mL), *Citrobacter youngae* (RSC-5, MIC of 10 mM, or 1,060 µg/mL), *Bacillus aryabhattai* (RSC-7, MIC of 6.0 mM, or 637 µg/ml), and *Bacillus anthracis* (RSC-9, MIC of 8.0 mM, or 850 µg/mL)—against fungi such as *Phytophthora capsici* (MIC of 8.0 mM or 850 µg/mL), *Rhizoctonia solani* (MIC of 10 mM or 1,060 µg/mL), and *Corynespora cassiicola* (MIC of 10 mM or 1,060 µg/mL).

On the other hand, Neto et al. (2021) reported that the antibacterial activity of benzaldehyde by broth microdilution method against *S. aureus* strains resulted in MIC ≥ 1,024 µg/mL without relevant activity. The synergistic effect of benzaldehyde in subinhibitory concentrations (MIC/8 = 128 µg/mL) was associated with the antibiotic norfloxacin (MIC of 256 µg/mL) against *S. aureus* strains, which reduced MIC to 128 µg/mL and with ciprofloxacin (MIC of 64 µg/mL) reduced MIC to 34 µg/mL. It was considered an antibacterial activity enhancer (Neto et al. 2021).

Benzyl alcohol is the primary aromatic alcohol recommended as a preservative and flavoring agent for foods, according to the directorate-general of the European Commission for Health and Consumer Protection (2002), and with antimicrobial activity by broth microdilution method with MIC of 25 µg/mL against *S. aureus*, 2,000 µg/mL against *E. coli* and *P. aeruginosa*, 2,500 µg/mL against *C. albicans*, and 5,000 against *A. niger*. It showed greater action against Gram-positive bacteria (Meyer et al. 2007).

Carvacrol is a monoterpene consisting of a phenolic ring with methyl and isopropyl substitutions (Wang and Wu 2021) and has been categorized as a generally recognized safe (GRAS) food additive that can be used as a flavoring in food products (Food and Drug Administration 2017). Carvacrol has been reported to show antimicrobial activity by broth microdilution method with MIC of 250 μg/mL against *E. coli*, *S. aureus*, *B. subtilis*, and *S. cerevisiae*, MIC of 1,000 μg/mL against *P. fluorescens*, and MIC > 3,000 μg/mL against *Lactobacillus plantarum* (Ben Arfa et al. 2006). Carvacrol was evaluated by broth microdilution method against *S. aureus* (ATCC 29213), *S. epidermidis* (ATCC 35894), *E. coli* (ATCC 8739), *P. aeruginosa* (ATCC 9027) with MIC from 300 to 5,000 μg/mL, and MBC from 1,250 to 10,000 μg/mL against *C. albicans* (ATCC 10231); it had MIC of 300 μg/mL and MBC of 600 μg/mL (Cacciatore et al. 2015). Carvacrol was also reported to have activity by broth microdilution method against cariogenic oral pathogens, such as *Streptococcus mutans*, *S. mitis*, *S. sanguis* and *S. salivarius* with MIC and MBC of 2,500 and 5,000 μg/mL, respectively; it was against *C. albicans* that had MIC of 1,250 and MBC of 2,500 μg/mL (Botelho et al. 2007). Carvacrol was evaluated against four strains of *Streptococcus pyogenes* (ATCC 19615 and ATCC 49399) and clinical isolates by broth microdilution method showed MIC of 125 μg/mL and MBC of 250 μg/mL for all (Wijesundara et al. 2021), demonstrating greater antimicrobial potential to control these microorganisms.

Stilbenes compounds represent a class of plant-derived secondary phenolic metabolites, characterized by a 1,2-diphenylethylene core, which plays an important role in protecting plants against pests and pathogens (Dubrovina and Kiselev 2017). Stilbenes have been reported MIC of 25 to 50 μg/mL by broth microdilution method against *Bacillus subtilis*, *Pseudomonas syringae*, *Botrytis cinerea*, *A. niger*, and *Cladosporium herbarum*, and their activity is associated to with the presence of a free hydroxyl in the molecule (Aslam et al. 2009).

Farnesol is an acyclic sesquiterpene alcohol isoprenoid, synthesized by the ergosterol pathway endogenously, characterized as a heat and pH stable molecule (Westwater et al. 2005, Rodrigues and Černáková 2020). Lopes et al. (2021) evaluated the antibacterial activity of farnesol by broth microdilution method against Gram-positive and Gram-negative bacteria and reported that it inhibited only Gram-positive bacteria, such as *S. aureus*, *S. epidermidis*, *Enterococcus faecium*, *E. faecalis*, *E. durans*, *Streptococcus agalactiae*, *S. pyogenes*, *S. mutans*, and *S. sobrinus* with MIC from 8 to > 1,024 μg/mL, of which *S. sobrinus* was the most sensitive one. The compound showed bactericidal activity with MBC ranging from 16 to > 512 μg/mL; however, it did not show relevant antibacterial activity (MIC > 1.024 μg/mL) against Gram-negative bacteria, such as *Acinetobacter baumannii*, *Stenotrophomonas maltophilia*, *Klebsiella pneumoniae*, *P. aeruginosa*, *Proteus mirabilis*, *Serratia marcescens*, *S. enterica*, *S. flexneri*, and *E. coli* (Lopes et al. 2021).

The synergistic effect of farnesol with the antibiotic gentamicin was carried out with static biofilms of methicillin-sensitive *S. aureus* (MSSA) and methicillin-resistant *S. aureus* (MRSA) exposed to several compound concentrations (Jabra-Rizk et al. 2006). The biofilm plate cell counts after zero, 4, and 24 h of the treatment showed that gentamicin (10 μg/mL) and 100 μM farnesol (22 μg/mL) combined reduced the count by more than 2 log bacterial units. The authors suggest that farnesol increased the permeability of the microorganisms to the antibiotic and it favors an adjuvant agent action for applications in preventing biofilm-related infections and contributing to potential drug resistance reversal (Jabra-Rizk et al. 2006).

Nagy et al. (2020) analyzed the effect of farnesol on *Candida auris* growth and biofilm formation at concentrations from 100 to 300 µM with a reduction of growth after 24 h. The ability to inhibit biofilm formation was concentration-dependent but without inhibition after 24 h.

Antimicrobial Mechanism of Action of Major Compounds in the Essential Oil

The mechanism of action of the antibacterial activity of essential oils has not yet been fully elucidated, but it is attributed to the ability of bioactive compounds to penetrate bacterial membranes and inhibit the cellular function properties inside of the cell. Generally, for most essential oils, Gram-positive bacteria are more susceptible than Gram-negative bacteria due to the complexity of the cell wall constitution and the presence of lipopolysaccharide, which hinder the penetration of the essential oil into the cell (Nazzaro et al. 2013).

The mechanism of action of the antifungal activity of the essential oil is related to cell wall disruption, sporulation inhibition, hyphae germination, and elongation (Nazzaro et al. 2017, Cavanagh 2007) that interfere with the mycotoxin production (Cai et al. 2021). Essential oils can impair mitochondrial function and inhibit the action of dehydrogenase enzymes involved in ATP biosynthesis, such as malate dehydrogenase, lactate dehydrogenase, and succinate dehydrogenase (Nazzaro et al. 2017).

Among the sulfur compounds, dimethyl sulfide has been reported to inhibit the expression of the invasion gene in *S. enterica*, which inhibits the expression of *hilA* as well as genes related to SPI-1 but other virulence genes, such as *ssrA* and *phoP*, are not affected (Antunes et al. 2010).

The antifungal activity and mechanism of action of sulfur compounds, including dimethyl trisulfide against *Colletotrichum gloeosporioides* to control mango postharvest, was reported by Tang et al. (2020). The conidia and mycelia of *C. gloeosporioides* were subjected to fumigation by the compound for 2, 4, 6, 8, or 10 h at 100 µL/L. The sulfur compounds caused damage to the integrity of plasma membranes, providing a reduction in the spore survival rate and alterations in the hyphae morphology. Dimethyl trisulfide also caused changes in cell wall structures, plasma membranes, Golgi complex, and mitochondria as well as protoplasm leakage with the formation of vacuoles. According to these authors, dimethyl trisulfide caused suppression in the expression of β-1, 3-D-glucan, chitin, and related genes in the biosynthesis of sterol and membrane proteins (Tang et al. 2020).

Phytol reduced 74–84% of *P. aeruginosa* biofilm formation and decreased the spasms and motility of the bacterium's flagella (Pejin et al. 2014). The presence of phytol increased the level of reactive oxygen species (ROS) in the *P. aeruginosa* cell, leading to a decrease in the glutathione defense system (GSH). In addition, this compound causes DNA damage in the bacteria (Lee et al. 2016). Inoue et al. (2005) also reported that 5 µg/mL and 20 µg/mL of phytol induced leakage of potassium ions.

The antifungal activity of citronellol may be related to its hydrophobic chemical characteristic that allows adhesion to cell membrane lipids and possible change in membrane permeability and integrity, causing microbial death (Kaur et al. 2011). Sharma et al. (2020) suggested that β-citronellol inhibits the morphological transition of *C. albicans* from yeast to hyphae and decreases the secretion of hydrolytic enzymes involved in the initial stage of infection as well as modulates the expression of associated genes. Altered expression of ERG genes also suggests that treatment with β-citronellol can lead to ergosterol depletion with cell membrane disruption (Sharma et al. 2020). The antibacterial activity of citronellol against *E. coli* (CECT 434) and *S. aureus* (CECT 976) evaluated by measuring the potential on the surface charge of cells showed that the compound caused a change for *E. coli* cells from –22.78 mV (control) to –8.13 mV and *S. aureus* cell from –27.10 mV (control) to –16.03 mV (Lopez-Romero et al. 2015). Membrane integrity results, assessed by propidium iodide uptake, suggested the ability of a compound to penetrate the cytoplasmic membrane (Lopez-Romero et al. 2015) with subsequent potassium ion leakage only for *S. aureus* (0.63 μg/mL of potassium ions).

The mechanism of action for antimicrobial activity of sesquiterpenes, including farnesol, occurs through the induction of ionic imbalance in the bacterial cell membrane, which induces potassium ion leakage and is directly proportional to the concentration of sesquiterpene in contact with the bacteria (Inoue et al. 2004). Kaneko et al. (2011) also reported that farnesol can inhibit mevalonate production in *S. aureus*. Farnesol, together with the antibiotics ampicillin and oxacillin, showed potentiation of the effect against *S. aureus* and the mechanism may be associated with the rupture of the cytoplasmic membrane by farnesol, promoting the leakage of potassium ions (Kim et al. 2018). Nagy et al. (2020) evaluated the action of farnesol in combination with echinocandin *in vitro* for *C. albicans* biofilm formation inhibition. According to these authors, the compound prevents *C. albicans* biofilm formation by inhibiting the Ras1-cAMP-PKA cascade and the synergism between the two substances used in the study caused simultaneous damage to cell membrane.

Trans-stilbene and *cis*-stilbene at 10 μg/mL inhibited human blood hemolysis by *S. aureus* (ATCC 6538). *Trans*-stilbene showed a dose-dependent reduction in hemolysis causing more than 90% inhibition. This compound also inhibited *S. aureus* biofilm formation and is dose-dependent at concentrations up to 200 μg/mL without affecting bacterial growth but at 1000 μg/mL, there was a reduction in the bacterial population (Lee et al. 2014). The authors suggest that the reductions in biofilm formation were due to the antibiotic activity of the compound in the biofilm instead of the antimicrobial activity (Lee et al. 2014). According to these authors, *trans*-stilbene also reduced the virulence *in vivo* of *S. aureus* when infected in *Caenorhabditis elegans*. Furthermore, the transcriptional analysis showed that *trans*-stilbene repressed the *hla* α-hemolysin gene and the intercellular adhesion locus (icaA and icaD), which is in agreement with the reported virulence and biofilm formation reductions (Lee et al. 2014).

The mechanism of action of the carvacrol antimicrobial activity has been related to cell membrane disruption, ATPase activity inhibition, membrane destabilization, cell ion leakage, membrane lipid fluidization, and reduced proton driving force (Di Pasqua et al. 2007, Ultee et al. 2007, Ultee et al. al. 2002). According to Wijesundara et al (2021), the antibacterial action mechanism of carvacrol against *S. pyogenes* caused changes in the cell membrane integrity, allowing leakage of cytoplasmic content, such as nucleic acids, and lactate dehydrogenase enzymes; they also suggest that the molecule is able to form membrane pores.

Benzaldehyde and derivatives have a broad spectrum of antimicrobial action and interact with the cell surface, inducing cell death, promoting cell membrane disintegration, and intracellular constituent release (Alamri et al. 2012, Park et al. 2001). Neto et al. (2021) reported that benzaldehyde had no potential to inhibit the efflux pump of *S. aureus*, but increased antimicrobial activity when in synergism with antibiotics (ciprofloxacin and norfloxacin) due to other resistance mechanism effects, such as the modification of antibiotic target or alteration of permeability to the compounds (Neto et al. 2021).

Conclusion

This chapter provided an overview of the chemical composition and antimicrobial activity of major compounds in the essential oil of *G. integrifolia*, *P. alliacea*, and *P. dodecandra* from the Phytolaccaceae and Petiveriaceae families. Few studies have reported the extraction and antimicrobial activity of essential oils from plants of these families, showing a vast unexplored research field in this area. The major compounds of the essential oil of the studied species showed antimicrobial activity *in vitro* against several bacteria and fungi but *in vivo* and toxicological studies are still needed to substantiate the potential action. Thus, essential oils and their respective major compounds in this chapter are promising for applications in the chemical, pharmaceutical, agricultural, and food industries.

References

Adams, R.P., K.R. Neisess, R.M. Parkhurst, L.P. Makhubu and L.W. Yohannes. 1989. *Phytolacca dodecandra* (Phytolaccaceae) in Africa: Geographical variation in morphology. Taxon. 38: 17–26.

Adelakun, O.E., O.J. Oyelade and B.F. Olanipekun. 2016. Use of essential oils in food preservation. pp. 71–84. *In*: V.R. Preedy (ed.). Essential Oils in Food Preservation, Flavor, and Safety. Academic Press is an imprint of Elsevier, 7.

Akisue, M.K., G. Akisue and F. De Oliveira. 1986. Caracterização farmacognóstica de Pau d'alho *Gallesia integrifolia* (Spreng.) Harms. Rev. Bras. Farmacogn. 1: 166–182.

Akitan, M.O. and J.O. Akinneye. 2020. Fumigant toxicity and phytochemical analysis of *Petiveria alliacea* (Linneaus) leaf and root bark oil on adult Culex quinquefasciatus. Bull. Natl. Res. Cent. 44: 1–29.

Alamri, A., M.H. El-Newehy and S.S. Al-Deyab. 2012. Biocidal polymers: Synthesis and antimicrobial properties of benzaldehyde derivatives immobilized onto amine-terminated polyacrylonitrile. Chem. Cent. J. 6: 1–13.

Almanza, L.V. 2012. Flora del valle de Tehuacán-Cuicatlán 105. Universidad Nacional Autónoma de México, Fascículo, 1–17.

Andrade, T.M., A.S. Melo, R.G.C. Dias, E.L.P. Varela, F.B. Oliveira, J.LF. Vieira et al. 2012. Potential behavioral and pro-oxidant effects of *Petiveria alliacea* L. extract in adult rats. J. Ethnopharmacol. 143: 604–610.

Antunes, L.C.M., M.M. Buckner, S.D. Auweter, R.B. Ferreira, P. Lolić and B.B. Finlay. 2010. Inhibition of *Salmonella* host cell invasion by dimethyl sulfide. Appl. Environ. Microbiol. 76: 5300–5304.

APG - Angiosperm Phylogeny Group. 2016. An update of the Angiosperm Phylogeny Group classification for the orders and families of flowering plants: APG IV. Bot. J. Linn. 181: 1–20.

Arunachalam, K., S.D. Ascencio, I.M. Soares, R.W.S. Aguiar, L.I. Silva, R.G. Oliveira et al. 2016. *Gallesia integrifolia* (Spreng.) Harms: *in vitro* and *in vivo* antibacterial activities and mode of action. J. Ethnopharmacol. 26: 128–137.

Arunachalam, K., S.O. Balogun, E.A. Pavan, G.V.B. Oliveira, R.G.T. Wagner, V.C. Filho et al. 2017. Chemical characterization, toxicology and mechanism of gastric antiulcer action of essential oil from *Gallesia integrifolia* (Spreng.) Harms in the *in vitro* and *in vivo* experimental models. Biomed. Pharmacother. 94: 292–306.

Aslam, S.N., P.C. Stevenson, T. Kokubun and D.R. Hall. 2009. Antibacterial and antifungal activity of cicerfuran and related 2-arylbenzofurans and stilbenes. Microbiol. Res. 164: 191–195.

Avato, P., F. Tursi, C. Vitali, V. Miccolis and V. Candido. 2000. Allylsulfide constituents of garlic volatile oil as antimicrobial agents. Phytomedicine 7: 239–243.

Ayedoun, M.A., M. Moudachirou, P.V. Sossou, F.X. Garneau, H. Gagnon and F.I. Jean. 1998. Volatile constituents of the root oil of *Petiveria alliacea* L. from Benin. J. Essent. Oil Res. 10: 645–646.

Balbach, A. 1993. As Plantas Curam [The Plants That Heal]. Missionary Publishing house, São Paulo, Brazil.

Barbosa, L.C.A., A.J. Demuner, R. Teixeira and M.S. Madruga. 1999. Chemical constituents of the bark of *Gallesia gorazema*. Fitoterapia 70: 152–156.

Ben Arfa, A., S. Combes, L. Preziosi-Belloy, N. Gontard and P. Chalier. 2006. Antimicrobial activity of carvacrol related to its chemical structure. Lett. Appl. Microbiol. 43: 149–154.

Benevides, P.J.C., M.C.M. Young, A.M. Giesbrecht, N.F. Roque and S.V. Bolzani. 2001. Antifungal polysulphides from *Petiveria alliacea* L. Phytochemistry 57: 743–747.

Beressa, T.B., O. Clemencio, E.L. Peter, H. Okella, P.E. Ogwang, W. Anke et al. 2020. Pharmacology, phytochemistry, and toxicity profiles of *Phytolacca dodecandra* L'Hér: A scoping review. Infect. Dis. Res. Treat. 13: 1–7.

Bieski, I.G.C., F.R. Santos, R.M. de Oliveira, M.M. Espinosa, M. Macedo, U.P. Albuquerque et al. 2012. Ethnopharmacology of medicinal plants of the pantanal region (Mato Grosso, Brazil) Evid.-Based Complement. Altern. Med. 2012: 272749.

Bortolucci, W.C., K.F. Raimundo, C.M.M. Fernandez, R.C. Calhelha, I.C.F.R. Ferreira, L. Barros et al. 2021. Cytotoxicity and anti-inflammatory activities of *Gallesia integrifolia* (Phytolaccaceae) fruit essential oil. Nat. Prod. Res. 18: 1–7.

Botelho, M.A., N.A.P. Nogueira, G.M. Bastos, S.G.C. Fonseca, T.L.G. Lemos, F.J.A. Matos et al. 2007. Antimicrobial activity of the essential oil from *Lippia sidoides*, carvacrol and thymol against oral pathogens. Braz. J. Med. Biol. Res. 40: 349–356.

Cacciatore, I., M. Di Giulio, E. Fornasari, A. Di Stefano, L.S Cerasa, L. Marinelli et al. 2015. Carvacrol codrugs: A new approach in the antimicrobial plan. PLoS One 10: 1–20.

Cai, J., R. Yan, J. Shi, J. Chen, M. Long, W. Wu et al. 2021. Antifungal and mycotoxin detoxification ability of essential oils: A review. Phytother. Res. 1–11.

Camargo, M.T.L.A. 2007. Contribuição etnofarmacobotânica ao estudo de *Petiveria alliacea* L. – Phytolacaceae – ("amansa-senhor") e a atividade hipoglicemiante relacionada a transtornos mentais. Dominguezia 23: 21–28.

Cardoso, G.N., K.V.S. Silva, M.I.O. Lima, J.M.M. Arrua and F.O. Pereira. 2020. Dermatophytes develop resistance to the monoterpenes geraniol and citronellol. Rev. Cuba. de Farm. 53: 1–12.

Castellar, A., R.F. Gagliardi, E. Mansur, H.R. Bizzo, A.M. Souza and S.G. Leitão. 2014. Volatile constituents from *in vitro* and *ex vitro* plants of *Petiveria alliacea* L. J. Essent. Oil Res. 26: 19–23.

Cavanagh, H.M. 2007. Antifungal activity of the volatile phase of essential oils: A brief review. Nat. Prod. Commun. 2: 1297–1302.

Chen, C.C. and C.T. Ho.1986. Identification of sulfurous compounds of shiitake mushroom (*Lentinus edodes* Sing.). J. Agric. Food Chem. 34: 830–833.

Chen, P.Y., Y.C. Chuang, U.I. Wu, H.Y. Sun, J.T. Wang, W.H. Sheng et al. 2021. Mechanisms of Azole resistance and trailing in *Candida tropicalis* bloodstream isolates. J. Fungi 7: 612.

Cheng, D., H.H. Ngo, W. Guo, S.W. Chang, D.D. Nguyen, Y. Liu et al. 2019. A critical review on antibiotics and hormones in swine wastewater: Water pollution problems and control approaches. J. Hazard. Mater. 387: 1–12.

Chouhan, S., K. Sharma and S. Guleria. 2017. Antimicrobial activity of some essential oils-present status and future perspectives. Medicines 4: 1–21.

Christenhusz, M.J.M. and J.W. Byng. 2016. The number of known plants species in the world and its annual increase. Phytotaxa. 261: 201–217.

Costa, D.C., S.H. Costa, T.A. Goncalves, F. Ramos, M.C. Castilho and A.S. Silva. 2015. Advances in phenolic compounds analysis of aromatic plants and their potential applications. Trends Food Sci. Technol. 44: 336–354.

Dhifi, W., S. Bellili, S. Jazi, N. Bahloul and W. Mnif. 2016. Essential Oils' chemical characterization and investigation of some biological activities: A critical review. Medicines 3: 1–16.

Di Pasqua, R., G. Betts, N. Hoskins, M. Edwards, D. Ercolini and G. Mauriello. 2007. Membrane toxicity of antimicrobial compounds from essential oils. J. Agric. Food Chem. 55: 4863–4870.

Dorman, H.D. and S.G. Deans. 2000. Antimicrobial agents from plants: antibacterial activity of plant volatile oils. J. Appl. Microbiol. 88: 308–316.

Duarte, M.R. and J.F. Lopes. 2005. Leaf and stem morphoanatomy of *Petiveria alliacea*. Fitoterapia 76: 599–607.

Dubrovina, A.S. and K.V. Kiselev. 2017. Regulation of stilbene biosynthesis in plants. Planta 246: 597–623.

El-Tarabily, K.A., M.T. El-Saadony, M. Alagawany, M. Arif, G.E. Batiha, A.F. Khafaga et al. 2021. Using essential oils to overcome bacterial biofilm formation and their antimicrobial resistance. Saudi J. Biol. Sci. 28: 5145–5156.

Filho, A.C.P.M., J.G.O. Filho and C.F.S. Castro. 2020. Avaliações antioxidante e antifúngica dos óleos essenciais de Hymenaea stigonocarpa Mart. ex Hayne e Hymenaea courbaril L. J. Biotechnol. Biodivers. 8: 1–11.

Food and Drug Administration - FDA. 2017. CFR-Code of federal regulations title 21. 3: 25–26.

Forzza, R.C., J.F.A. Baumgratz, C.E.M. Bicudo, D.A. Canhos, A.A. Carvalho Jr, M.A.N. Coelho et al. 2012. New Brazilian floristic list highlights conservation challenges. Bioscience 62: 39–45.

European commission for health and consumer protection. SCF. Opinion of the Scientific Committee on Food on benzyl alcohol (expressed on 26 September 2002). Scientific Committee on Food. SCF/CS/ ADD/FLAV/78 Final. 17 Sept, 2002, 2002.

Guarim Neto, G. and R.G. de Morais. 2003. Recursos medicinais de espécies do Cerrado de Mato Grosso: Um estudo bibliográfico. Acta Bot. Brasilica 17: 561–584.

Gomes, P.B., E.C. Noronha, C.T.V. Melo, J.N.S. Bezerra, M.A. Neto, C.S. Lino et al. 2008. Central effects of isolated fraction from the root of *Petiveria alliacea* L. (tipi) in mice. J. Ethnopharmacol. 120: 209–214.

Gomes, P.B., M.M.S. Oliveira, C.R.A. Nogueira, E.C. Noronha, L.M.V. Carneiro, J.N.S. Bezerra et al. 2005. Study of antinociceptive effect of isolated fractions from *Petiveria alliacea* L. (tipi) in mice Biol. Pharm. Bull. 28: 42–46.

Ghaneian, M.T., M.H. Ehrampoush, A. Jebali, S. Hekmatimoghaddam and M. Mahmoudi. 2015. Antimicrobial activity, toxicity and stability of phytol as a novel surface disinfectant. Environ. Eng. Manag. J. 2: 13–16.

Grandtner, M.M. and J. Chevrette. 2013. Dictionary of Trees, Volume 2_ South America. Elsevier.

Inoue, Y., A. Shiraishi, T. Hada, K. Hirose, H. Hamashima and J. Shimada. 2004. The antibacterial effects of terpene alcohols on *Staphylococcus aureus* and their mode of action. FEMS Microbiol. Lett. 237: 325–331.

Inoue, Y., H. Toshiko, A. Shiraishi, K. Hirose, H. Hamashima and S. Kobayashi. 2005. Biphasic effects of geranylgeraniol, teprenone, and phytol on the growth of *Staphylococcus aureus*. Antimicrob. Agents Chemother. 49: 1770–1774.

Jabra-Rizk, M.A., T.F. Meiller, C.E. James and M.E. Shirtliff. 2006. Effect of farnesol on *Staphylococcus aureus* biofilm formation and antimicrobial susceptibility. Antimicrob. Agents Chemother. 50: 1463–1469.

Kaneko, M., N. Togashi, H. Hamashima, M. Hirohara and Y. Inoue. 2011. Effect of farnesol on mevalonate pathway of *Staphylococcus aureus*. J. Antibiot. 64: 547–549.

Kaur, S., S. Rana, H.P. Singh, D.R. Batish and R.K. Kohli. 2011. Citronellol disrupts membrane integrity by inducing free radical generation. Z. Naturforsch. C. J. Biosci. 66: 260–266.

Kerdudo, A., V. Gonnot, E.N. Ellong, L. Boyer, T. Michel, S. Adenet et al. 2015. Essential oil composition and biological activities of *Petiveria alliacea* L. from Martinique. J. Essent. Oil Res. 27: 186–196.

Kim, C., D. Hesek, M. Lee and S. Mobashery. 2018. Potentiation of the activity of b-lactam antibiotics by farnesol and its derivatives. Bioorganic Med. Chem. Lett. 28: 642–645.

Kim, J.W., J.E. Huh, S.H. Kyung and K.H. Kyung. 2004. Antimicrobial activity of alk (en) yl sulfides found in essential oils of garlic and onion. Food Sci. Biotechnol. 13: 235–239.

Kim, S., R. Kubec and R.A. Musah. 2006. Antibacterial and antifungal activity of sulfur-containing compounds from *Petiveria alliacea* L. J. Ethnopharmacol. 104: 188–192.

Kotan, R., S. Kordali and A. Cakirk. 2007. Screening of antibacterial activities of twenty-one oxygenated monoterpenes. Z. Naturforsch. C Biosci. 62: 507–513.

Kupcová, K., I. Stefanova, Z. Plavcova, J. Hosek, P. Hrouzek and R. Kubec. 2018. Antimicrobial, cytotoxic, anti-inflammatory, and antioxidant activity of culinary processed shiitake medicinal mushroom (*Lentinus edodes*, agaricomycetes) and its major sulfur sensory-active compound–lenthionine. Int. J. Med. Mushrooms 20: 165–175.

Lee, K., J.H. Lee, S.Y. Ryu, M.H. Cho and J. Lee. 2014. Stilbenes reduce *Staphylococcus aureus* hemolysis, biofilm formation, and virulence. Foodborne Path. Dis. 11: 710–717.

Lee, W., E.R. Woo and D.G. Lee. 2016. Phytol has antibacterial property by inducing oxidative stress response in *Pseudomonas aeruginosa*. Free Radic. Res. 50: 1309–1318.

Legère, K. 2009. Plant names in the Tanzanian Bantu language Vidunda: Structure and (some) etymology. pp. 217–228. *In*: M. Matondo, F. McLaughlin and E. Potsdam (eds.). Selected Proceedings of the 38th Annual Conference on African Linguistics: Linguistic Theory and African Language Documentation. Cascadilla Proceedings Project, Somerville, MA.

Lemma, A. 1970. Laboratory and field evaluation of the molluscicidal properties of *Phytolacca dodecandra*. Bull. World Health Organ. 42: 597–612.

Leontiev, R., N. Hohaus, C. Jacob, M.C. Gruhlke and A.J. Slusarenko. 2018. A comparison of the antibacterial and antifungal activities of thiosulfinate analogues of allicin. Sci. Rep. 8: 1–19.

Lima, D.S., M.A.N. Pontes, F.P. Andrade Junior, B.H.C. Santos, W.A. Oliveira and I.O. Lima. 2020. Atividade antibacteriana de citronelal e citronelol contra cepas de *Escherichia coli* produtoras de ESBL. Arch. Health Invest. 9: 238–241.

Lima, I.L., E.L. Longui, I.M. Andrade, J.N. Garcia, A.C.S. Zanatto, E. Morais et al. 2010. Efeito da procedência em algumas propriedades da madeira de *Gallesia integrifolia* (Spreng.) Harms. Revista do Instituto Florestal. 22: 61–69.

Lima, T.C.M., G.S. Morato and R.N. Takahashi. 1991. Evaluation of antinociceptive effect of *Petiveria alliacea* (Guine) in animals. Mem. Inst. Oswaldo Cruz 86: 153–158.

Lopes, A.P., R.R.D.O.C. Branco, F.A.A. Oliveira, M.A.S. Campos, B.C. Sousa, I.R.C. Agostinho et al. 2021. Antimicrobial, modulatory, and antibiofilm activity of tt-farnesol on bacterial and fungal strains of importance to human health. Bioorg. Med. Chem. Lett. 47: 1–8.

Lopez-Romero, J.C., H. González-Ríos, A. Borges and M. Simões. 2015. Antibacterial effects and mode of action of selected essential oils components against *Escherichia coli* and *Staphylococcus aureus*. Evid. Based Complement. Altern. Med. 1–9.

Lorenzi, H. 2002. Árvores brasileiras: Manual de identificação e cultivo de plantas arbóreas nativas do Brasil. 4th ed São Paulo: Instituto Plantarum de Estudos da Flora Ltda.

Luz, D.A., A.M. Pinheiro, M.L. Silva, M.C. Monteiro, R.D. Prediger, C.S.F. Maia et al. 2016. Ethnobotany, phytochemistry and neuropharmacological effects of *Petiveria alliacea* L. (Phytolaccaceae): A review. J. Ethnopharmacol. 185: 182–201.

Matebie, W.A., W. Zhang and G. Xie. 2019. Chemical composition and antimicrobial activity of essential oil from *Phytolacca dodecandra* collected in ethiopia. Molecules 24: 342–350.

McCullough, F., P. Gayral, J. Duncan and J. Christie. 1980. Molluscicides in schistosomiasis control. Bull. World Health Organ. 58: 681–689.

Meharie, B.G. and T.A. Tunta. 2021. *Phytolacca dodecandra* (Phytolaccaceae) root extract exhibits antioxidant and hepatoprotective activities in mice with CCl4-induced acute liver damage. Clin. Exp. Gastroenterol. 14: 59–70.

Meyer, B.K., A. Ni, B. Hu and L. Shi. 2007. Antimicrobial preservative use in parenteral products: Past and present. J. Pharm. Sci. 96: 3155–3167.

Morais, L.A.S. 2009. Influência dos fatores abióticos na composição química dos óleos essenciais. Hortic. Bras. 27: 4050–4063.

Morita, K. and S. Kobayashi. 1967. Isolation, structure, and synthesis of lenthionine and its analogs. Chem. Pharm. Bull. 15: 988–993.

Mulyaningsih, S., F. Sporer, J. Reichling and M. Wink. 2011. Antibacterial activity of essential oils from *Eucalyptus* and of selected components against multidrug-resistant bacterial pathogens. Pharmaceutical Biology 49: 893–899.

Muñoz, V., M. Sauvain, G. Bourdy, S. Arrázola, J. Callapa, G. Ruiz et al. 2000. A search for natural bioactive compounds in Bolivia through a multidisciplinary approach. Part III. Evaluation of the antimalarial activity of plants used by Altenos Indians. J. Ethnopharmacol. 71: 123–131.

Nascimento, A.M.A., M.G.L. Brandão, G.B. Oliveira, I.C.P. Fortes and E. Chartone-Souza. 2007. Synergistic bactericidal activity of *Eremanthus erythropappus* oil or β-bisabolene with ampicillin against *Staphylococcus aureus*. Antonie Van Leeuwenhoek 92: 95–100.

Nagy, F., Z. Toth, L. Daroczi, A. Szekely, A.M. Borman, L. Majoros et al. 2020. Farnesol increases the activity of echinocandins against *Candida auris* biofilms. Med. Mycol. J. 58: 404–407.

Nazzaro, F., F. Fratianni, L.D. Martino, R. Coppola and D.D. Feo. 2013. Effect of essential oils on pathogenic bacteria. Pharmaceuticals 6: 1451–1474.

Nazzaro, F., F. Fratianni, R. Coppola and V.D. Feo. 2017. Essential oils and antifungal activity. Pharmaceuticals 10: 1–20.

Neto, L.J.L., A.G.B. Ramos, T.S. Freitas, C.R.S. Barbosa, D.L.S. Júnior and A. Siyadatpanah. 2021. Evaluation of Benzaldehyde as an antibiotic modulator and its toxic effect against drosophila melanogaster. Molecules 26: 1–14.

Neves, I.I., C.A.G. Camara, J.C.S. Oliveira and A.V. Almeida. 2011. Acaricidal activity and essential oil composition of *Petiveria alliacea* L. from Pernambuco (Northeast Brazil). J. Essent. Oil Res. 23: 23–26.

Olomieja, A.O., I.O. Olanrewajul, J.I. Ayo-Ajayi, E.G. Jolayemi, U.O. Daniel and R.C. Mordi. Antimicrobial and antioxidant properties of *Petiveria alliacea*. 2021 4th International Conference on Science and Sustainable Development. Conf. Ser. Earth Environ. Sci. 655: 1–10.

Olomieja, A.O., G.E. Jolayemi, F.E. Owolabi and R.C. Mordi. 2020. Chemical components and antimicrobial activity of essential oils of *Petiveria alliacea* leaves etracts. Arch. Ecotoxicol. 2: 47–50.

Oluwa, A.A., O.N. Avoseh, O. Omikored, I.A. Ogunwander and O.A. Lawal. 2017. Study on the chemical constituents and anti-inflammatory activity of essential oil of *Petiveria alliacea* L. Br. J. Pharm. Res. 15: 1–8.

Park, E.S., W.S. Moon, M.J. Song, M.N. Kim, K.H. Chung and J.S. Yoon. 2001. Antimicrobial activity of phenol and benzoic acid derivatives. Int. Biodeter. Biodegradation 47: 209–214.

Pavitra, P.S., A. Mehta and S. Verma. 2019. Essential oils: From prevention to treatment of skin cancer. Drug Discov. Today 24: 644–655.

Pejin, B., A. Savic, M. Sokovic, J. Glamoclija, A. Ciric, M. Nikolic et al. 2014. Further *in vitro* evaluation of antiradical and antimicrobial activities of phytol. Nat. Prod. Res. 28: 372–376.

Pereira, F.D.O., J.M. Mendes, I.O. Lima, K.S.D.L. Mota, W.A.D. Oliveira and E.D.O Lima. 2015. Antifungal activity of geraniol and citronellol, two monoterpenes alcohols, against *Trichophyton rubrum* involves inhibition of ergosterol biosynthesis. Pharm. Biol. 53: 228–234.

Pérez-Cantero, A., L. López-Fernández, J. Guarro and J. Capilla. 2020. Azole resistance mechanisms in *Aspergillus*: update and recent advances. Int. J. Antimicrob. Agents 55: 1–41.

Powo. Plants of the World Online. 2021. Facilitated by the Royal Botanic Gardens, Kew. http://www.plantsoftheworldonline.org/ Accessed 15 dec. 2021.

Raimundo, K.F., W.C. Bortollucci, J. Glamoclija, M. Sokovic, J.E. Gonçalves, G.A. Linde et al. 2018. Antifungal activity of *Gallesia integrifolia* fruit essential oil. Braz. J. Microbiol. 49: 229–235.

Raimundo, K.F., W.C. Bortollucci, I.L. Rahal, H.L. Marko de Oliveira, R. Piau Júnior, A.C. Araújo et al. 2021. Insecticidal activity of *Gallesia integrifolia* (Phytolaccaceae) essential oil. Bol. Latinoam. Caribe Plantas Med. Aromat. 20: 38–50.

Ribeiro-Santos, R., M. Andrade, A. Sanches-Silva and N.R. Melo. 2018. Essential oils for food application: Natural substances with established biological activities. Food Bioproc. Tech. 11: 43–71.

Rocha, L.D., L.T. Maranho and K.H. Preussler. 2006. Organização estrutural do caule e lâmina foliar de *Petiveria alliacea* L., Phytolaccacea. Rev. Bras. Farm. 3: 98–101.

Rodrigues, C.F. and L. Černáková. 2020. Farnesol and tyrosol: Secondary metabolites with a crucial quorum-sensing role in *Candida* biofilm development. Genes 11: 1–15.

Rodrigues, J.S. and P. Lindberg. 2021. Metabolic engineering of Synechocystis sp. PCC 6803 for improved bisabolene production. Metab. Eng. Commun. 12: e00159.

Rontani, J.F. and J.K. Volkman. 2003. Phytol degradation products as biogeochemical tracers in aquatic environments. Org. Geochem. 34: 1–35.

Ross, Z.M., E.A. O'Gara, D.J. Hill, H.V. Sleightholme and D.J. Maslin. 2001. Antimicrobial properties of garlic oil against human enteric bacteria: Evaluation of methodologies and comparisons with garlic oil sulfides and garlic powder. Appl. Environ. Microbiol. 67: 475–480.

Rzedowski, J. and G.C. de Rzedowski. 2000. Flora del Baíjo y de Regiones Adyacentes-Phytolaccacea. Instituto de Ecologia, Fascículo 91, Pátzcuaro, México.

Santos, A.S., G.S. Silva, K. Silva, V.S.O. Lima, M.I. Arrua, J.M.M.O. Lima et al. 2017. Antifungal activity of geraniol and citronellol against food-relevant dematiaceous fungi *Cladosporium* spp. Revista Do Instituto Adolfo Lutz 76: 1–8.

Santos, P.L., J.P.S. Matos, L. Picot, J.R. Almeida, J.S. Quintans and L.J. Quintans-Júnior. 2019. Citronellol, a monoterpene alcohol with promising pharmacological activities—A systematic review. Food Chem. Toxicol. 123: 459–469.

Savolainen, V., M.W. Chase, S.B. Hoot, C.M. Morton, D.E. Soltis, C. Bayer et al. 2000. Phylogenetics of flowering plants based on combined analysis of plastid atpB and rbcL gene sequences. Syst. Biol. 49: 306–362.

Schäferhoff, B., K.F. Müller and T. Borsch. 2009. Caryophyllales phylogenetics: Disentangling Phytolaccaceae and Molluginaceae and description of Microteaceae as a new isolated family. Willdenowia 39: 209–228.

Schmelzer, G and A. Gurib-Fakim. 2008. Plant Resources of Tropical Africa: Medicinal Plants 1. PROTA Foundation, Wageningen.

Seol, G.H. and Y.K. Shin. 2018. Essential Oils and factors related to cardiovascular diseases. Potential of Essential Oils 7: 129–144.

Sharma, Y., S.K. Rastogi, S. Ahmedi and N. Manzoor. 2020. Antifungal activity of β-citronellol against two non-albicans *Candida* species. J. Essent. Oil Res. 32: 198–208.

Shin, S. 2003. Anti-Aspergillus activities of plant essential oils and their combination effects with ketoconazole or amphotericin B. Arch. Pharm. Res. 26: 389–393.

Shin, S. and S. Lim. 2004. Antifungal effects of herbal essential oils alone and in combination with ketoconazole against *Trichophyton* spp. J. Appl. Microbiol. 97: 1289–1296.

Silva, J.P.B., S.C.M. Nascimento, D.H. Okabe, A.C.G. Pinto, F.R. Oliveira, T.P. Paixão et al. 2018. Antimicrobial and anticancer potential of *Petiveria alliacea* L. (Herb to "Tame the Master"): A review. Pharmacogn. Rev. 12: 85–93.

Steinmann, V.W. 2010. Neotropical phytolaccaceae. *In*: W. Milliken, B. Klitgård and A. Baracat. Neotropikey - Interactive Key and Information Resources for Flowering Plants of the Neotropics. Disponível em <http://www.kew.org/science/tropamerica/neotropikey/families/Phytolaccaceae.htm>. Accessed 15 dec. 2021.

Stevens, P.F. 2001 [onwards]. Angiosperm Phylogeny Website. Version 12, July 2012 [and more or less continuously updated since]. Available at <http://www.mobot.org/MOBOT/research/APweb/>. Accessed 15 dec. 2021.

Tang, L., J. Mo, T. Guo, S. Huang, Q. Li, P. Ning et al. 2020. *In vitro* antifungal activity of dimethyl trisulfide against *Colletotrichum gloeosporioides* from mango. World J. Microbiol. Biotechnol. 36: 1–15.

Tropicos.org. Missouri Botanical Garden. *Gallesia integrifolia* (Spreng.) Harms – synonyms. Available: ⟨http://legacy.tropicos.org/Name/24800087?tab=synonyms/ Accessed 23 dec. 2021.

Tropicos.org. Missouri Botanical Garden. *Petiveria alliacea* – synonyms. Available: ⟨http://www.tropicos.org/Name/24800061?tab=synonyms/ Accessed 19 dec. 2021.

Tropicos.org. Missouri Botanical Garden. *Phytolacca dodecandra* (L'Herit) – synonyms. Available: ⟨http://legacy.tropicos.org/Name/24800056?tab=synonyms/ Accessed 23 dec. 2021.

Ullah, I., A.L. Khan, L. Ali, A.R. Khan, M. Waqas, J. Hussain et al. 2015. Benzaldehyde as an insecticidal, antimicrobial, and antioxidant compound produced by *Photorhabdus temperata* M1021. J. Microbiol. 53: 127–133.

Ultee, A., M.H.J. Bennik and R.J.A.E.M. Moezelaar. 2002. The phenolic hydroxyl group of carvacrol is essential for action against the food-borne pathogen *Bacillus cereus*. App. Environm. Microbiol. 68: 1561–1568.

Urru, I., M.C. Stensmyr and B.S. Hansson. 2011. Pollination by brood-site deception. Phytochemistry 72: 1655–1666.

Utama, I.M.S., R.B.H. Wills, S. Ben-Yehoshua and C. Kuek. 2002. *In vitro* efficacy of plant volatiles for inhibiting the growth of fruit and vegetable decay microorganisms. J. Agric. Food Chem. 50: 6371–6377.

Verma, R.S., R.C. Padalia, V.R. Singh, P. Goswami, A. Chauhan and B. Bhukya. 2017. Natural benzaldehyde from *Prunus persica* (L.) Batsch. Int. J. Food Prop. 20: 1259–1263.

Xu, Z. and M. Deng. 2017. Phytolaccaceae. *In*: Identification and Control of Common Weeds: Volume 2. Springer, Dordrecht. https://doi.org/10.1007/978-94-024-1157-7_26.

Wang, P. and Y. Wu. 2021. A review on colloidal delivery vehicles using carvacrol as a model bioactive compound. Food Hydrocoll. 120: 1–15.

Webbe, G. and J. Lambert. 1983. Schistosomiasis: Plants that kill snails and prospects for disease control. Nature 302: 754.

Westwater, C., E. Balish and D.A. Schofield. 2005. Candida albicans-conditioned medium protects yeast cells from oxidative stress: A possible link between quorum sensing and oxidative stress resistance. Eukaryotic Cell 4: 1654–1661.

Wijesundara, N.M., S.F. Lee, Z. Cheng, R. Davidson and H.V. Upasinghe. 2021. Carvacrol exhibits rapid bactericidal activity against *Streptococcus pyogenes* through cell membrane damage. Sci. Rep. 11: 1–14.

Zoghbi, M.D.G.B., E.H.A. Andrade and J.G.S. Maia. 2002. Volatile constituents from *Adenocalymma alliaceum* Miers and *Petiveria alliacea* L., two medicinal herbs of the Amazon. Flavour Fragr. J. 17: 133–135.

9

The Immunomodulatory and Antimicrobial Effect of Essential Oil in Animals

Rao Zahid Abbas,[1]* Faisal Siddique[2] and Sara Omer Swar[3]

Introduction

The medicinal plant is gaining popularity, and in recent decades it has been used more in the treatment of diseases and the promotion of health and production of animals as well as human beings. Although there are numerous studies showing the efficacy of herbal products, there is still a lack of evidence in many cases. Humans and animals have been using essential oils for centuries. Essential oils have been documented separately from plants for over 5,000 years. Aromatic essential oils were also utilized to treat a variety of diseases throughout ancient times in Greece, the Arab world, and Central Asia (Giannenas et al. 2020). The discovery of essential oils and chemical ingredients at the end of the nineteenth century aided the development of the pharmaceutical industry. Unfortunately, when antimicrobial medicines were established as the most efficient treatment effect for antibacterial microorganisms in the twentieth century, further scientific studies on essential oils were ceased (Horváth and Ács 2015). There has recently been a resurgence of curiosity in the usage of essential oils. Some of the reasons for this are concerns about the growth of multidrug-resistant bacteria as well as the use of antibiotics in animal feed, which showed a vital part in the production of antibacterial bacteria and is serious for humans and lead to medical risk (Yap et al. 2014).

[1] Department of Parasitology, University of Agriculture Faisalabad.
[2] Department of Microbiology, Cholistan University of Veterinary and Animal Sciences, Bahawalpur.
[3] Department of Food Technology, College of Agricultural Engineering Sciences, Salahaddin University, Kurdistan, Iraq.
* Corresponding author: raouaf@hotmail.com

Ruminants, including cattle and buffalo, form a symbiotic relationship with the bacteria found in the rumen. These ruminal microorganisms break down animal feed, which contains carbohydrates, proteins, and fibers consumed by both animals and germs. However, there are significant disadvantages in terms of energy losses and ammonia production. These losses did not only affect production efficiency but also frequently led to global pollution (Calsamiglia et al. 2007). Ruminant researchers are particularly interested in modifying interactions between distinct microbial communities in order to maximize the efficacy of energy and peptide use in the ruminant. This is performed through dietary changes and the use of dietary supplements that have an effect on the environment and either help or hinder specific bacterial species. Antibiotic ionophores have been demonstrated to be especially effective in lowering ruminant protein and energy losses. The EU stopped the use of antimicrobial medicines in animal feed by the end of 2006 (Cuong et al. 2021).

Researchers were encouraged by the prohibition to hunt for other antibacterial substances, such as herbal plant extract, probiotics, and essential oils. Furthermore, in response to shifting consumer preferences and an increasing preference for organic or low-cost foods, researchers have concentrated on plants, herbal products, and related compounds. Essential oil is a condensed mixture of hydrophobic bioactive molecules that are derived in liquid volatile form from plants via a distillation process. In the sixteen century, Paracelsus von Hohenheim invented the term 'essential oils'. They are also known as ethereal oils, volatile oils, and aetheroleum. It is simply defined as an oil extract derived from plants, such as clove oils. The term essential is derived from the word 'essence', which means scent or plant aroma and taste, and refers to the ability of these chemicals to provide distinct flavors to various plants (Prajapati et al. 2021).

The Physicochemical Characteristics of Essential Oils

Essential oils derived from different plants are increasing in popularity due to their remarkable biological characteristics and applications in the food and pharmaceutical sectors. Volatile oils are obtained from a variety of plant components by various techniques. The chemical structure and purity of volatile oil differ based on genes and environmental influences. It is feasible to increase the efficacy of volatile oil from botanicals by using several biotechnological approaches. The chemical structure and natural and biological properties of essential oils have been used to classify them (Alam and Singh 2021). They are divided into two main groups: phenylpropanoids and terpenoids. Both families contain phenolic composites, which are regarded to be principally responsible for essential oils' antioxidant effect. The relative qualities and composition of essential oils are determined by a variety of parameters, including the environment, geographic location, and a variety of other factors (Şanli and Karadoğan 2016).

The composition of volatile oil comprises approximately 60 different bioactive compounds. Terpenes, the primary ingredients of essential oils, are created and released by the unique cells of plants and are thus obtained from secondary isoprenoid cascades. These chemicals are distinguished by their derivation from a fundamental structure of five carbon chains known as an isoprene unit (C5–C40) and are subdivided according to the quantity of these components in their structure. Terpenoids could have a range of chemical functional structures comprising phenolic, ether, alcoholic, ketonic, aldehyde, ketone and hydrocarbon (Fongang and Bankeu 2020).

Phenylpropanoids might not be the most frequent essential oil molecules, although they are present in substantial amounts in several plants. The word "phenylpropanoid" denotes molecules with a three-carbon chain linked to a six-carbon aromatic loop. Phenylpropanoids are mostly derived from aromatic amino acids, such as phenylalanine, through the shikimate metabolic process that is exclusively found in microbes and plant structures (Burčul et al. 2020). There are different well-known high-yielding botanical families, such as Liliaceae (garlic), Poaceae, Asteraceae, Alliaceae, Piperaceae (black peper), Lauraceae (cinnamon), Myrtaceae (thyme), Apiaceae (Anise), Lamiaceae (oregano), and Poaceae, has been stated to produce essential oils having biomedical and pharmaceutical significance (Spada et al. 2021, McCaskill 2021, Mucha and Witkowska 2021).

The Role of Essential Oils in the Performance of Poultry Birds

The EU 2006 banned the application of antimicrobial in animal feed due to rising numbers of multidrug-resistant microorganisms and consumer demand for antibiotic-free chicken meat, eggs, and byproducts. Alternative antimicrobials have become important. Essential oils can be offered before and after broiler harvesting as an alternative to conventional antibacterial treatments (Adewole et al. 2021). In 2016, the essential oil industry's market share in the United States was $6.6 billion, an increase of 286% over its 2004 market profit of $2.3 billion. They are often used in food and perfume flavors as well as in the treatment of emulsification, anti-inflammatory, and anesthesia (Sharmeen et al. 2021). Plant-based essential oils (EOs), such as cinnamaldehyde, thymol, carvacrol, and eugenol, have been studied and assessed for antibacterial activities against a variety of diseases, especially campylobacteriosis. These are now also potentially appealing as substitute antimicrobial drugs, even though they are appropriate for organic and non-traditional uses. They may be effective not just in reducing preventable health-related disorders but also in boosting the bird's productivity (Micciche et al. 2019).

The consumption of EOs as well as other natural remedies as productivity boosters have yielded some encouraging outcomes. Body weight gain, growth development, nutrient digestibility, feed conversion efficiency, and egg production are all common and widely accepted indicators for chicken farming. In broiler farming, essential oils can be employed as a single or mixed growth booster (Su et al. 2021). Several studies have found that consuming EO has a significant effect on body weight gain. If nutritional EOs had been used, poultry production would have increased as well. When thyme, oregano, lemon, anise, and okes essential oils were blended with normal poultry feed, nutritional digestibility increased considerably (Prajapati et al. 2021). A study was conducted to compare the administration of the nutritional antimicrobial medicine avilamycin with a mixture of Eos, including pepper, oregano, and cayenne pepper. When essential oils were administered in feed, the body weight increased (Abd El-Hack et al. 2021). According to certain research, curcumin supplementation improves birds' blood antioxidant capacity and immunological health (Oladokun et al. 2021). For sustaining and improving egg output and productivity, appropriate nourishment, sustainable strategies, and maintenance are deemed required for the development of the poultry sector. Several kinds of research have been accompanied to examine the potential impacts of garlic on egg production and its quality (Mohammed and Al-Hameed 2021). Phytoconstituents and herbs, whether used alone or in combination, could support livestock function great way to stay healthy. When garlic powder was added to layer feed at a rate of 24 mg/kg, egg production, egg shell quality, and feed efficiency increased. Similar advantages were reported when essential oils were given to nesting quails (Çabuk et al. 2014). Summer season stress causes a decrease in egg laying, increased egg cracking, and even death. During extreme heat, volatile oils blend and organic acid addition in commercial layer diets are helpful to egg development and immunological performance (Özek 2011).

The Significance of EOs in Bacterial Bovine Fermentation as Food Supplements

Antimicrobial medicines are routinely used in cattle feed to promote development, minimize disease outbreaks and digestive issues, and improve nutrient utilization. However, global outrage over the widespread use of antimicrobials in cattle feeding has surged in recent decades, owing to the rise of antibacterial drugs resistance microbes which might endanger global health (Tian et al. 2021).

As a result, much effort has been expended in the search for antibiotic alternatives. Phytoconstituents provide a unique potential in this area, even though many plants release bioactive components with antibacterial characteristics, such as terpenoids and tannins. These chemicals were proven to alter ruminal fermentation in order to increase ruminant nutrition use. Furthermore, the well-documented antibacterial action of volatile oil and other bioactive ingredients has prompted a number of studies to look into the possibility of all of these bioactive chemicals modulating gut bacterial metabolism to raise cow production levels (Lillehoj et al. 2018). Essential oils possess antimicrobial activity against Gram-positive and Gram-negative bacteria, protozoa, and fungi. Essential oils have also been used to cure a number of food-borne ailments (Basavegowda and Baek 2021). Thymol and carvacrol derived from oregano oil inhibited the growth of *E. coli* strain O157:H7. The well-known antibacterial effects of essential oils have sparked a number of researchers to study how they might affect ruminal fermentation to boost nutritional intake and digestibility in ruminants. The limited numbers of bioactive chemicals found in essential oils that have been studied so far seem promising throughout the process. However, there is a wide range of volatile oils and their molecules, and research is underway on most of these substances. Furthermore, most of the study has been done in the laboratory, and more research is needed to examine the effects in animals, the mode of action of various essential oils and their associated components, and the amount which affects ruminal fermentation. The ruminal microbial fermentation depends upon many factors, such as the nature and chemical structure of essential oils, dose rate, etc. Researchers discovered the importance of EOs in ruminal microbial fermentation for the first time in 1968. *In vitro*, they tested a crude extract of the sagebrush species Artemisia tridentate (Basavegowda and Baek 2021).

Ruminant microflora symbiosis provides domesticated animals with the distinctive benefit of being able to use non-protein forms of nitrogen as nutrition. The bacterial polypeptide that travels from the rumen to the gastrointestinal tract provides a good source of amino acids for the animal's production of meat and milk proteins. However, the bacterial proteins produced in the rumen are insufficient to provide the amino acid requirements of high-producing domesticated animals. Diets are typically supplemented with feed protein sources; however, such procedures might raise feed prices. Furthermore, ruminants' inadequate nitrogen absorption results in the release of N-rich feces into the atmosphere. They estimated that 0.3 percent of the nitrogen absorbed by dairy cows is emitted in urine. As a result, animals are consuming more nitrogen, which benefits both cattle production rates and the ecosystem (Kholif and Olafadehan 2021).

Essential oils and their active components have been discovered to have a significant impact on ruminal nitrogen metabolism in studies conducted all over the world. Thymol supplementation at 1 g/l in casein-containing feed resulted in the amino acid buildup and a decrease in ammonia nitrogen content, indicating that ruminal microorganisms are blocked from deamination amino acids. Essential oils subdued the development of hyper ammonia-producing bacteria, such as *Cl. Sticklandii, Clostridium aminophilum*, and *P. anaerobius* in the rumen. These microorganisms, which have strong deamination activity, constitute around 1% of the ruminal microbiota. The impact of essential oils on hyper ammonia generating bacteria is primarily dosage-dependent, although the low concentration of essential oils can specifically suppress HAP bacteria and excessive amounts harm many microbes found in rumens (Simitzis 2017). Despite having no effect on fiber breakdown, essential oils have the capacity to limit colonization and breakdown of quickly fermentable compounds. Essential oils have the most profound effects on the rumen due to their specialized activity on certain ruminal bacteria, inhibiting peptide, glucose, and amino acid degradation (Kim et al. 2015). Cows that were given 250–750 gm of oregano leaves per day showed decreased ruminal concentration in an *in vitro* research (Hristov et al. 2013). Different essential oils supplemented with dairy cows, such as cinnamaldehyde (50 mg/kg), eugenol (75 mg/kg), and cinnamon oil (50 mg/day), enhanced milk production (Benchaar et al. 2016). Feeding dairy cows juniper berry (2 g/day) and garlic (5 g/day) oils, on the other hand, would have no effect on milk yield or quality (Yang et al. 2010a).

The Immunomodulatory Role of EOs in Animals

The increasing global population of 7.8 billion people puts a significant burden on the livestock sector, particularly in cattle farming. Cattle, goats, and sheep are the major ruminants utilized internationally as sources of animal protein, each serving significantly to the national economy. Acute inflammatory diseases produce considerable damage to the animal business, resulting in both health and socioeconomic ramifications. Increased risk of infectious disease outbreak dissemination and worsening follows improvement and growth of milk and beef production. This suggests that a better understanding of the role of essential oils in bovine immune function is required to provide the best strategies for combating current and future diseases and improving food and nutritional security. High-yielding farm animal and their progeny are much more susceptible to a variety of diseases, leading to shorter life spans and worse climatic adaptation (Vlasova and Saif 2021).

Understanding animal immune systems, particularly cow immune systems, is important to determine the beneficial effect of essential oils during feed supplementation. Physical (skin and mucous membrane) and mechanical (defensin and antimicrobial peptides) first-line defensive barriers occur in cattle, as well as other species, and constitute an important aspect of the innate immune system. In cattle, macrophages, natural killer cells, mucosa-associated invariant T cells, gamma delta T cells, and granulocytes are the most important cells in the innate immune system (Levings and Roth 2013). Pathogen pattern recognition (PPR) receptors in the cell membranes of these fighting cells attach pathogens during the initial stage of the immune system. Toll-like receptors are the best example of PPR (Levings 2013).

Mast cells in cattle are another vital constituent of the non-specific immune system, and they have a significant part in allergy and inflammatory responses. Complement is another important innate immune defense mechanism. It is made up of a collection of inactive proteins (C1–C9) found in serum. Complement encounters infections via the classical, lectin, and alternative pathways. A new investigation of rumen epithelial cells of the goat was undertaken. Subacute rumen acidosis (SARA) is a common goat disease that causes severe economic losses. It is caused by a high carbohydrate intake and a lack of fiber in the diet of ruminants. The accumulation of fatty acids and organic acids in the rumen promotes the manufacture of lipopolysaccharides in Gram-Negative bacteria, e.g., *Escherichia coli* disrupts the integrity of the rumen's epithelial cells and results in SARA. This lipopolysaccharide is captured by Toll-like receptors (TLR-2, TLR-3, and TLR-4). NF-κB signaling is an important mechanism by which host cells produce a variety of pro-inflammatory cytokines, including IL-10, IL-1, and TNF-α in response to LPS. As a result, measures to minimize ruminant epithelial inflammation are critical. The medications that are now utilized to decrease inflammation can have unwanted adverse effects. As a result, finding essential oils that reduce epithelium inflammatory response, while being non-toxic is critical for animal protection.

Tea tree oil is extracted from tea leaf tissue and is used to treat inflammatory diseases. This tree contains around a hundred bioactive chemicals, one of which is terpinene-4-ol, and it can inhibit the release of cytokines, such as TNF-α, IL-1 and IL-10. The study discovered that employing tea tree oil as a ruminant feed additive suppresses the NF-κB signaling pathway by reducing the production of proinflammatory cytokines (Hu et al. 2021). Several studies have shown that thyme oil, garlic oil, and cinnamon oil can improve with SARA treatment (Castillejos et al. 2006). Carvacrol was used as a feed supplement in broiler diets for up to 15 days. It reduced the production of proinflammatory cytokines like IL-6, TNF-, and IL-1 as well as the expression of TLR-4 and the NF-B p65 gene (Liu et al. 2019).

Bovine adaptive immune system has two types' humoral immune response (B cells, plasma cells, and memory cells) and cell-mediated immune response (CD4 and CD8 T). Immunization can strengthen the adaptive immune system towards becoming faster, better, and stronger in response to a particular disease following constant exposure. Bovine plasma cells may synthesize five types of immunoglobulins: IgM, IgG, IgA, IgD, and IgE. IgG is further classified as IgG1, IgG2, and IgG3. IgM is further subdivided into IgM1 and IgGM2. Through complement stimulation, Ab-dependent cellular cytotoxicity, and receptor-mediated endocytosis, these immunoglobulins neutralize the foreign antigen (Shao et al. 2020).

The concentration of immunoglobulins differs amongst cattle breeds. IgG, particularly IgG1 and IgG2, is plentiful in cow colostrum and is more powerful against viral and bacterial illnesses. IgA is found in minimal amounts in bovine serum but is prevalent in various fluids and is vital for antiviral response in the respiratory tracts and digestive tracts (Maunsell et al. 2019). The eucalyptus oil has positive effects on the stimulation of CD8 cells in lower doses as these cells play fundamental roles in immune memory during primary respiratory virus infections (Shao et al. 2020). Cinnamaldehyde was provided as a supplement to a barley grain diet for cannulated cattle bulls. However, there was no increase in lymphocyte cells, implying that it does not affect the adaptive immune system (Yang et al 2010a).

Although, different results have been observed when eugenol is used as feed additives. These indicate that eugenol administration may boost specific immunity (Yang et al. 2010b). The blood metabolites of milking goats were studied after three essential oils were added to the goat's diet, such as thyme, clove, and anise. There are still no increases in immunoglobulin percentages among groups, indicating that essential oils had no influence on adaptive immunity in goats (El-Essawy et al. 2021).

The Antimicrobial Role of Essential Oil in Animals

Essential oils have a wide range of impacts on animal health, production, and performance due to their antimicrobial characteristics, but the precise mechanism of action is unknown. The primary mechanism of action of volatile oil is thought to be cell damage initiation via apoptosis, cell cycle progression, rupturing of the cellular membrane, and impaired function of critical organs inside a cell. Such processes are mediated by cellular membrane contacts and damage produced by the volatile oils' lipid-soluble properties (Sharifi-Rad et al. 2017). Volatile oils were shown to have various degrees of antibacterial activity depending on the kind and chemical composition of the oil and the type of bacterium. Significant differences have been discovered between bacterium species and within specific species and subspecies.

The exploration of essential oils' antibacterial capabilities is gaining popularity, as new medicines are needed to address diseases caused by multidrug bacterial resistant strains that are a serious problem in both humans and animals (Shao et al. 2020).

Gram-positive bacteria, as compared to Gram-negative bacteria, are more susceptible to essential oils presented in Table 1. The cell wall of Gram-positive bacteria is composed of peptidoglycan, teichoic acid, and polypeptides. Gram-negative bacteria's outermost membrane is made up of hydrophilic lipopolysaccharides, which operate as a barrier between biomolecules and hydrophobic substances, making Gram-negative bacteria more resistant to hydrophobic antibacterial compounds found in volatile oil (Nazzaro et al. 2013). *Staphylococcus aureus* infections can occur in humans, animals, birds, and cold-blooded animals. *Staph aureus* causes mastitis in farm animals. Lemongrass (*C. citratus*), oregano (*Origanum Vulgare*), savoury (*Satureja montana*), and litsia (*L. cubeba*) have methicillin-resistant and anti-staphylococcal activity due to bioactive chemicals contained in essential oils, such as thymol and carvacrol (Ehsani et al. 2017, Kot et al. 2019, Vitanza et al. 2019, Hu et al. 2019).

The essential oils derived from *Helichrysum pandurifolium*, *Leptospermum scoparium*, *Helichrysum araxinum*, *C. citratus*, and *Helichrysum trilineatum* have an antibacterial effect against *Staphylococcus pseudintermedius* (Najar et al. 2020, Nocera et al. 2020). Using these essential oils intramammary lowers the bacterial count significantly during mastitis. Enterococcus spp., such as *E. faecalis* and *E. faecium*, are still the most common pathogenic bacteria in urinary infections, endocarditis, skin infections, diarrhea, and infant illnesses. Enterococcus species can also induce mastitis and diarrhea in cows, endocarditis, sepsis, arthritis, and amyloidosis in chickens. There is a scarcity of information on the activity of volatile oil against *E. faecalis* strains from animal sources (Hammerum 2012, Šeputienė et al. 2012).

Essential oils derived from *S. officinalis*, *S. sclarea*, and *I. verum* have antibacterial activity against *E. faecalis* (Benmalek et al. 2014, Ghorbani 2017). However, oral administration must be discouraged or curtailed since EOs may interfere with the presence of inhabitant enterococci species in the intestinal microflora, reducing the beneficial impact of these microbes. *Mycobacterium tuberculosis* and *M. bovis* are two mycobacterium species that cause tuberculosis in humans and animals. The use of essential oils as alternative tuberculosis therapies is not recommended due to the dangers of this deadly disease. Essential oils, on the other hand, could have been used as alternative therapies to prevent infection caused by non-mycobacterial species. *M. avium* is characterized as a non-mycobacterial species, despite the fact that it is associated with tuberculosis in the chicken species (Peruč et al. 2018, Baldin et al. 2019). *Escherichia coli* have been linked to a variety of human and animal diseases. It is a critical threat in veterinary medicine since it can cause illness in all kinds of animals, resulting in large economic damages. It causes avian colibacillosis, which is a systemic illness in birds. The essential oils of *Cymbopogon citratus*, *Mentha piperita*, *Syzygium aromaticum*, and *Cinnamomum*

Table 1: List of common essential oils used against microbes.

Sr. No.	Essential Oils (EO)	Plant Origin	Action Against Bacteria	Reference
1.	Cinnamon EO	*Cinnamomum zeylanicum*	• *E. coli* • *Salmonella* spp. • *Pseudomonas aeruginosa*	Zhang et al. 2016
2.	Clove EO	*Syzygium aromaticum*	• *E. coli* • *Pseudomonas aeruginosa* • *Staphylococcus aureus*	Ebani et al. 2020
3.	Peppermint EO	*Mentha piperita*	• *Klebsiella pneumoniae* • *Pseudomonas aeruginosa* • *Acinetobacter baumannii*	Elmenawey et al. 2019
4.	Basil EO	*Ocimum basilicum*	• *Pseudomonas aeruginosa,* • *Shigella* sp. • *Listeria monocytogenes* • *Staphylococcus aureus*	Stefan et al. 2013
5.	Geranium EO	*Pelargonium graveolens*	• *Proteus mirabilis* • *Escherichia coli* • *Enterococcus faecalis*	Carmen and Hancu 2014
6.	Oregano EO	*Origanum vulgare*	• *Escherichia coli* • *Klebsiella oxytoca* • *Klebsiella pneumoniae*	Mugnaini et al. 2012
7.	Tea tree EO	*Melaleuca alternifolia*	• *Pseudomonas aeruginosa* • *Campylobacter jejuni* • *Trichophyton equinum*	Kačániová et al. 2017
8.	Aromatic Litsea EO	*Litsea cubeba*	• *Staphylococcus aureus*	Hu et al. 2019
9.	Thyme EO	*Thymus vulgaris*	• *Staphylococcus aureus* • *Pseudomonas aeruginosa* • *Klebsiella pneumoniae* • *Enterobacter cloacae*	Messaoudi et al. 2015
10.	Lavender EO	*Lavandula officinalis*	• *Aeromonas hydrophila* • *Aeromonas hydrophila* • *Citrobacter freundii* • *Proteus mirabilis* • *Salmonella enterica* • *Pseudomonas aeruginosa*	Hossain et al. 2017
11.	Lemon grass EO	*Cymbopogon citratus*	• *Pseudogymnoascus destructans*	Gabriel et al. 2018
12.	Citronella EO	*Cymbopogon citratus*	• *C. freundii* • *S. enterica* • *E. tarda* • *P. aeruginosa* • *P. mirabilis*	De Silva et al. 2017
13.	Olive EO	*Olea europaea* L.	• *B. cereus* • *E. faecalis* • *M. catarrhalis* • *S. typhi*	Yipel et al. 2016

Table 1 contd. ...

...Table 1 contd.

Sr. No.	Essential Oils (EO)	Plant Origin	Action Against Bacteria	Reference
14.	Anise EO	*Pimpinella anisum*	• *Staphylococcus aereus* • *E. coli*	Fatma et al. 2013
15.	Eucalyptus EO	*Eucalyptus globulus*	• *E tarda* • *S. iniae* • *S. parauberis* • *L. garviae* • *V. harveyi*	Park et al. 2016

zeylanicum were found to have higher antibacterial activity against *E. coli*. The external therapy of otitis externa and rashes multidrug-resistant *E. coli* strains could have been recommended using these bioactive ingredients (Marchese et al. 2017, Ebani et al. 2018).

The use of volatile oil to cure salmonella illnesses has mostly been investigated in the field of food microbiology. Furthermore, foodborne infections caused by *Salmonella enterica sub species enterica* are becoming a rising public health concern; changes in human culture as a result of modern food production and manufacturing practices involving live animals contribute to zoonotic disease epidemics. The essential oils derived from *C. zeylanicum*, *M. piperita*, *Aloysia triphylla*, and *L. cubeba* showed antibacterial activity against *Salmonella enterica sub species enterica* and *S. typhimurium*. Such essential oils appear to increase the quality of meat and egg production when used as food supplements in chicken feed (Krishan and Narang 2014, Zhang et al. 2019).

Pseudomonas aeruginosa is a significant source of nosocomial infection in humans and animal in hospitals causing septicemia, bronchitis, urogenital, and dermatitis. Several species of the Pseudomonas genera, although less commonly, are associated with diseases of a variety of body parts. It has evolved resistance mechanisms to a wide range of antimicrobial drugs. The unique composition of the cell membrane, which protects the bacteria from hazardous substances and enables it to form a biofilm, is responsible for its innate tolerance to a wide range of treatments. There are no specific treatments for *P. aeruginosa* due to the existence of broad-spectrum beta-lactamases (Potron et al. 2015). Essential extracts of *Melaleuca alternifolia*, *Cinnamomum aromaticum*, *Pinus eldarica*, *Myroxylon balsamum*, and *S. aromaticum* were shown to have antibacterial action against *P. aeruginosa* (Kavanaugh and Ribbeck 2012, Sadeghi et al. 2016). Campylobacter species are the leading cause of gastroenteritis worldwide. The prevalence of *C. jejuni* in broiler and layer birds has been documented. Orange bitter oils, jasmine, cedarwood, spikenard, tea tree oil, patchouli, carrot seed, celery seed, ginger root, marigold, gardenia, and mugwort essential oils showed antibacterial activity against campylobacter species (Kurekci et al. 2015).

Fungal infections, often known as mycoses, are a major impediment to veterinary medicine's goal of increasing meat, milk, and egg production. Antifungal medications are frequently used off-label in animals, notably pets. These drugs are not permitted in food-producing animals due to the unknown withdrawal period. In such cases, an alternative method of preventing fungal infections may be beneficial (Leite et al. 2015).

Dermatophytes are fungi that cause dermatitis and pseudomycetoma. These fungal illnesses can be treated locally or systemically with antifungal medications. Iitraconazole is the only veterinary drug approved for dermatophyte treatment. Furthermore, it is solely used to treat microsporiasis in cats. All further antifungal therapeutic systemic medicines have been used off-label as a result of these reasons. The discovery of increasing azole resistance in virulence isolates indicated that searching for other drugs for therapy may be advantageous. *T. vulgaris, Coriandrum sativum, L. angustifolia, S. montana, R. officinal, Foeniculum vulgare, Clinopodium nepeta, Thymus numidicus, O. vulgare, M. alternifolia*, and *Croton argyrophylloides* essential oils exhibit antifungal action and are used as a topical treatment worldwide (Mugnaini et al. 2013, Leite et al. 2015, Nardoni et al. 2017, Debbabi et al. 2020).

Malassezia pachydermatis is zoonotic yeast caused otitis and seborrhoic dermatitis in atopic dogs. *M. alternifolia, Zataria multiflora, C. citratus, M. piperita, Satureja* species, *C. limon, O. vulgare, Chamaecyparis obtuse, Lavandula officinalis, Salvia sclarea* essential oils have been reported antifungal activity against *Malassezia pachydermatis* (Pistelli et al. 2012, Bismarck et al. 2020). *Aspergillus* species are responsible for abortion, otitis externa and mycotic rhinitis in bovines and dogs respectively. *O. vulgare, C. citratus, I. verum, L. cubeba, T. vulgaris, A. triphylla* essential oil showed antifungal activity against aspergillosis (Nardoni et al. 2018). *Candida albicans* is a major zoonotic pathogen that affects both humans and animals. Essential oils of *Santolina chamaecyparissus, Mentha suaveolens*, and *Anethum graveolens* have been shown to be effective against candidiasis (Pietrella et al. 2011, Tian et al. 2017).

Conclusion

The plant extract is increasing in popularity, and it has been increasingly used in the treatment of various diseases, and the stimulation of human and animal wellness and productivity in the past few decades. Humans and animals have used essential oils for millennia. Antibiotics have adverse complications and acquired resistance, thus there appears to be increasing interest in the usage of essential oils these days. Essential oils inhibited the growth of hyper ammonia producing bacteria, such as *Clostridium sticklandii, Clostridium aminophilum*, and *Peptostreptococcus anaerobius* in rumen. These microorganisms, which have strong deamination activity, constitute around 1% of the ruminal microbiota. The essential oils predominantly stimulate the innate immune response by activation of proinflammatory cytokine. The primary antimicrobial mode of action of volatile oil is thought to be cell damage initiation via apoptosis, cell cycle progression, rupturing of the cellular membrane, and impaired function of critical organs inside a cell. Gram-positive bacteria, as compared to Gram-negative bacteria, are more susceptible to essential oils.

References

Abd El-Hack, M.E., M.T. El-Saadony, A.M. Saad, H.M. Salem, N.M. Ashry, M.M.A. Ghanima and K.A. El-Tarabily. 2021. Essential oils and their nanoemulsions as green alternatives to antibiotics in poultry nutrition: A comprehensive review. Poult. Sci. 101584.

Adem, S.R., A.S. Ayangbenro and R.E. Gopane. 2020. Phytochemical screening and antimicrobial activity of *Olea europaea subsp. africana* against pathogenic microorganisms. Scie. Afric. 10: e00548.

Adewole, D.I., S. Oladokun and E. Santin. 2021. Effect of organic acids-essential oils blend and oat fiber combination on broiler chicken growth performance, blood parameters, and intestinal health. Anim. Nutrit. 7(4): 1039–1051.

Alam, A. and V. Singh. 2021. Composition and pharmacological activity of essential oils from two imported Amomum subulatum fruit samples. J. Taib. Univers. Medic. Sci. 16(2): 231–239.

Baldin, V.P., D.L.S.R. Bertin, F.C.M. Mariano, A.L. Ieque, K.R. Caleffi-Ferracioli, S.V.L. Dias, A.L. de Almeida, J.E. Gonçalves, C.D.A. Garcia, R.F. Cardoso and G. Ginger. 2019. Esential oil and fractions against *Mycobacterium* spp. J. Ethnopharmacol. 244: 112095.

Basavegowda, N. and K.H. Baek. 2021. Synergistic antioxidant and antibacterial advantages of essential oils for food packaging applications. Biomolec. 11(9): 1267.

Baser, K., C. Hüsnü and F. Chlodwig. 2010. Essential oils used in veterinary medicine. Essential 10: 881–890.

Benchaar, C. 2016. Diet supplementation with cinnamon oil, cinnamaldehyde, or monensin does not reduce enteric methane production of dairy cows. Animal 10(3): 418–425.

Benmalek, Y., O.A. Yahia, A. Belkebir and M.L. Fardeau. 2014. Anti-microbial and anti-oxidant activities of *Illicium verum*, *Crataegus oxyacantha* ssp. *monogyna* and *Allium cepa red* and white varieties. Bioengin. 4: 244–248.

Bismarck, D., A. Dusold, A. Heusinger and E. Müller. 2020. Antifungal *in vitro* activity of essential oils against clinical isolates of *Malassezi pachydermatis* from canine ears: A report from a practice laboratory. Complement. Med. Res. 27: 143–154.

Burčul, F., B. Ivica, R. Mila and P. Olivera. 2020. Terpenes, phenylpropanoids, sulfur and other essential oil constituents as inhibitors of cholinesterases. Curr. Medic. Chemist. 27(26).

Çabuk, M., S. Eratak, A. Alçicek and M. Bozkurt. 2014. Effects of herbal essential oil mixture as a dietary supplement on egg production in quail. Sci. World J. 2014.

Calsamiglia, S., M. Busquet, P.W. Cardozo, L. Castillejos and A. Ferret. 2007. Invited review: Essential oils as modifiers of rumen microbial fermentation. J. Dairy Sci. 90(6): 2580–2595.

Carmen, G. and G. Hancu. 2014. Antimicrobial and antifungal activity of *Pelargonium roseum* essential oils. Advan. Pharma. Bullet. 4: 511–520.

Castillejos, L., S. Calsamiglia and A. Ferret. 2006. Effect of essential oil active compounds on rumen microbial fermentation and nutrient flow in *in vitro* systems. J. Dairy Sci. 89: 2649–2658.

Chouhan, S., K. Sharma and S. Guleria. 2017. Antimicrobial activity of some essential oils-present status and future perspectives. Medic. 4(3): 58.

Cuong, N.V., B.T. Kiet, V.B. Hien, B.D. Truong, D.H. Phu, G. Thwaites and M.J. Carrique. 2021. Antimicrobial use through consumption of medicated feeds in chicken flocks in the Mekong Delta of Vietnam: A three-year study before a ban on antimicrobial growth promoters. Plos One 16(4): e0250082.

De Silva, B.C.J., W.G. Jung, S. Hossain, S.H.M.P. Wimalasena, H.N.K.S. Pathirana and G.J. Heo. 2017. Antimicrobial property of lemongrass (*Cymbopogon citratus*) oil against pathogenic bacteria isolated from pet turtles. Lab. Anim. Res. 33(2): 84–91.

Debbabi, H., R.E. Mokni, I. Chaieb, S. Nardoni, F. Maggi, G. Caprioli and S. Hammami. 2020. Chemical composition, antifungal and insecticidal activities of the essential oils from Tunisian *Clinopodium Nepeta* Subsp. *nepeta and Clinopodium Nepeta* Subsp. *Glandulosum*. Molecules 25: 2137.

Ebani, V.V. and F. Mancianti. 2020. Use of essential oils in veterinary medicine to combat bacterial and fungal infections. Vet. Sci. 7: 193.

Ebani, V.V., B. Najar, F. Bertelloni, L. Pistelli, F. Mancianti and S. Nardoni. 2018. Chemical composition and *in vitro* antimicrobial efficacy of sixteen essential oils against *Escherichia coli* and *Aspergillus fumigatus* isolated from poultry. Vet. Sci. 5: 62.

Ehsani, A., O. Alizadeh, M. Hashemi, A. Afshari and M. Aminzare. 2017. Phytochemical, antioxidant and antibacterial properties of *Melissa officinalis* and *Dracocephalum moldavica* essential oils. Vet. Res. Forum. 8: 223–229.

El-Essawy, A.M., U.Y. Anele, A.M. Abdel-Wahed, A.R. Abdou and I.M. Khattab. 2021. Effects of anise, clove and thyme essential oils supplementation on rumen fermentation, blood metabolites, milk yield and milk composition in lactating goats. Anim. Feed Sci. Technol. 271: 114760.

Elmenawey, M.A., F.A. Mohammed, E.A. Morsy, G.A. Abdel-Alim and M.H.H. Awaad. 2019. The impact of essential oils blend on experimental colisepticemia in broiler chickens. Int. J. Vet. Sci. 8: 294–299.

Fatma, M., H. Abdallah, H. Sobhy and G. Enan. 2013. Evaluation of antiviral activity of selected anise oil as an essential oil against bovine herpes virus. Type-1 *In Vitro* 10(5): 496–499.

Fongang, F.Y.S. and K.J.J. Bankeu. 2020. Terpenoids as Important Bioactive Constituents of Essential Oils, Essential Oils Bioactive Compounds, New Perspectives and Applications, Mozaniel Santana de Oliveira, Wanessa Almeida da Costa and Sebastião Gomes Silva, IntechOpen. Doi: 10.5772/intechopen.91426.

Fournomiti, M., A. Kimbaris, I. Mantzourani, S. Plessas, I. Theodoridou, V. Papaemmanouil, I. Kapsiotis, M. Panopoulou, E. Stavropoulou, E.E. Bezirtzoglou and A. Alexopoulos. 2015. Antimicrobial activity of essential oils of cultivated oregano (*Origanum vulgare*), sage (*Salvia officinalis*), and thyme (*Thymus vulgaris*) against clinical isolates of *Escherichia coli*, *Klebsiella oxytoca*, and *Klebsiella pneumoniae*. Microb. Ecol. Health Dis. 26: 23289.

Gabriel, K.T., L. Kartforosh, S.A. Crow and C.T. Cornelison. 2018. Antimicrobial activity of essential oils against the fungal pathogens *Ascosphaera apis* and *Pseudogymnoascus destructans*. Mycopathol. 183: 921–934.

Ghorbani, A. and M. Esmaeilizadeh. 2017. Pharmacological properties of *Salvia officinalis* and its components. J. Tradit. Complement. Med. 7: 433–440.

Giannenas, I., E. Sidiropoulou, E. Bonos, E. Christaki and P. Florou-Paneri. 2020. The history of herbs, medicinal and aromatic plants, and their extracts: Past, current situation and future perspectives. In. Feed Addit. 2: 1–18.

Hammerum, A.M. 2012. Enterococci of animal origin and their significance for public health. Clin. Microbiol. Infect. 18: 619–625.

Horváth, G. and K. Ács. 2015. Essential oils in the treatment of respiratory tract diseases highlighting their role in bacterial infections and their anti-inflammatory action: A review. Flav. Fragr. J. 30(5): 331–341.

Hossain, S., H. Heo, B. De Silva, S. Wimalasena, H. Pathirana and G.J. Heo. 2017. Antibacterial activity of essential oil from lavender (*Lavandula angustifolia*) against pet turtle-borne pathogenic bacteria. Laborat. Anim. Res. 33(3): 195–201.

Hu, W., C. Li, J. Dai, H. Cui and L. Lin. 2019. Antibacterial activity and mechanism of *Litsea cubeba* essential oil against methicillin-resistant *Staphylococcus aureus* (MRSA). Ind. Crop. Prod. 130: 34–41.

Hu, Z., M. Lin, X. Ma, G. Zhao and K. Zhan. 2021. Effect of tea tree oil on the expression of genes involved in the innate immune system in goat rumen epithelial cells. Anim. 11(8): 2460.

Kačániová, M., M. Terentjeva, N. Vukovic, C. Puchalski, S. Roychoudhury, S. Kunová, A. Klūga, M. Tokár, M. Kluz and E. Ivanišová. 2017. The antioxidant and antimicrobial activity of essential oils against *Pseudomonas* spp. isolated from fish. Saudi Pharma. J. 25: 1108–1116.

Kavanaugh, N.L. and K. Ribbeck. 2012. Selected antimicrobial essential oils eradicate *Pseudomonas* spp. and *Staphylococcus aureus* biofilms. Appl. Environ. Microbiol. 78: 4057–4061.

Kholif, A.E. and O.A. Olafadehan. 2021. Essential oils and phytogenic feed additives in ruminant diet: Chemistry, ruminal microbiota and fermentation, feed utilization and productive performance. Phytoch. Revi. 2: 1–22.

Kim, B.S., S.J. Park, M.K. Kim, Y.H. Kim, S.B. Lee, K.H. Lee, N.Y. Choi, Y.R. Lee, Y.E. Lee and Y.O. You. 2015. Inhibitory effects of Chrysanthemum boreale essential oil on biofilm formation and virulence factor expression of Streptococcus mutans. Evid. Based Complement. Alternat. Med. 616309.

Kot, B., K. Wierzchowska, M. Piechota, P. Czerniewicz and G.C. Chrznowski. 2018. Antimicrobial activity of five essential oils from lamiaceae against multidrug-resistant *Staphylococcus aureus*. Nat. Prod. Res. 24: 3587–3591.

Krishan, G. and A. Narang. 2014. Use of essential oils in poultry nutrition: A new approach. J. Adv. Vet. Anim. Res. 1: 156–162.

Kurekci, C., J. Padmanabha, S.L. Bishop-Hurley, E. Hassan, R.A. Al-Jassim and C.S. McSweeney. 2013. Antimicrobial activity of essential oils and five terpenoid compounds against *Campylobacter jejuni* in pure and mixed culture experiments. Int. J. Food Microbiol. 166: 450–457.

Leite, M.C., A.P. de Brito Bezerra, J.P. de Sousa and L.E. de Oliveira. 2015. Investigating the antifungal activity and mechanism(s) of geraniol against *Candida albicans* strains. Med. Mycol. 53: 275–284.

Levings, R.L. and J.A. Roth. 2013. Immunity to bovine herpesvirus 1: I. Viral lifecycle innate immunity. Anim. Health Res. Rev. 14(1): 88–102.

Lillehoj, H., Y. Liu, S. Calsamiglia et al. 2018. Phytochemicals as antibiotic alternatives to promote growth and enhance host health. Vet. Res. 49: 76.

Liu, S.D., M.H. Song, W. Yun, J.H. Lee, H.B. Kim and J.H. Cho. 2019. Effect of carvacrol essential oils on immune response and inflammation-related genes expression in broilers challenged by lipopolysaccharide. Poult. Sci. 98(5): 2026–2033.

Marchese, A., R. Barbieri, E. Coppo, I.E. Orhan, M. Daglia, S.F. Nabavi, M. Izadi, M. Abdollahi, S.M. Nabavi and M. Ajami. 2017. Antimicrobial activity of eugenol and essential oils containing eugenol: A mechanistic viewpoint. Crit. Rev. Microbiol. 43: 668–689.

Maunsell, F.P. and C. Chase. 2019. *Mycoplasma bovis*: Interactions with the immune system and failure to generate an effective immune response. Vet. Clin. North Am. Food Anim. Pract. 35(3): 471–83.

McCaskill, L.D. 2021. The use of essential oils in traditional chinese veterinary medicine: Small animal practice. Am. J. Tradit. Chin. Vet. Medi. 67: 5–10.

Messaoudi, M., M. Benreguieg, M. Merah and Z.A. Messaoudi. 2019. Antibacterial effects of Thymus algeriensis extracts on some pathogenic bacteria. Act. Scie. Biol. Sci. 41: e48548–e48548.

Micciche, A., M.J. Rothrock, Y. Yang and S.C. Ricke. 2019. Essential oils as an intervention strategy to reduce *Campylobacter* in poultry production: A review. Front. Microbiol. 10: 1058.

Mohammed, H.A. and S.A. Al-Hameed. 2021. Effect of dietary black Cumin seeds (*Nigella Stavia*), garlic (*Allium Sativum*) and lettuce leaves (*Lactuca Sativa*) on performance and egg quality traits of native layer hens. IOP. Conf. Ser. Earth Envi. Sci. 910: 012001.

Mucha, W. and D. Witkowska. 2021. The applicability of essential oils in different stages of production of animal-based foods. Molecules 26(13): 3798.

Mugnaini, L., S. Nardoni, L. Pistelli, M. Leonardi, L. Giuliotti, M.N. Benvenuti, F. Pisseri and F. Mancianti. 2013. An herbal antifungal formulation of *Thymus serpillum*, *Origanum vulgare* and *Rosmarinus officinalis* for treating ovine dermatophytosis due to *Trichophyton mentagrophytes*. Mycoses 56: 333–337.

Najar, B., V. Nardi, C. Cervelli, G. Mecacci, F. Mancianti, V.V. Ebani, S. Nardoni and L. Pistelli. 2020. Volatilome analyses and *in vitro* antimicrobial activity of the essential oils from five South Africa *Helichrysum* species. Molecules 25: 3196.

Nardoni, S., C.D. Ascenzi, G. Rocchigiani, R.A. Papini, L. Pistelli, G. Formato, B. Najar and F. Mancianti. 2018. Stonebrood and chalkbrood in *Apis mellifera* causing fungi: *In vitro* sensitivity to some essential oils. Nat. Prod. Res. 32: 385–390.

Nardoni, S., S. Giovanelli, L. Pistelli, L. Mugnaini, G. Profili, F. Pisseri and F. Mancianti. 2015. *In vitro* activity of twenty commercially available, plant-derived essential oils against selected dermatophyte species. Nat. Prod. Commun. 10: 1473–1478.

Nazzaro, F., F. Fratianni, L. De Martino, R. Coppola and V. De Feo. 2013. Effect of essential oils on pathogenic bacteria. Pharmaceut. 6(12): 1451–1474.

Nocera, F.P., S. Mancini, B. Najar, F. Bertelloni, L. Pistelli, A. De Filippis, F. Fiorito, L. De-Martino and F. Fratini. 2020. Antimicrobial activity of some essential oils against Methicillin-susceptible and Methicillin-resistant *Staphylococcus pseudintermedius*-associated Pyoderma in dogs. Animals 10: 1782.

Novak, K. 2013. Functional polymorphisms in toll-like receptor genes for innate immunity in farm animals. Vet. Immunol. Immunopathol. 157: 1–11.

Oladokun, S., J. MacIsaac, B. Rathgeber and D. Adewole. 2021. Essential oil delivery route: Effect on broiler chicken's growth performance, blood biochemistry, intestinal morphology, immune, and antioxidant status. Animals 11(12): 3386.

Özek, K., K.T. Wellmann, B. Ertekin and B. Tarım. 2011. Effects of dietary herbal essential oil mixture and organic acid preparation on laying traits, gastrointestinal tract characteristics, blood parameters and immune response of laying hens in a hot summer season. J. Anim. Feed Sci. 20(4): 575–586.

Park, J.W., M. Wendt and G.J. Heo. 2016. Antimicrobial activity of essential oil of *Eucalyptus globulus* against fish pathogenic bacteria. Lab. Anim. Res. 32: 87–90.

Peruč, D., I. Gobin, M. Abram, D. Broznić, T. Svalina, S. Štifter, M.M. Staver and B. Tićac. 2018. Antimycobacterial potential of the juniper berry essential oil in tap water. Archives of industrial hygiene and toxicology. Arch. Hig. Rada Toksikol. 69: 46–54.

Pietrella, D., L. Angiolella, E. Vavala, A. Rachini, F. Mondello, R. Ragno, F. Bistoni and A. Vecchiarelli. 2011. Beneficial effect of *Mentha suaveolens* essential oil in the treatment of vaginal candidiasis assessed by real-time monitoring of infection. BMC Complement. Altern. Med. 11: 18–25.

Pistelli, L., F. Mancianti, A. Bertoli, P.L. Cioni, L. Leonardi, F. Pisseri, L. Mugnaini and S. Nardoni. 2012. Antimycotic activity of some aromatic plants essential oils against canine isolates of *Malassezia pachydermatis*: An *in vitro* assay. Open Mycol. J. 6: 17.

Potron, A., L. Poirel and P. Nordmann. 2015. Emerging broad-spectrum resistance in *Pseudomonas aeruginosa* and *Acinetobacter baumannii*: Mechanisms and epidemiology. Int. J. Antimicrob. Agents 24: 568–585.

Prajapati, D.R., V.R. Patel, A.P. Raval, A.B. Parmar, A. Londhe and S.S. Patel. 2021. Essential Oils: Alternative to improve production, health and immunity in poultry. J. Trop. Anim. Res. 1(3): 92–98.

Sadeghi, M., B. Zolfaghari, A. Jahanian-Najafabadi and S.R. Abtahi. 2016. Anti-pseudomonas activity of essential oil, total extract, and proanthocyanidins of *Pinus eldarica* Medw. bark. Res. Pharm. Sci. 11: 58–64.

Şanli, A. and T. Karadoğan. 2016. Geographical impact on essential oil composition of endemic *Kundmannia Anatolica* Hub. Mor. (Apiaceae). Afr. J. Tradit. Complement Altern. Med. 14(1): 131–137.

Šeputienė, V., A. Bogdaitė, M. Ružauskas and E. Sužiedėlienė. 2012. Antibiotic resistance genes and virulence factors in *Enterococcus faecium* and *Enterococcus faecalis* from diseased farm animals: Pigs, cattle and poultry. Pol. J. Vet. Sci. 3: 431–438.

Shao, J., Z. Yin, Y. Wang, Y. Yang, Q. Zhang and J. Lu. 2020. Effects of different doses of Eucalyptus oil from Eucalyptus globulus Labill on respiratory tract immunity and immune function in healthy Rats. Front. Pharmacol. 11: 1287.

Sharifi-Rad, J., A. Sureda, G.C. Tenore, M. Daglia, M. Sharifi-Rad, M. Valussi, R. Tundis, M. Sharifi-Rad, M.R. Loizzo, A.O. Ademiluyi and R. Sharifi-Rad. 2017. Biological activities of essential oils: From plant chemoecology to traditional healing systems. Molecules 22(1): 70–75.

Sharmeen, J.B., F.M. Mahomoodally, G. Zengin and F. Maggi. 2021. Essential oils as natural sources of fragrance compounds for cosmetics and cosmeceuticals. Molecules 26: 666.

Simitzis, P.E. 2017. Enrichment of animal diets with essential oils—A great perspective on improving animal performance and quality characteristics of the derived products. Medicines 4(2): 35–40.

Spada, M., O.A. Cuzman, I. Tosini, M. Galeotti and F. Sorella. 2021. Essential oils mixtures as an eco-friendly biocidal solution for a marble statue restoration. Int. Biodet. Biodegrad. 163: 105280.

Stanfield, R.L., J. Haakenson, T.C. Deiss, M.F. Criscitiello, I.A. Wilson and V.V. Smider. 2018. The unusual genetics and biochemistry of bovine immunoglobulins. Adv. Immunol. 137: 135–64.

Stefan, M., M.M. Zamfirache, C. Padurariu, E. Trută and I. Gostin. 2013. The composition and antibacterial activity of essential oils in three Ocimum species growing in Romania. Cent. Euro. J. Biol. 8(6): 600–608.

Su, G., L. Wang, X. Zhou, X. Wu, D. Chen, B. Yu and J. He. 2021. Effects of essential oil on growth performance, digestibility, immunity and intestinal health in broilers. Poult. Sci. 101242.

Tian J., Z. Lu, Y. Wang, M. Zhang, X. Wang, X. Tang, X. Peng and H. Zeng. 2017. Nerol triggers mitochondrial dysfunction and disruption via elevation of Ca2+ and ROS in *Candida albicans*. Int. J. Biochem. Cell Biol. 85: 114–122.

Tian, M., X. He, Y. Feng, W. Wang, H. Chen, M. Gong and A. Van-Eerde. 2021. Pollution by antibiotics and antimicrobial resistance in livestock and poultry manure in China, and countermeasures. Antibiotics 10(5): 539.

Vitanza, L., A. Maccelli, M. Marazzato, F. Scazzocchio, A. Comanducci, S. Fornarini, M.E. Crestoni, A. Filippi, C. Fraschetti, F. Rinaldi et al. 2019. *Satureja montana* L. essential oil and its antimicrobial activity alone or in combination with gentamicin. Microb. Pathog. 126: 323–331.

Vlasova, A.N. and L.J. Saif. 2021. Bovine immunology: Implications for dairy cattle. Front. Immunol. 12.

Yang, W.Z., B.N. Ametaj, C. Benchaar and K.A. Beauchemin. 2010a. Dose response to cinnamaldehyde supplementation in growing beef heifers: Ruminal and intestinal digestion. J. Anim. Sci. 88: 680–688.

Yang, W.Z., B.N. Ametaj, C. Benchaar, M.L. He and K.A. Beauchemin. 2010b. Cinnamaldehyde in feedlot cattle diets: Intake, growth performance, carcass characteristics, and blood metabolites. J. Anim. Sci. 88: 1082–1092.

Yap, P., S.X. Yiap, B.C. Ping, H.C. and S.H.E. Lim. 2014. Essential oils, a new horizon in combating bacterial antibiotic resistance. Open Microbio. J. 8: 6.

Yipel, F.A., A. Acar and M. Yipel. 2016. Effect of some essential oils (*Allium sativum L., Origanum majorana* L.) and ozonated olive oil on the treatment of ear mites (Otodectes cynotis) in cats. Turk. J. Vet. Anim. Sci. 40: 782–787.

Zhang, S., Y.R. Shen, S. Wu, Y.Q. Xiao, Q. He and S.R. Shi. 2019. The dietary combination of essential oils and organic acids reduces *Salmonella enteritidis* in challenged chicks. Poult. Sci. 98: 6349–6355.

Zhang, Y., X. Liu, Y. Wang, P. Jiang and S. Quek. 2016. Antibacterial activity and mechanism of cinnamon essential oil against *Escherichia coli* and *Staphylococcus aureus*. Food Cont. 59: 282–289.

Index

For Product Safety Concerns and Information please contact our EU
representative GPSR@taylorandfrancis.com
Taylor & Francis Verlag GmbH, Kaufingerstraße 24, 80331 München, Germany

www.ingramcontent.com/pod-product-compliance
Lightning Source LLC
Chambersburg PA
CBHW060359220326
41598CB00023B/2966

9 781032 128160